Grundlehren der mathematischen Wissenschaften 269

A Series of Comprehensive Studies in Mathematics

Pierre Schapira

Microdifferential Systems in the Complex Domain

Springer-Verlag
Berlin Heidelberg NewYork Tokyo
1985

Pierre Schapira
Université Paris Nord
Département de Mathématiques
Av. J.-B. Clément
93430 Villetaneuse
France

AMS Subject Classification (1980): 58 G 05, 58 G 15, 58 G 17

Library of Congress Cataloging in Publication Data
Schapira, Pierre, 1943–
Microdifferential systems in the complex domain.
(Grundlehren der mathematischen Wissenschaften; 269)
Bibliography: p.
Includes index.
1. Differential equations, Partial. 2. Differential operators. 3. Cauchy problem. I. Title. II.
Series. QA377.S35 1985 515.3'53 84-13981

ISBN-13: 978-3-642-64904-2   e-ISBN-13: 978-3-642-61665-5
DOI: 10.1007/978-3-642-61665-5

© Springer-Verlag Berlin Heidelberg 1985
Softcover reprint of the hardcover 1st edition 1985

Typesetting and printing: Zechnersche Buchdruckerei, Speyer
Bookbinding: J. Schäffer OHG, Grünstadt
2141/3140-543210

Preface

The words "microdifferential systems in the complex domain" refer to several branches of mathematics: microlocal analysis, linear partial differential equations, algebra, and complex analysis.

The microlocal point of view first appeared in the study of propagation of singularities of differential equations, and is spreading now to other fields of mathematics such as algebraic geometry or algebraic topology. However it seems that many analysts neglect very elementary tools of algebra, which forces them to confine themselves to the study of a single equation or particular square matrices, or to carry on heavy and non-intrinsic formulations when studying more general systems. On the other hand, many algebraists ignore everything about partial differential equations, such as for example the "Cauchy problem", although it is a very natural and geometrical setting of "inverse image".

Our aim will be to present to the analyst the algebraic methods which naturally appear in such problems, and to make available to the algebraist some topics from the theory of partial differential equations stressing its geometrical aspects.

Keeping this goal in mind, one can only remain at an elementary level. Actually the algebra we use is rather naive, and there are no fine results on partial differential equations. In that sense, the only theorem we prove here is the (microdifferential) Cauchy-Kowalewski theorem, but we show how algebraic techniques allow us to give a meaning to this theorem for systems, and using geometrical arguments and sheaf theory, we show how it leads to deep results such as the Kashiwara constructibility theorem and its various generalizations.

As should be clear to the reader, much of the material of this book is due to Masaki Kashiwara. It is a pleasure to thank him here for the numerous discussions we had together. I also would like to thank Emmanuel Andronikof, Christian Houzel, Yves Laurent and Teresa Monteiro-Fernandès for their comments and advice on various parts of the book.

Finally I want to express my gratitude to Mrs Catherine Simon for her patience in typing the manuscripts.

Paris, September 1984   P. Schapira

Contents

Introduction

In this book we shall treat systems of differential or microdifferential equations, that is, modules (or sheaves of modules) over various rings of differential or microdifferential operators on a complex manifold.

The idea of regarding a system of linear equations as a module over a ring is basic in algebraic geometry, but was only initiated around 1960 in the case of differential operators with constant coefficients by B. Malgrange [1] (cf. also S. Matsuura [1], L. Ehrenpreis [1], V. P. Palamodov [1]). Then D. Quillen [1], and later I. N. Bernstein [1], [2] and M. Kashiwara [1] considered rings of differential operators with variable coefficients and overcame the difficulty of the non commutativity by introducing filtrations over those rings and using the commutativity of the associated graded rings. In particular, I. N. Bernstein for the case of polynomial coefficients, and M. Kashiwara for the case of analytical coefficients, defined the characteristic variety of a system, and the operations on systems.

But the theory took its full importance with the paper of M. Sato, M. Kashiwara, T. Kawaï [1] in 1971, who introduced (after M. Sato [1], [2]) the microlocal point of view, and thus obtained deep results, such as the involutivity of the characteristic variety, or general structure theorems. Their theory is now a basic tool which plays an important role in many parts of mathematics, such as algebraic geometry, singularity theory, Lie algebras and even Feynman integrals.

The aim of this book is to give (in Chapters I and II) an elementary and self-contained introduction to this theory, then to illustrate it: in Chapter III we discuss the Cauchy problem for general systems, and, as a by-product, we prove propagation and constructibility theorems for various sheaves of solutions of microdifferential equations.

In more details the contents of the book are as follows.

Chapter one gives the basic facts about microdifferential operators. We construct the sheaf $\mathscr{E}_X$ of microdifferential operators on T^*X, the cotangent bundle of a complex manifold X, and prove the division and the preparation theorems for $\mathscr{E}_X$. Then, after recalling the abstract Cauchy-Kowalewski theorem in scales of Banach spaces, we prove a microdifferential Cauchy-Kowalewski theorem which will become a basic tool in Chapter III. Next we construct carefully the sheaf $\mathscr{B}_{Z|X}$ associated to a submanifold Z of X, and its microlocalization $\mathscr{C}_{Z|X}$, the difficulty being in the proof of the invariance of the construction. We apply it, together with the division theorem, to obtain an elementary proof of the fact that every contact transformation can be locally "quantized", that is, extended as a ring sheaf isomorphism of $\mathscr{E}_X$. We finally give the structure theorem for systems with simple characteristics, in particular for simple holonomic systems, an easy application of the preceding results.

Chapter two studies the algebraic structure of $\mathscr{E}_X$. It begins with a paragraph where we prove all the results on filtered rings that we need later on, (except for Gabber's theorem on the involutivity that we admit without proof). Thus, after having proved that the filtration on $\mathscr{E}_X$ is "zariskian", we obtain the main properties of this sheaf: it is coherent, noetherian, flat over the sheaf of rings $\mathscr{D}_X$ of differential operators, its homological dimension is equal to the dimension of X, the support of a coherent $\mathscr{E}_X$-module $\mathscr{M}$ (denoted char($\mathscr{M}$) and often called the characteristic variety of $\mathscr{M}$) is an involutive analytic subset of T^*X, etc. ... We then construct the main operations on $\mathscr{E}_X$-modules, beginning first with the sheaves $\mathscr{C}_{Z|X}$, and define direct or inverse images of $\mathscr{E}_X$-modules. We prove for example the formula on the multiplicities of the induced system in the non characteristic case.

Chapter three deals with Cauchy problems and propagation and constructibility results that can be deduced from the refined Cauchy-Kowalewski theorem. We first recall various notions of "microcharacteristic varieties" of an $\mathscr{E}_X$-module $\mathscr{M}$ along an involutive submanifold V. The first one, $C_V(\mathscr{M})$, is purely geometric: it is the normal cone to the characteristic variety of $\mathscr{M}$ along V. But in many problems $C_V(\mathscr{M})$ has to be replaced by a bigger variety, $C_V^1(\mathscr{M})$, which takes into account not only the principal symbols involved in the system, but also the lower order terms. We make a special study of microcharacteristic varieties associated to a submersion or an embedding. The first case is useful in the study of direct images of coherent $\mathscr{D}_X$-modules and the second one is a tool to calculate the characteristic variety of the systems induced on a submanifold and to give a criterion of coherency of these induced systems. Then we prove Cauchy-Kowalewski and propagation theorems for a system $\mathscr{M}$ with values in a system $\mathscr{S}$ with simple characteristics along an involutive manifold V, under suitable assumptions about the 1-microcharacteristic variety of $\mathscr{M}$ along V. As an application we obtain a generalization of Y. Hamada's theorem, which solves the Cauchy problem when the data possess meromorphic or logarithmic singularities along hypersurfaces.

Next we recall the microlocal interpretation of Whitney's conditions and the notion of the micro-support of a complex of sheaves recently introduced by M. Kashiwara and P. Schapira. Then we are ready to prove the Kashiwara constructibility theorem for sheaves of holomorphic solutions of holonomic systems, and various extension of this theorem.

We shall assume the reader has a basic knowledge in real and complex differential calculus (manifolds, tangent and cotangent bundles, etc. ...) and a basic knowledge in algebra (e. g. Atiyah-Macdonald [1]). As the theory we present here may be considered as a non-commutative generalization of analytic geometry, it is clear that some familiarity with this last subject would be welcome (e. g. Cartan [1]), but that is not really necessary.

For the reader's convenience we have included four appendices.

In Appendix A we recall some elementary notions on symplectic geometry, especially concerning involutive and Lagrangean manifolds.

Appendix B is mainly devoted to the construction of derived functors in abelian categories, as for example the functors $Ext^j(\cdot,\cdot)$ and $Tor_j(\cdot,\cdot)$. To remain at an elementary level we have avoided the sophisticated techniques of "derived categories" as well as the techniques of spectral sequences, even if it happens (at the end of the book) that we make use of double complexes.

In Appendix C we recall basic notions on sheaves and sheaf cohomology. Finally in Appendix D we recall some well known results of analytic geometry, such as for example the notion of multiplicity.

These appendices may be helpful for the reader, but they can certainly not replace the excellent books which are available on these sujects, such as Abraham-Marsden [1], Cartan-Eilenberg [1], Godement [1], Hörmander [2], Serre [1], etc. ...

To conclude, let us emphasize the aspects of the subject which are not treated in this book. We do not discuss the real case, that is, microfunction solutions of microdifferential systems, nor infinite order systems. The reason is that the sheaf of microfunction solutions may be obtained from the sheaf of holomorphic solutions by a purely geometric method, namely Sato's microlocalization, and this theory has now been generalized by the "microlocal theory of sheaves" of Kashiwara-Schapira [4]. This paper deals with sheaves on real manifolds, from the viewpoint of their micro-support. When applying these general results to the sheaf of holomorphic solutions of (micro)-differential systems, it is in fact possible to recover many results on microfuntion solutions of microdifferential systems, and on microdifferential systems of infinite order. But one cannot obtain in this way results on coherent $\mathscr{E}_X$-modules (as for example in the case of meromorphic singularities), since these purely geometric methods do not take into account the "growth conditions". In other words, the theory of microdifferential systems in the real domain is more closely related to the general theory of sheaves than to the (essentially algebraic) theory we have developped here.

Another important feature we have not treated is the theory of "regular holonomic systems" of M. Kashiwara and T. Kawaï [1], but it would have made this book considerably more extensive, and we do believe that anyone having mastered the material presented here will be able to understand their paper. Moreover their theory might not have acquired its final form yet, since it appears that regular holonomic systems can be investigated by purely topological means, through the so-called "Riemann-Hilbert correspondence" (cf. Kashiwara [7]).

The book has three chapters and four appendices (A, B, C, D). Each chapter is divided into paragraphs and each paragraph into sections. Each appendix consists also of several sections. A statement (Theorem, Example, Remark, etc ...) is referred to as in the following example: (II, Theorem 3.1.2) means the second statement of Section 1 of § 3 of Chapter II. If this reference is made in Chapter II, then "II" is omitted. Similarly (Appendix, Theorem C.1.2) means the second statement of Section 1 of Appendix C. The formulas have a similar, but independent enumeration.

Each chapter is preceded by a short introduction. At the end of each paragraph we give some exercises, of various interest and difficulties. Some exercises are just particular cases of theorems stated in the course of the text, but some others may indicate a possible developement of the theory.

We have avoided, as much as possible, to give bibliographical references inside the text, and have prefered to end each chapter by a few historical and bibliographical comments. The reason is that many important results have a long and complicated history, and it would be tedious to quote each time every one who contributed to a result. On the other hand it seems incorrect to quote only the person who has initiated the subject, or who has put it into final form.

Chapter I. Microdifferential Operators

Summary

After recalling the principal notions concerning the Ring[1] $\mathscr{D}_X$ of holomorphic differential operators on a complex manifold X, we introduce the Ring $\hat{\mathscr{E}}_X$ of formal microdifferential operators on the cotangent bundle T^*X, assuming for a while that X is an open set in a complex (finite dimensional) vector space. This Ring is defined as a sheaf of formal series of homogeneous holomorphic functions, the composition law being obtained by extending the Leibniz formula. Then by means of the quasi-norms of L. Boutet de Monvel and P. Kree [1], we construct the sub-Ring $\mathscr{E}_X$ of $\hat{\mathscr{E}}_X$ of microdifferential operators of M. Sato, M. Kashiwara, T. Kawaï [1], and prove the invertibility of operators whose principal symbols do not vanish.

We construct various Banach algebras of zero order microdifferential operators, and use these algebras to obtain the extension to microdifferential operators of the classical division theorems of Späth and Weïerstrass, as stated in Sato-Kashiwara-Kawaï [1], and also an extension of the Cauchy-Kowalewski theorem.

The division theorem asserts that if the principal symbol of an operator P has a zero exactly of order p in some direction $\dfrac{\partial}{\partial \xi_1}$, then any operator Q may be divided by P with a remainder which is a polynomial of order at most $p-1$ in D_{x_1}, having as coefficients microdifferential operators independent of D_{x_1}.

Roughly speaking the Cauchy-Kowalewski theorem asserts that if $(x)=(x_1,\ldots,x_n)$ are coordinates on X, the Cauchy Problem with values in the sheaf of operators which do not depend on $D_{x_1},\ldots,D_{x_p}$, is well posed for operators P which are polynomials in $D_{x_1},\ldots,D_{x_p}$, with zero order microdifferential operators as coefficients (and of course, with data on a non characteristic hypersurface). The proof is an application of the abstract Cauchy-Kowalewski theorem in scales of Banach spaces. We recall here this theorem with proof, following F. Trèves [1].

---

[1] In the sequel, we shall write Ring, Ideal, Module, etc. ... with a capital letter, instead of sheaf of rings, sheaf of ideals, sheaf of modules, etc. ... (cf. Appendix C.2).

The left $\mathscr{D}_X$-module $\mathscr{B}_{Z|X}$, associated to a submanifold Z of X, and its microlocalisation $\mathscr{C}_{Z|X}$, are constructed in § 4. We avoid the cohomological tools, by associating a left Ideal of $\mathscr{D}_X$ to a system of functions defining Z, and proving the intrinsic character of the constructions.

Then, following Sato-Kashiwara-Kawaï [1] and Kashiwara [5], we are ready to prove that given a complex contact transformation ϕ from an open set $U \subset T^*X$ to an open set $U' \subset T^*X'$, ϕ can be locally "quantized", that is extended to an isomorphism $\hat{\phi}$ of filtered Rings from $\mathscr{E}_X|_U$ to $\phi^{-1}(\mathscr{E}_{X'}|_{U'})$. In fact if we set $\Lambda_\phi^a = \{(x, x'; \xi, \xi'); (x'; -\xi') = \phi(x, \xi)\}$, we have to find an Ideal $\mathscr{I}$ of $\mathscr{E}_{X \times X'}$, whose symbol Ideal coincides with the defining Ideal of Λ_ϕ^a. Then we must prove, by successive applications of the division theorem, that given a section P of $\mathscr{E}_X$ there exists a unique section Q of $\mathscr{E}_{X'}$ such that $P - Q$ belongs to $\mathscr{I}$. When ϕ is the identity, $\mathscr{I}$ is naturally associated to a volume element dx on X, and the anti-isomorphism $P \mapsto Q$ is nothing but the adjoint with respect to dx. In the general case $\hat{\phi}$ is the composite of an anti-isomorphism associated to Λ_ϕ^a and of the adjoint, for a volume element. We discuss some examples of quantized contact transformations, and in particular the action on $\mathscr{E}_X$ of a complex change of coordinates. This allows us to define now $\mathscr{E}_X$ when X is a complex manifold.

Once we are able to make use of quantized contact transformations, the theory of systems with "simple characteristics" becomes transparent, as shown in Sato-Kashiwara-Kawaï [1]. Such a system $\mathscr{M}$ is a left $\mathscr{E}_X$-module endowed with a generator u such that, if $\mathscr{I}$ denotes the left Ideal of $\mathscr{E}_X$ annihilating u, the symbol Ideal $\mathscr{\bar{I}}$ of $\mathscr{I}$ coincides with $\mathscr{I}_V$, the defining Ideal of a smooth conic involutive manifold V. When V is regular involutive, V is isomorphic by a contact transformation to the manifold $V' = \{(x, \xi) \in T^*\mathbb{C}^n$; $\xi_1 = \ldots = \xi_p = 0$, $\xi_n \neq 0\}$, and successive applications of the Cauchy-Kowalewski theorem permit us to prove that $\mathscr{M}$ is then locally isomorphic to a module $\mathscr{E}_X/\mathscr{I}$, where $\mathscr{I}$ is the left Ideal generated by $D_{x_1}, \ldots, D_{x_p}$. When V is Lagrangean, it is isomorphic to $V' = \{(x, \xi) \in T^*\mathbb{C}^n$; $x_1 = \xi_2 = \ldots = \xi_n = 0$, $\xi_1 \neq 0\}$, the conormal bundle to the hypersurface $\{x_1 = 0\}$, and there exists a complex number α, unique modulo $\mathbb{Z}$, such that $\mathscr{M} = \mathscr{E}_X/\mathscr{I}$, the Ideal $\mathscr{I}$ being generated by $x_1 D_{x_1} - \alpha$, $D_{x_2}, \ldots, D_{x_n}$.

The notions of symplectic geometry used in this Chapter are recalled in Appendix A.

§ 1. Construction of the Ring $\mathscr{E}_X$

1.1. Differential Operators

We briefly recall some classical constructions on differential operators, (cf. Hörmander [1], Wells [1]).

Let E be a complex vector space of finite dimension n, X an open set in E. Let $\tau\colon TX \to X$ be the tangent bundle to X, and $\pi\colon T^*X \to X$ the cotangent bundle. Thus $TX \cong X \times E$ and $T^*X \cong X \times E^*$, where E^* denotes the dual vector space to E. Let $\langle\,,\,\rangle$ be the natural pairing on $TX \underset{X}{\times} T^*X$.

We denote by $\mathscr{O}_X$ the sheaf of holomorphic functions on X, and by d the differential on X.

To a holomorphic vector field u (i. e. a section of the bundle TX) and a holomorphic 1-form ω (i. e. a section of the bundle T^*X) we may associate the holomorphic function $\langle u, \omega \rangle$ on X. The bracket of two vector fields u and v is defined by the formula (where ω is a 1-form):

$$\langle [u, v], \omega \rangle = \langle u, d\langle v, \omega \rangle \rangle - \langle v, d\langle u, \omega \rangle \rangle .$$

Vector fields operate on sections of $\mathscr{O}_X$ by:

$$u(f) = \langle u, df \rangle$$

and we get:

$$(u \circ v - v \circ u)(f) = [u, v](f) .$$

The sheaf $\mathscr{D}_X$ of (holomorphic, finite order) differential operators on X is the sub-Algebra of $\mathscr{H}om_{\mathbb{C}}(\mathscr{O}_X, \mathscr{O}_X)$ generated by $\mathscr{O}_X$ and by the vector fields.

Let P be a section of $\mathscr{D}_X$. To P we associate $P(x, \xi)$, its "total symbol", a function on T^*X, by setting:

(1.1.1) $$P(x, \xi) = e^{-\langle x, \xi \rangle} P(e^{\langle x, \xi \rangle}) .$$

It is immediately verified that the total symbol only depends on the affine structure of E and not on the vector space structure.

The "Leibniz formula" calculates the total symbol $R(x, \xi)$ of the composite $P \circ Q$ of two differential operators P and Q:

(1.1.2) $$R(x, \xi) = (e^{d_\xi d_y}(P(x, \xi) Q(y, \eta)))|_{x=y, \xi=\eta} ,$$

(cf. Boutet de Monvel [1] for this formulation, and cf. Formula (1.1.9) below for a formulation with coordinates).

The total symbol $P(x, \xi)$ is a polynomial in the ξ-variable. The degree of this polynomial is called the order of P, and denoted $\mathrm{ord}(P)$, with the convention that $\mathrm{ord}(0) = -\infty$. The homogeneous part of higher degree is called the "principal symbol" of P, and denoted $\sigma(P)$. The Leibniz formula gives:

(1.1.3) $$\sigma(P \circ Q) = \sigma(P) \cdot \sigma(Q) .$$

If P is a differential operator of order at most m we set:

(1.1.4) $$\sigma_m(P) = \begin{cases} \sigma(P) & \text{if } \mathrm{ord}(P) = m \\ 0 & \text{if } \mathrm{ord}(P) < m \end{cases} .$$

Remark that $\sigma_m(\cdot)$ is additive: if P an Q are of order at most m, $\sigma_m(P + Q) = \sigma_m(P) + \sigma_m(Q)$, but $\sigma(\cdot)$ is not additive.

If P is of order m and Q of order m', the bracket $[P, Q] = P \circ Q - Q \circ P$ is of order at most $m + m' - 1$, and one checks (cf. Hörmander [1]):

$$(1.1.5) \qquad \sigma_{m+m'-1}([P, Q]) = \{\sigma_m(P), \sigma_{m'}(Q)\},$$

where $\{\cdot, \cdot\}$ is the Poisson bracket on T^*X (cf. Appendix A).

The Ring $\mathscr{D}_X$ is filtered by the subsheaves $\mathscr{D}_X(m)$ of operators of order at most m, and the graded Ring (cf. Ch. II, § 1) is:

$$(1.1.6) \qquad \mathrm{gr}(\mathscr{D}_X) = \bigoplus_m \mathscr{D}_X(m) / \mathscr{D}_X(m-1)$$

We can naturally identify $\mathrm{gr}(\mathscr{D}_X)$ with the subsheaf of $\pi_*(\mathscr{O}_{T^*X})$ of holomorphic functions on $T^*X = X \times E^*$, polynomial with respect to the variable of E^*, by the map $\sigma(\cdot)$.

Let χ be a bi-holomorphic map between two open sets X and X' of E. Then χ induces a natural isomorphism $\chi_*(\mathscr{O}_X) \cong \mathscr{O}_{X'}$ which extends to an isomorphism $\chi_*(\mathscr{D}_X) \cong \mathscr{D}_{X'}$ by sending a vector field u on X to the vector field $X' \circ u$ on χ', where χ' denotes the tangent may to χ.

Let ϕ be the bi-holomorphic map between T^*X and T^*X' associated to χ, that is $\phi = (\chi, {}'\chi'^{-1})$. One checks that if P is a section of $\mathscr{D}_X$, $\hat{\phi}(P)$ the image of P in $\mathscr{D}_{X'}$, we have $\mathrm{ord}(P) = \mathrm{ord}(\hat{\phi}(P))$ and:

$$(1.1.7) \qquad \sigma(\hat{\phi}(P)) = \sigma(P) \circ \phi^{-1}.$$

Thus the principal symbol $\sigma(P)$ is a function on T^*X which only depends on the analytic structure of X, not on the embedding $X \hookrightarrow E$. In fact the principal symbol may be defined "intrinsically" as follows. Let $(x, \xi) \in T^*X$, and let f be a section of $\mathscr{O}_X$ in a neighborhood of x, such that $df(x) = \xi$. Consider:

$$Q(x, f, \tau) = e^{-\tau f} P(e^{\tau f}).$$

This is a polynomial in the τ-variable of degree $\mathrm{ord}(P)$, and the coefficient of $\tau^{\mathrm{ord}(P)}$ is $\sigma(P)$. But we emphasize that the total symbol is not intrinsically defined on T^*X.

Finally let us choose a system of local coordinates $x = (x_1, \ldots, x_n)$ on X. Then $(dx_1, \ldots, dx_n)$ define, at each $x^0 \in X$, a basis of $T^*_{x^0}X$. A vector of T^*X may be decomposed as $\left(x; \sum_j \xi_j \, dx_j\right) = (x_1, \ldots, x_n; \xi_1, \ldots, \xi_n)$, and we shall say that (x, ξ) are the associated coordinates on T^*X. Let $\left(\dfrac{\partial}{\partial x_1}, \ldots, \dfrac{\partial}{\partial x_n}\right)$ be the dual basis to $(dx_1, \ldots, dx_n)$, (at each point x^0). Then every differential operator of order m may be written in a unique way as:

$$(1.1.8) \qquad P = \sum_{|\alpha| \le m} a_\alpha \frac{\partial^\alpha}{\partial x^\alpha},$$

where a_α are sections of $\mathscr{O}_X$, $\alpha = (\alpha_1, \ldots, \alpha_n) \in \mathbb{N}^n$, $|\alpha| = \alpha_1 + \ldots + \alpha_n$,
$$\frac{\partial^\alpha}{\partial x^\alpha} = \frac{\partial^{\alpha_1}}{\partial x_1^{\alpha_1}} \cdots \frac{\partial^{\alpha_n}}{\partial x_n^{\alpha_n}}.$$

We shall often write D_{x_i} or even D_i instead of $\dfrac{\partial}{\partial x_i}$. Remark that with such a system of coordinates, the Leibniz formula becomes:

$$(1.1.9) \qquad R(x,\xi) = \sum_{\alpha \in \mathbb{N}^n} \frac{1}{\alpha!} \frac{\partial^\alpha}{\partial \xi^\alpha} P(x,\xi) \frac{\partial^\alpha}{\partial x^\alpha} Q(x,\xi).$$

We shall denote by 1 or I the unit in $\mathscr{D}_X$.

1.2. Formal Microdifferential Operators

Our aim is to construct a sheaf on T^*X such that if the principal symbol of a differential operator P is invertible on an open set $U \subset T^*X$, P itself is invertible in this new sheaf (Proposition 1.3.5). We do this by extending the class of total symbols on T^*X.

The sheaf $\hat{\mathscr{E}}_X$ we construct in this section will appear as too big for application (cf. Proposition 4.2.1) and will only be used as a tool for the construction of $\mathscr{E}_X$, although many properties of $\mathscr{E}_X$ could be proved for $\hat{\mathscr{E}}_X$, (even with less work).

Let X be open in E as in 1.1 (in fact we may take for X the underlying affine space of E), π the projection $T^*X \cong X \times E^* \to X$.

Let e be the Euler operator on E^*, that is the radial vector field. For a system of linear coordinates $(\xi_1, \ldots, \xi_n)$ on E^*:

$$(1.2.1) \qquad e = \sum_{j=1}^n \xi_j \frac{\partial}{\partial \xi_j}.$$

Let $\mathscr{O}_{T^*X}(j)$ be the subsheaf of $\mathscr{O}_{T^*X}$ of homogeneous functions of degree $j \in \mathbb{Z}$, that is of sections f of $\mathscr{O}_{T^*X}$ which satisfy:

$$(1.2.2) \qquad e(f) = jf.$$

Let U be an open set in T^*X. We denote by $\hat{\mathscr{E}}_X(m)(U)$ the space of formal series:

$$P = \sum_{-\infty < j \leqslant m} p_j,$$

where p_j is a section of $\mathscr{O}_{T^*X}(j)$ on U. The correspondence $U \to \hat{\mathscr{E}}_X(m)(U)$ clearly defines a sheaf, denoted $\hat{\mathscr{E}}_X(m)$, and we set (cf. II, § 1.4):

$$\hat{\mathscr{E}}_X = \bigcup_m \hat{\mathscr{E}}_X(m).$$

We endow the sheaf (of $\mathbb{C}$-vector spaces) $\hat{\mathscr{E}}_X$ with a structure of a ringed sheaf by extending the Leibniz composition formula. For two sections P and Q of $\hat{\mathscr{E}}_X$ we define the new section $R = P \circ Q$ by:

$$(1.2.3) \qquad R(x,\xi) = (e^{d_\xi d_y}(P(x,\xi)Q(y,\eta)))_{x=y, \xi=\eta}$$

or equivalently if we have a basis on E for which $x = (x_1, \ldots, x_n)$:

$$(1.2.4) \qquad R(x, \xi) = \sum_{\alpha \in \mathbb{N}^n} \frac{1}{\alpha!} \left(\frac{\partial^\alpha}{\partial \xi^\alpha} P(x, \xi) \frac{\partial^\alpha}{\partial x^\alpha} Q(x, \xi) \right).$$

If $P = \sum_{j \leq m} p_j$, $Q = \sum_{j' \leq m'} q_{j'}$, then $R = \sum_{l \leq m + m'} r_l$, with:

$$(1.2.5) \qquad r_l(x, \xi) = \sum_{l = j + j' - |\alpha|} \frac{1}{\alpha!} \left(\frac{\partial^\alpha}{\partial \xi^\alpha} p_j \frac{\partial^\alpha}{\partial x^\alpha} q_{j'} \right).$$

The order of a section P of $\hat{\mathscr{E}}_X$ on a connected open set $U \subset T^* X$ is the smallest integer $m \in \mathbb{Z} \cup \{-\infty\}$ such that P belongs to $\hat{\mathscr{E}}_X(m)$, and is denoted by $\mathrm{ord}(P)$.

If P is of order at most m, we define $\sigma_m(P)$ as the homogeneous part of order m of $\sum_j p_j$. If P is of order exactly m (on a connected open set) we write $\sigma(P)$ instead of $\sigma_m(P)$, and say that $\sigma(P)$ is the "principal symbol" of P, (with $\sigma(0) = 0$). By contrast we call $\sum_j p_j$ the "total symbol" of P.

The sheaf $\hat{\mathscr{E}}_X$ is filtered by the subsheaves $\hat{\mathscr{E}}_X(j)$, and the associate graded Ring

$$\mathrm{gr}(\hat{\mathscr{E}}_X) = \bigoplus_{j \in \mathbb{Z}} \hat{\mathscr{E}}_X(j) / \hat{\mathscr{E}}_X(j - 1)$$

is the subsheaf $\bigoplus_{j \in \mathbb{Z}} \mathscr{O}_{T^* X}(j)$ of $\mathscr{O}_{T^* X}$. The principal symbol is thus nothing but the natural map from the filtered Ring $\hat{\mathscr{E}}_X$ to its graded Ring.

Remark that σ is multiplicative:

$$(1.2.6) \qquad \sigma(P \circ Q) = \sigma(P) \cdot \sigma(Q),$$

but not additive, and we have, as for differential operators:

$$(1.2.7) \qquad \sigma_{m + m' - 1}([P, Q]) = \{\sigma_m(P), \sigma_{m'}(Q)\}$$

for $P \in \hat{\mathscr{E}}_X(m)$, $Q \in \hat{\mathscr{E}}_X(m')$.
Remark also that $\hat{\mathscr{E}}_X(0)$ is a sub-Ring of $\hat{\mathscr{E}}_X$.

We may send $\pi^{-1}(\mathscr{D}_X)$ into $\hat{\mathscr{E}}_X$ by assigning to a section P of $\mathscr{D}_X$ its total symbol. This is a morphism of Rings, since the composition law on $\hat{\mathscr{E}}_X$ is given by the Leibniz formula and this morphism is clearly injective. Thus we are led to identify a function f of $\mathscr{O}_X$ with the operator of order zero of total symbol f, and the operator $D_{x_i} = \frac{\partial}{\partial x_i}$ with the section of $\hat{\mathscr{E}}_X$ of order one, of total symbol ξ_i. One should take care however, that the total symbol, and, a fortiori the Leibniz formula ist not invariant by a non affine change of coordinates.

By analogy with the differential case we sometime write $P(x, D)$ for a section of $\hat{\mathscr{E}}_X$, but untill we define the action of a bi-holomorphic map (from X to X') on $\mathscr{E}_X$ in § 5, microdifferential operators should be identified with their symbols.

Example 1.2.1. Take $X = \mathbb{C}$, $U = \{(x, \xi) \in T^*X : x \neq 0, \xi \neq 0\}$, and let P and Q be the sections of $\hat{\mathscr{E}}_X$ of total symbol x^{-1} and ξ^{-1} respectively. Then: $(P \circ Q)(x, \xi) = x^{-1}\xi^{-1}$ but $(Q \circ P)(x, \xi) = \sum_{\alpha \geqslant 1} \alpha! x^{-\alpha} \xi^{-\alpha}$.

Remark 1.2.2. Let $p_k(x, \xi)$ be a section of $\mathscr{O}_{T^*X}(k)$, defined in a neighborhood of $(x^0; \xi^0) = (x^0; 1, 0, \ldots, 0) \in T^*\mathbb{C}^n$. We may develop p_k in a Taylor series of ξ'/ξ_1, where $\xi = (\xi_1, \xi')$:

$$p_k(x; \xi) = \xi_1^k p_k(x; 1, \xi'/\xi_1),$$

$$= \sum_{\alpha' \in \mathbb{N}^{n-1}} a_{k, \alpha'}(x)(\xi'/\xi_1)^{\alpha'} \xi_1^k.$$

Now if P is a section of $\hat{\mathscr{E}}_X(m)$ defined in a neighborhood of (x^0, ξ^0) we obtain:

$$P = \sum_{k \leqslant m} p_k,$$

$$= \sum_{\substack{\alpha' \in \mathbb{N}^{n-1}, \alpha_1 \in \mathbb{Z} \\ |\alpha'| + \alpha_1 \leqslant m}} b_\alpha(x) \xi^\alpha.$$

All the p_k are defined and holomorphic in a common neighborhood of $(x^0; \xi^0)$, and thus all the $b_\alpha(x)$ are holomorphic in a common neighborhood of x^0.

Remark 1.2.3. Let $T_X^* X$ be the zero section of T^*X, identified with X. Since the only homogeneous functions defined in a neighborhood of $T_X^* X$ are polynomials, we find that:

(1.2.8) $\hat{\mathscr{E}}_X|_{T_X^* X} \cong \mathscr{D}_X$

For the same reason, if $\dim E > 1$, microdifferential operators globally defined all over $T^*X \setminus T_X^* X$ are differential operators, since a holomorphic function on $T^*X \setminus T_X^* X$ extends to T^*X, ($\operatorname{codim}_{T^*X} T_X^* X \geqslant 2$).

1.3. Microdifferential Operators

Definition 1.3.1. Let U be an open set in T^*X. A microdifferential operator on U is a section P of $\mathscr{E}_X(U)$ which satisfies the following condition:

Let $\sum_{-\infty < j \leqslant m} p_j$ be the total symbol of P. Then for every compact set $K \subset U$ there exists $\varepsilon > 0$ such that:

$$\sum_{j \leqslant 0} |p_j|_K \varepsilon^{-j}/(-j)! < \infty .$$

where we put $|p_j|_K = \sup_K |p_j|$.

We denote by $\mathscr{E}_X(U)$ the set of microdifferential operators on U, by $\mathscr{E}_X$ the sheaf it defines, and let $\mathscr{E}_X(m) = \hat{\mathscr{E}}_X(m) \cap \mathscr{E}_X$.

The main tool in the study of $\mathscr{E}_X$ is the quasi-norm introduced by L. Boutet de Monvel and P. Kree [1]. Assume we have choosen a basis on E. Let P be a section of $\hat{\mathscr{E}}_X(m)(U)$, K a compact set in U, and t a real number, $t > 0$. We set:

$$(1.3.1) \quad N_m(P, K, t) = \sum_{\substack{k \geqslant 0 \\ \alpha \in \mathbb{N}^n, \beta \in \mathbb{N}^n}} \frac{2(2n)^{-k} k!}{(|\alpha| + k)!(|\beta| + k)!} |D_x^\alpha D_\xi^\beta p_{m-k}|_K t^{2k + |\alpha| + |\beta|}.$$

Proposition 1.3.2. *The following assertions are equivalent:*
a) *P belongs to $\mathscr{E}(m)(U)$.*
b) *For all compact set K in U there exists $t > 0$ such that $N_m(P, K, t)$ is finite.*

Recall now that for two series $f(t)$ and $g(t)$, the notation $f(t) \ll g(t)$ means that g is a "majorant series" of f, that is, if $f(t) = \sum_{n \geqslant 0} a_n t^n$, $a_n \geqslant 0$, $g(t) = \sum_{n \geqslant 0} b_n t^n$, $b_n \geqslant 0$, then $a_n \leqslant b_n \ \forall n$.

Proposition 1.3.3. *For P in $\hat{\mathscr{E}}_X(m)(U)$, Q in $\hat{\mathscr{E}}_X(m')(U)$, we have:*

$$N_{m+m'}(P \circ Q, K, t) \ll N_m(P, K, t) \cdot N_{m'}(Q, K, t),$$

and for $m' = m$:

$$N_m(P + Q, K, t) \ll N_m(P, K, t) + N_m(Q, K, t).$$

We shall not repeat here the proofs of Propositions 1.3.2 and 1.3.3, based on Cauchy inequalities and refer to Boutet de Monvel-Kree [1] or Björk [2].

Corollary 1.3.4. *The sheaf $\mathscr{E}_X$ is a sub-Ring of $\hat{\mathscr{E}}_X$.*

Remark that $\mathscr{E}_X(0)$ is itself a sub-Ring of $\hat{\mathscr{E}}_X(0)$.

Proposition 1.3.5. *Let P be a section of $\mathscr{E}_X$ on U and assume $\sigma(P)$ invertible on U (i.e. $\sigma(P)(x, \xi) \neq 0 \ \forall (x, \xi) \in U$). Then there exists a unique section Q of $\mathscr{E}_X$ on U such that:*

$$P \circ Q = Q \circ P = I.$$

Proof. It is enough to prove the existence of Q_1 and Q_2 such that $P \circ Q_1 = I$ and $Q_2 \circ P = I$. We shall only give the proof for the left inverse.

We may assume U connected. Let m be the order of P, and let Q_{-m} be the section of $\hat{\mathscr{E}}_X(U)$ of total symbol $\dfrac{1}{\sigma(P)}$. Then clearly Q_{-m} belongs to $\mathscr{E}_X(U)$, and it is sufficient to find a left inverse to $Q_{-m} \circ P$. But:

$$\sigma(Q_{-m} \circ P) = \sigma(Q_{-m}) \circ \sigma(P) = 1$$

and we may write:

$$Q_{-m} \circ P = I - R, \quad R \in \mathscr{E}_X(-1)(U).$$

Now the serie $\sum\limits_{k \geqslant 0} R^k$ defines an inverse to $I - R$ in $\hat{\mathscr{E}}_X(U)$, and it converges in $\mathscr{E}_X(U)$ since for every compact set K in U there exists $C \geqslant 0$ with $N_0(R, K, t) \leqslant t^2 C$. This completes the proof. $\quad\square$

Exercises to I.1

Ex. 1.1. On $\mathbb{C}^n$ with coordinates $(x_1, \ldots, x_n)$, let $\phi(x) = x_1 + \sum\limits_{j=2}^{n} x_j^2$. Find n linearly independant vector fields $X_1, \ldots, X_n$ such that $[X_i, X_j] = 0 \ \forall i, j$, $X_j(\phi) = 0 \ \forall j > 1$, $X_1(\phi) \equiv 1$.

Ex. 1.2. Let P and Q be two microdifferential operators "with constant coefficients", that is operators whose total symbols only depend on ξ. Check the Leibniz formula for $P \circ Q$.

Ex. 1.3. Let X be a connected and simply connected neighborhood of 0 in $\mathbb{C}$, and define for f holomorphic on X, $D^{-1}(f)$ as the primitive of f on X which is 0 at 0.

Let $P = \sum\limits_{-\infty < j \leqslant m} a_j(x) \xi^j$ be a section of $\hat{\mathscr{E}}_X$ defined for $\xi \neq 0$. Show that if P belongs to $\mathscr{E}_X$, the operator:

$$P(x, D) = \sum\limits_{-\infty < j \leqslant m} a_j(x) D^j$$

acts on $\Gamma(X, \mathscr{O}_\mathbb{C})$.

Show that if P and Q are of order $\leqslant 0$, then:

$$(P \circ Q)(x, D)(f) = P(x, D)(Q(x, D)(f)),$$

(cf. Bony-Schapira [?]).

Ex. 1.4. Extend Proposition 1.3.5. to the case where P is a square matrix of operators of order at most zero, such that the determinant of the symbols of order zero of the entries is invertible.

Ex. 1.5. Let $X_1, \ldots, X_n$ be n vector fields on $\mathbb{C}^n$ which satisfy:

$$[X_i, X_j] = 0, \quad [x_j, X_k] = -\delta_k^j.$$

Calculate the X_j's in the basis $\left(\dfrac{\partial}{\partial x_1}, \ldots, \dfrac{\partial}{\partial x_n} \right)$.

§ 2. Division Theorems

2.1. A Banach Algebra of Operators

For a compact set K in a metric space Y, let K_t denote the set $\{y \in Y; d(y, K) < t\}$, and $\overline{K}_t$ its closure. In this section we shall take $E = \mathbb{C}^n$, endowed with the hermitian norm $|x| = \langle x \cdot \overline{x} \rangle^{1/2}$.

We denote by $\{x_n\}_n$ a sequence, and $x_n \xrightarrow[n]{} 0$ means that x_n tends to 0. Let X be an open set in E and K a compact set in T^*X. We set $\hat{\mathscr{E}}_0(K) = \varinjlim_{U \supset K} \hat{\mathscr{E}}_X(0)(U)$, and similarly for $\mathscr{E}_0(K)$.

Definition 2.1.1. We denote by $X(K, t)$ the vector space of sections P of $\hat{\mathscr{E}}_0(K)$ such that $N_0(P, K, t)$ is finite, and we endow this space with the norm $N_0(\cdot, K, t)$.

Proposition 2.1.2. *For all positive t, t', t'' with $2(t' + t'') < t$ we have a natural continuous injection:*

$$X(K, t) \to X(\overline{K}_{t'}, t'').$$

Proof. We use the inequalities:

$$\alpha! \leqslant |\alpha|! \leqslant \alpha! \, 2^{(n-1)|\alpha|}, \quad \alpha \in \mathbb{N}^n$$

$$|\alpha|! \, |\beta|! \, k! \leqslant \frac{(|\alpha|+k)! \, (|\beta|+k)!}{k!} \leqslant |\alpha|! \, |\beta|! \, k! \, 2^{|\alpha|+|\beta|+2k}$$

Let h be a holomorphic function in a neighborhood of $\overline{K}_{\rho+s}$. We have:

$$\frac{1}{\alpha! \beta!} (|D_x^\alpha D_\xi^\beta h|_{K_\rho}) (s/\sqrt{2})^{|\alpha|+|\beta|} \leqslant |h|_{\overline{K}_{\rho+s}} \leqslant \sum_{\alpha, \beta} \frac{1}{\alpha! \beta!} (|D_x^\alpha D_\xi^\beta h|)_K (\rho+s)^{|\alpha|+|\beta|}$$

We obtain for a constant $C > 0$:

$$N_0(P, K_\rho, s/2) \leqslant C \sum_{\alpha, \beta, k} \frac{2(2n)^{-k}}{k! \, \alpha! \, \beta!} (\rho+s)^{|\alpha|+|\beta|} s^{2k} (|D_x^\alpha D_\xi^\beta p_{-k}|_K)$$

$$\leqslant C N_0(P, K, 2^n (\rho+s)). \quad \square$$

Then Propositions 1.3.2 and 1.3.3 imply that $X(K, t)$ is a normed algebra and

$$(2.1.1) \qquad \mathscr{E}_0(K) = \varinjlim_{t > 0} X(K, t).$$

Proposition 2.1.3. *The space $X(K, t)$ is a Banach algebra.*

Proof. Let $\{P^n\}_n$ be a Cauchy sequence in $X(K, t)$, where $P^n = \sum_n p_j^n$. There

exists $t' > 0$ such that each p_j^n belongs to $\mathcal{O}_{T^*X}(j)(\overline{K}_{t'})$, and the sequences $\{p_j^n\}_n$ are Cauchy sequences in these spaces.

Let $p_j = \lim_n p_j^n$, $P = \sum_j p_j$.

Since for all α, β

$$|D_x^\alpha D_\xi^\beta (p_j^n - p_j)|_K \xrightarrow[n]{} 0,$$

we get:

$$N_0(P^n - P^m, K, t) \leqslant \varepsilon \Rightarrow N_0(P^n - P, K, t) \leqslant \varepsilon,$$

thus P^n converges to P in $X(K, t)$.   □

To a Banach space A and to a strictly positive number ρ we associate the space $A(\rho)$ of all formal series $\{g = \sum_{j \geqslant 0} a_j \tau^j; a_j \in A\}$ such that the quasi-norm:

(2.1.2)
$$\|g\|_\rho = \sum_{j \geqslant 0} \|a_j\| \rho^j$$

is finite. The space $A(\rho)$ endowed with the norm $\| \cdot \|_\rho$ is clearly a Banach space.

Let K be a compact set of T^*X contained in the hyperplane $\{(x, \xi); x_1 = x_1 = 0\}$, where $(x_1, \ldots, x_n)$ is a system of coordinates on X. We set:

(2.1.3)
$${}^1X(K, t) = \{P \in X(K, t); [P, D_1] = 0\}.$$

Remark that P commutes with D_1 if and only if the total symbol $\sum_j p_j$ of P does not depend on the x_1-variable.

We endow the closed sub-algebra ${}^1X(K, t)$ with the induced norm $N_0(\cdot, K, t)$.

Proposition 2.1.4. *Let B_ε be the open ball $\{x_1 \in \mathbb{C}; |x_1| < \varepsilon\}$, $\overline{B}_\varepsilon$ its closure. Then:*

$$\mathcal{E}_0(\overline{B}_\varepsilon \times K) = \varinjlim_{t > 0} {}^1X(K, t)(\varepsilon + t)$$

Proof. Let P belong to $\mathcal{E}_0(\overline{B}_\varepsilon \times K)$. Then for some $\varepsilon'' > \varepsilon' > \varepsilon > 0$, $t > 0$, $N_0(P, \overline{B}_{\varepsilon''} \times K, 2t)$ is finite. Thus:

$$\sum_{\alpha, \beta, k} \frac{2(2n)^{-k} t^k}{k! |\alpha|! |\beta|!} |D_x^\alpha D_\xi^\beta p_{-k}|_{\overline{B}_{\varepsilon''} \times K} t^{|\alpha| + |\beta|} < \infty$$

By the Cauchy inequalities, we know that if h is a holomorphic function in a neighborhood of $\overline{B}_{\varepsilon''} \times K$:

$$\sum_{j \geqslant 0} \frac{\varepsilon'^j}{j!} |D_{x_1}^j h|_{\{0\} \times K} \leqslant \frac{\varepsilon''}{\varepsilon'' - \varepsilon'} |h|_{\overline{B}_{\varepsilon''} \times K}.$$

To $P = \sum_k p_k$ we associate the sequence $\{P^j\}_j$, where $P^j = \sum_k p_{-k}^j$, p_{-k}^j

$= \frac{1}{j!} (D_{x_1}^j p_{-k})|_{\{0\} \times K}.$

Thus if we take $\alpha_1 = 0$ in the preceding inequality we get:

$$\sum_j N_0(P^j, \{0\} \times K, 2^n t)\varepsilon^{tj} < \infty .$$

Let us prove the converse inclusion.

Let $\{P^j\}_j$ be a sequence in ${}^1X(K, t)$ such that $\sum_j N_0(P^j, \{0\} \times K, t)(\varepsilon + t_1)^j$ is finite for some $t_1 > 0$. We associate $P = \sum_j x_1^j P^j$, and remark that if h is holomorphic in a neighborhood of $\overline{B}_\varepsilon \times K$ we have:

$$\sum_{\alpha, \beta} \frac{1}{\alpha! \beta!} (|D_x^\alpha D_\xi^\beta h|_{\overline{B}_\varepsilon \times K}) (t/2)^{|\alpha|+|\beta|} \leqslant 2 \sum_{\alpha, \beta} \frac{1}{\alpha! \beta!} (|D_x^\alpha D_\xi^\beta h|_{\{0\} \times K}) (\varepsilon + t)^{|\alpha|+|\beta|} .$$

Thus:

$$\sum_{\alpha, \beta, k} \frac{2(2n)^{-k}}{\alpha! \beta! k!} (|D_x^\alpha D_\xi^\beta p_{-k}|_{\overline{B}_\varepsilon \times K}) (t/2)^{|\alpha|+|\beta|+2k}$$

$$\leqslant 2 \sum_{\alpha_1} \frac{(\varepsilon + t)^{\alpha_1}}{\alpha_1!} \sum_{\alpha', \beta, k} \frac{2(2n)^{-k}}{\alpha'! \beta! k!} (|D_{x'}^{\alpha'} D_\xi^\beta D_{x_1}^{\alpha_1} p_{-k}|_{\{0\} \times K}) (\varepsilon + t)^{|\alpha'|+|\beta|+2k}$$

$$\leqslant 2 \sum_j (\varepsilon + t)^j N_0(P^j, \{0\} \times K, 2^n (\varepsilon + t))$$

which completes the proof. $\square$

Remark that Proposition 2.1.4 asserts that we may regard a microdifferential operator on $\overline{B}_\varepsilon \times K$ as a holomorphic function defined for $|x_1| \leqslant \varepsilon$ with value in the space of microdifferential operators not depending on x_1. If K is a compact set contained in $\{(x, \xi); \xi_1 = \xi_1^0, \xi_n \neq 0\}$, the same construction holds with x_1 replaced by ξ_1/ξ_n. In that case we set:

$$(2.1.4) \qquad X^1(K, t) = \{P \in X(K, t); [P, x_1] = 0\}$$

That is $X^1(K, t)$ is the set of sections of $X(K, t)$ whose total symbol does not depend on the ξ_1-variable.

2.2. The Späth and Weierstrass Theorems

To an operator P of $\mathscr{E}_X$ one associates the $\mathbb{C}$-linear endomorphism ad_P of $\mathscr{E}_X$:

$$(2.2.1) \qquad \mathrm{ad}_P(Q) = [P, Q] .$$

One checks immediately that:

$$(2.2.2) \qquad \begin{cases} \mathrm{ad}_{x_1}^k(Q) = 0 \Leftrightarrow \exists Q_j, \quad 0 \leqslant j < k, \text{ such that} \\ \mathrm{ad}_{x_1}(Q_j) = 0, \quad Q = \sum_{j=0}^{k-1} Q_j D_1^j , \end{cases}$$

and similarly with x_1 and D_1 interchanged. One can also write D_1 on the left (for other Q_j's).

Theorem 2.2.1 (Späth). *Let P be a microdifferential operator defined in a neighborhood of $(x^0, \xi^0) \in T^* X$. Assume that $\dfrac{\partial^j}{\partial \xi_1^j}(\sigma(P))$ is zero at (x_0, ξ_0) for $0 \leqslant j < p$ and different from zero for $j = p$. Then for any section Q of $\mathscr{E}_X$ defined in a neighborhood of (x^0, ξ^0) there exist unique S and R such that:*

$$\begin{cases} Q = SP + R, \\ \operatorname{ord}(R) \leqslant \operatorname{ord}(Q), \quad \operatorname{ad}_{x_1}^p(R) = 0. \end{cases}$$

Other Formulations of the Preceding Theorem and Remarks

With the same hypothesis, there exist $\tilde{S}$ and $\tilde{R}$, with $\operatorname{ord}(\tilde{R}) \leqslant \operatorname{ord}(Q)$, $\operatorname{ad}_{x_1}^p(\tilde{R}) = 0$, and $Q = P\tilde{S} + \tilde{R}$.

We may also interchange x_1 and D_1 in Theorem 2.2.1. Thus if $\sigma(P)$ has a zero of order p at (x_0, ξ_0) in the x_1-variable, we may write uniquely:

$$Q = S'P + R'$$

with $\operatorname{ord}(R') \leqslant \operatorname{ord}(Q)$, $\operatorname{ad}_{D_1}^p(R') = 0$.

It is useful to notice that if P and Q commute with an operator A not depending on D_1, then S and R will commute with A since:

$$0 = [A, Q] = [A, S]P + [A, R],$$

$$\operatorname{ad}_{x_1}^p([A, R]) = [A, \operatorname{ad}_{x_1}^p(R)] = 0.$$

(use the unicity of the division).

Moreover if P and Q do not depend on x_1, the same will hold for R and S since $\operatorname{ad}_{x_1}^p(R) = 0$ implies $\operatorname{ad}_{x_1}^p([D_1, R]) = 0$.

Proof of the Theorem. a) Uniqueness follows from the uniqueness of the division for symbols. In fact assume $0 = SP + R$, $\operatorname{ord}(R) \leqslant \operatorname{ord}(SP)$, and let $m = \operatorname{ord}(SP)$.

$$0 = \sigma(S)\sigma(P) + \sigma_m(R)$$

and $\sigma_m(R)$ is a polynomial with degree less than p in the ξ_1 variable. Thus $\sigma(S) = \sigma_m(R) = 0$ (cf. Gunning-Rossi [1]).

b) We may assume $n > 1$, $\xi^0 = (\xi_1^0, \ldots, \xi_n^0)$ with $\xi_n^0 \neq 0$. In fact let us denote by $(x, t; \xi, \tau)$ the coordinates on $T^*(X \times \mathbb{C})$, and let us consider P and Q as operators in $\mathscr{E}_{X \times \mathbb{C}}$ defined in a neighborhood of $(x^0, 0; \xi^0, 1)$, commuting with t and D_t. Then S and R will commute with t and D_t, thus will be sections of $\mathscr{E}_X$ in a neighborhood of (x^0, ξ^0).

Let m be the order of P, and let D_n^{-m} be the microdifferential operator of total symbol ξ_n^{-m}. Replacing P by $D_n^{-m} P$ we may assume P of order zero. Likewise we may assume Q of order zero. Now we shall use a division procedure for Banach algebras following an idea by C. Houzel.

Let A be a Banach space. Consider the linear map u:

(2.2.3)
$$\begin{cases} A(\rho) \times A^p \to A(\rho), \\ (g, h) \mapsto g\tau^p + \sum_{j=0}^{p-1} h_j \tau^j, \end{cases}$$

where $h = (h_0, \ldots, h_{p-1})$.

Then clearly u is an isomorphism and $\|u^{-1}\| \leqslant \rho^{-p}$. Thus $u + v$ will remain invertible as soon as $\|v\| < \rho^p$.

Assume now that we have a family of Banach algebras with unit, $(A_t)_{0 < t \leqslant t_0}$, with continuous injections of norm 1 of A_t in $A_{t'}$ for $t' \leqslant t$.

Lemma 2.2.2. *Let $f(\tau)$ belong to $A_{t_0}(\rho_0)$, and assume that for $0 \leqslant j < p$,*
$$\left\| \frac{\partial^j}{\partial \tau^j} f(0) \right\|_t$$
tends to zero for $t \xrightarrow{>} 0$, and $\dfrac{\partial^p}{\partial \tau^p} f(0)$ is invertible in A_{t_0}. Then there exists $\rho_1 > 0$, such that for each $\rho < \rho_1$, $\rho > 0$ there exists $t' > 0$ such that for $0 < t \leqslant t'$, the map:
$$\begin{cases} A_t(\rho) \times A_t^p \to A_t(\rho), \\ (g; (h_0, h_1, \ldots, h_{p-1})) \mapsto gf + \sum_{j=0}^{p-1} h_j \tau^j \end{cases}$$
is an isomorphism.

Proof of the Lemma. We write:
$$f(\tau) = \sum_{j=0}^{p-1} c_j \tau^j + c_p \tau^p + r(\tau) \tau^{p+1}.$$

Since c_p is invertible in A_{t_0}, we may assume $c_p = 1$. Then the map defined in the lemma is the sum of two map u and v with $u(g, h) = g\tau^p + \sum_{j=0}^{p-1} h_j \tau^j$, $v(g, h) = \sum_{j=0}^{p-1} g c_j \tau^j + g r(\tau) \tau^{p+1}$. The norm of v in $L(A_t(\rho), A_t(\rho))$ is bounded by ρ^p for ρ small enough and $t \leqslant \rho$ since $\|r(\tau) \tau^{p+1}\|_{A_t(\rho)} \leqslant C \cdot \rho^{p+1}$ for a constant C depending neither on t nor on ρ, and $\|c_j\|_t$ tends to zero when $t \to 0$.   □

End of the Proof of Theorem 2.2.1

Let L_t be the compact set:
$$L_t = \{(x, \xi); \xi_1 = \xi_1^0, |(x, \xi) - (x^0, \xi^0)| \leqslant t\}$$
and let $^1X(L_t, t) = A_t$ be the Banach algebra constructed in § 2.1. The section P defines an element $f(\tau) \in A_{t_0}(\rho_0)$ for some $t_0 > 0$, $\rho_0 > 0$ such $\dfrac{\partial^p f}{\partial \tau^p}(0)$ is inver-

tible in A_{t_0}, if we have choosen t_0 small enough. Moreover $\left\|\dfrac{\partial^j f}{\partial \tau^j}(0)\right\|_{A_t}$ tends to zero when $t \xrightarrow{\geq} 0$ for $j<p$, since $\dfrac{\partial^j}{\partial \xi_1^j}\sigma(P)(x^0,\xi^0)=0$, and P is of order zero. $\quad\square$

Corollary 2.2.3 (Weierstrass). *With the same hypothesis on P, there exists an invertible operator A and an operator W such that:*

$$\begin{cases} P=AW, \quad W=D_1^p+\sum_{j=0}^{p-1}R_j D_1^j, \\ [x_1, R_j]=0, \quad \operatorname{ord}(W)\leqslant p. \end{cases}$$

There is a similar statement with x_1 and D_1 interchanged. There also exists a similar statement where we write $P=\tilde{W}\tilde{E}$. We shall say that W is a "Weierstrass polynomial" (of degree p, with respect to D_1).

Proof. We apply the division theorem with $Q=D_1^p$. We get:

$$\begin{cases} D_1^p=EP+R, \\ \operatorname{ord}(R)\leqslant p, \quad R=\sum_{j=0}^{p-1}R_j D_1^j, \quad \operatorname{ad}_{x_1}(R_j)=0. \end{cases}$$

One has to check that E is invertible. But:

$$\sigma(E)\sigma(P)=\sigma(D_1^p-R),$$

$$=\xi_1^p-\sum_{j=0}^{p-1}\sigma_{p-j}(R_j)\xi_1^j.$$

Since $\dfrac{\partial^j}{\partial\xi_1^j}\sigma(P)(x^0,\xi^0)=0, j<p$, we get:

$$\sigma(E)(x^0,\xi^0)\left(\dfrac{\partial^p}{\partial\xi_1^p}\sigma(P)(x^0,\xi^0)\right)=p!$$

thus $\sigma(E)\neq 0$ in a neighborhood of (x^0,ξ^0) and we may apply Proposition 1.3.5. $\quad\square$

Exercises to I.2

Ex. 2.1. Prove directly the division theorem for differential operators (Hint: proceed by induction on $\operatorname{ord}(P)$, and use the division theorem for principal symbols).

Ex. 2.2. Prove the division theorem for formal microdifferential operators (with the same method as for Ex. 2.1. Note however that, for $P=\sum_j p_j$ to define a section of $\hat{\mathscr{E}}_X$, we need all the p_j's to be holomorphic in a common neighborhood of (x^0, ξ^0)).

Ex. 2.3. Let $(x^0, \xi^0) = (0; 0, \ldots, 0,1) \in T^*\mathbb{C}^n$. Prove that every microdifferential operator P in a neighborhood of (x^0, ξ^0) may be written in a unique way:

$$\begin{cases} P = x_1 \circ Q \circ D_1 + R, \\ \mathrm{ord}(R) \leqslant \mathrm{ord}(P), \quad [x_1, [D_1, R]] = 0. \end{cases}$$

§ 3. Refined Microdifferential Cauchy-Kowalewski Theorem

3.1. Statement of the Theorem

Let $x = (x_1, \ldots, x_n)$ be a system of linear coordinates on E, (x, ξ) the associate coordinates on T^*E. We set:

$$x = (x_1, x', x''), \quad \text{where } x' = (x_2, \ldots, x_p)$$

and similarly for $\xi = (\xi_1, \xi', \xi'')$. Let X be an open set in E and let Y be the hypersurface $Y = \{x \in X; \ x_1 = 0\}$. Let U be an open set in T^*X, V the manifold:

$$V = \{(x, \xi) \in U; \ \xi_1 = \ldots = \xi_p = 0\}$$

Let ρ be the natural projection $Y \underset{X}{\times} T^*X \to T^*Y$, $\rho(0, x', x''; \xi_1, \xi', \xi'')$ $= (x', x''; \xi', \xi'')$, and let $V' = \rho(V)$, the manifold of T^*Y with equations $\{\xi' = 0\}$.

Following Kashiwara-Oshima [1] we introduce the sub-Algebra $\mathscr{E}_V$ of $(\mathscr{E}_X)|_V$ generated over $(\mathscr{E}_X(0))|_V$ by $D_1, \ldots, D_p$. A section P of $\mathscr{E}_V$ may thus be written:

$$P = \sum_{0 \leqslant \alpha \leqslant m} A_\alpha D^\alpha,$$

where $\alpha = (\alpha_1, \ldots, \alpha_p) \in \mathbb{N}^p$, $A_\alpha \in (\mathscr{E}_X(0))|_V$.

We introduce the left $\mathscr{E}_V$-module:

(3.1.1) $$\mathscr{L}_V = \mathscr{E}_V / \mathscr{E}_V (D_1, \ldots, D_p),$$

where $\mathscr{E}_V(D_1, \ldots, D_p)$ is the left Ideal $\sum_{i=1}^{p} \mathscr{E}_V D_i$. Then $\mathscr{L}_V$ is clearly isomorphic, as an $\mathscr{E}_X(0)$-module, to

$$\mathscr{E}_X(0)\Big/\Big(\sum_{i=1}^{p}\mathscr{E}_X(-1)D_i\Big).$$

A section f of $\mathscr{L}_V$ may be considered as a microdifferential operator of order zero, not depending on $D_1,\dots,D_p$ (i.e.: commuting with $x_1,\dots,x_p$), the action of D_j $(1\leqslant j\leqslant p)$ on P being described by:

$$D_j(f)=D_j\circ f\mod\Big(\sum_{i=1}^{p}\mathscr{E}_X(-1)D_i\Big).$$

If we write $f=f(x,D'')$ to express that we may represent f by a total symbol $\sum_j f_j(x,\xi'')$ independant of ξ_1,ξ', then:

$$D_i(f)=\frac{\partial}{\partial x_i}f\quad 1\leqslant i\leqslant p,$$

which means that $\sum_j\dfrac{\partial}{\partial x_i}(f_j(x,\xi''))$ is the total symbol of $D_i(f)$.

As a particular case, take $p=n$. We get:

(3.1.2) $\mathscr{O}_X\cong\mathscr{D}_X/\mathscr{D}_X(D_1,\dots,D_n)$

and the action of $\mathscr{D}_X$ on $\mathscr{O}_X$ is given by:

$$Q(f)=Q\circ f\mod\Big(\sum_{i=1}^{n}\mathscr{D}_X D_i\Big).$$

We shall also use the sub-Algebra $\mathscr{E}_{V'}$ of $\mathscr{E}_Y$ and the left $\mathscr{E}_{V'}$-module $\mathscr{L}_{V'}$, constructed by the same method. There exists a natural "restriction" morphism from $(\mathscr{L}_V)|_{Y\times_X T^*X}$ to $\rho^{-1}\mathscr{L}_{V'}$. To a section $f(x,D'')$ of $\mathscr{L}_V$ we just associate the section of $\mathscr{L}_{V'}$, $f(0,x',x'',D'')$, i.e. the class of $f(\mathrm{mod}\,x_1\mathscr{L}_V)$, by means of the isomorphism (of $\mathscr{E}_{V'}$-module):

$$\mathscr{L}_{V'}\cong\rho_*(\mathscr{L}_V/x_1\mathscr{L}_V)$$

(We shall make a systematic study of this problem in Chapters II and III).

In this whole section, P denotes a section of $\mathscr{E}_V$ of order $m\geqslant0$ defined in a neighborhood of V. We assume that:

(3.1.3) $P=\displaystyle\sum_{0\leqslant|\alpha|\leqslant m}A_\alpha(x;D',D'')D_1^{\alpha_1}D_{x'}^{\alpha'},$

where

$\alpha=(\alpha_1,\alpha')=(\alpha_1,\dots,\alpha_p)\in\mathbb{N}^p,\quad A_\alpha\in\mathscr{E}_X(0),\quad[A_\alpha,x_1]=0,\quad A_{(m,0,\dots,0)}\equiv1.$

Let $(x^0,\xi^0)=(0,x_2^0,\dots,x_n^0;\ 0,0,\xi_{p+1}^0,\dots,\xi_n^0)=(0,x^{0'},x^{0''};\ 0,0,\xi^{0''})$ belonging to $V\cap(Y\underset{X}{\times}T^*X)$. Let:

$$K(r)=\{(x;\,0,\xi'')\in V;\ |x_j^0-x_j|\leqslant r,j=1,\dots,p,x^{0''}=x'',\xi^{0''}=\xi''\}$$

and let $K'(r)=\rho(K(r)\cap(Y\underset{X}{\times}T^*X))$ (that is, the compact in V' given by the same equations, with $x_1=0$).

Theorem 3.1.1. *There exists $\rho_0>0$, $\delta>0$, $\lambda_0>0$ such that for all $r<\rho_0$, $g\in\mathscr{S}_V(K(r))$, $(h)=(h_0,\ldots,h_{m-1})\in\mathscr{S}_V^m(K'(r))$, there exists a unique solution $f\in\mathscr{S}_V(K(\delta r))$ of the Cauchy problem:*

$$(3.1.4)\qquad Pf=g,\quad f|_Y=h_0,\ldots,\left.\left(\frac{\partial^{m-1}}{\partial x_1^{m-1}}f\right)\right|_Y=h_{m-1}.$$

Moreover for $|(\tilde{x};\,0,\tilde{\xi}'')|\leqslant\lambda_0$, we may replace $K(r)$ by $(\tilde{x};\,0,\tilde{\xi}'')+K(r)$ (thus Y by $\tilde{x}_1+Y$, $K'(r)$ by $(\tilde{x};\,0,\tilde{\xi}'')+K'(r)$).

Remark 3.1.2. By taking $V=T_X^*X$, the zero section of T^*X, we recover the classical refined Cauchy-Kowalewski theorem, (cf. Leray [1]).

In order to prove this theorem in § 3.3, we recall an abstract version of it.

3.2. The Abstract Cauchy-Kowalewski Theorem in Scales of Banach Spaces

Let $(X_s)_{s\in[0,\,1]}$ be a scale of Banach spaces. That is, we have continuous injections with norms $\leqslant 1$ from X_s to $X_{s'}$ for $s'\leqslant s$.

Let $A(x_1)$ be a holomorphic map from $\overline{D}_{\varepsilon_0}=\{x_1\in\mathbb{C};\,|x_1|\leqslant\varepsilon_0\}$ with values in $L(X_s,X_{s'})$ for all $s'<s$ such that:

$$(3.2.1)\qquad\qquad \|A(x_1)\|_{s,\,s'}\leqslant\frac{C}{s-s'}$$

for a constant C not depending on s,s'.

Theorem 3.2.1. *Let $v(x_1)$ be a holomorphic function on $\overline{D}_{\varepsilon_0}$ with values in X_1. Then for all $s<1$ there exists a holomorphic function $u(x_1)$ defined on the set $\{x_1\in\mathbb{C};\,|x_1|<\delta_0(1-s)\}$ with values in X_s which is a solution of:*

$$(3.2.2)\qquad\begin{cases}\dfrac{du}{dx_1}=A(x_1)u(x_1)+v(x_1),\\[2mm]u(0)=0,\end{cases}$$

where $\delta_0=\inf(\varepsilon_0,(Ce)^{-1})$.
   Moreover for all $s<1$ the solution is unique.

Proof. We shall follow Trèves [1].
   We define by induction:

$$u_k(x_1)=\int_0^{x_1}v(t)\,dt+\int_0^{x_1}A(t)u_{k-1}(t)\,dt$$

with $u_{-1}=0$.

If u_{k-1} is holomorphic for $|x_1| < \varepsilon_0$ with values in X_s, u_{k+1} will be holomorphic on the same open set with values in $X_{s'}$, for all $s' < s$. Thus all u_k will be defined for $|x_1| < \varepsilon_0$, with values in $\bigcup_{s<1} X_s$. Set $w_0 = u_0$, $w_k = u_k - u_{k-1}$

$$w_{k+1}(x_1) = \int_0^{x_1} A(t) w_k(t) dt, \quad k \geqslant 0.$$

Lemma 3.2.2. *We have:*

$$\|w_k(x_1)\|_s \leqslant M(x_1) \left(\frac{Ce|x_1|}{1-s} \right)^k$$

for $|x_1| < \varepsilon_0$, with $M(x_1) = \int_0^{x_1} \|v(t)\|_1 dt$.

Proof of the Lemma. The assertion is true for $k = 0$.

Assume it is true up to order k.

Take s, s' with $0 \leqslant s' < s < 1$

$$\|w_{k+1}(x_1)\|_{s'} \leqslant \frac{C}{s-s'} \int_0^{x_1} \|w_k(t)\|_s dt \leqslant M(x_1) \frac{C}{s-s'} \frac{(Ce)^k}{(1-s)^k} \frac{|x_1|^{k+1}}{(k+1)}.$$

Take $s = s' + (1-s')/k + 1$

$$1 - s = \frac{k}{k+1} (1-s') \quad \|w_{k+1}(x_1)\|_{s'} \leqslant M(x_1) \frac{(C|x_1|)^{k+1}}{(1-s')^{k+1}} e^k \left(1 + \frac{1}{k} \right)^k.$$

and $\left(1 + \frac{1}{k} \right)^k < e$. □

End of the Proof of Theorem 3.2.1

Let M be a constant, with $M(x_1) \leqslant M$. The series $\sum_k w_k$ converges in X_s uniformly on every compact set in $\{x_1 \in \mathbb{C}; |x_1| < \varepsilon_0\}$ such that $|x_1| < (Ce)^{-1}(1-s)$ on this compact set.

Let $u(x_1)$ be the sum $\sum_k w_k$. Then $u(x_1) = \lim_k u_k(x_1)$, thus:

$$u(x_1) = \int_0^{x_1} v(t) dt + \int_0^{x_1} A(t) u(t) dt,$$

which proves the existence of the solution.

To prove the uniqueness, assume $\dfrac{du}{dx_1} = A(x_1) u(x_1)$, and let us define w_k as above by induction, with $w_0(x_1) = u(x_1)$. Then $w_k(x_1) = w_{k-1}(x_1)$, and since the series $\sum_k w_k$ is convergent, we get $w_k = 0$. □

3.3. Proof of Theorem 3.1.1

a) It is enough to solve the Cauchy problem (3.1.4) where $(h)=0$. In fact let $\tilde{f}$ be a solution of:

$$\begin{cases} P\tilde{f}=g-P\left(\sum_{j=0}^{m-1} x_1^j h_j\right), \\ \gamma_Y(\tilde{f})=0, \end{cases}$$

where $\gamma_Y(\tilde{f})$ denotes the m first traces of $\tilde{f}$ on Y. Then $f=\tilde{f}+\sum_{j=0}^{m-1} x_1^j h_j$ will solve (3.1.4).

b) Instead of solving (3.1.4), we first solve a Cauchy problem for square matrices of rank N of first order microdifferential operators.

Let $\tilde{A}_j$ $(j=2,\ldots,p)$ and B be $(N \times N)$-matrices whose entries are in $\mathscr{E}_X(0)$ and do not depend on D_1. Let I be the unit $(N \times N)$-matrix. Let:

$$\tilde{P}=ID_1+\sum_{j=2}^{p} \tilde{A}_j D_j+\tilde{B}.$$

We shall solve for $\vec{g} \in \mathscr{S}_V^N(K(r))$

(3.3.1) $\tilde{P}\vec{f}=\vec{g}, \quad \vec{f}|_Y=0$

in $\mathscr{S}_V^N(K(\delta r))$, and prove the uniqueness of $\vec{f}$.

We denote by X_s' the Banach space of sections f of $\mathscr{E}_0^N(K(s))$ not depending on $(x_1, D_1, \ldots, D_p)$, such that $N_0(f, K(s), t)$ is finite.

Since $K(s)$ is contained in V, the norm $N_0(\cdot, K(s), t)$ on X_s' coincides with the quotient norm:

(3.3.2) $N_0^1(f, K(s), t)=\inf N_0\left(f+\sum_{j=1}^{p} Q_j D_j, K(s), t\right),$

where the infimum is taken for $Q_j \in \mathscr{E}_X(-1)(K(s))$.

Then by Proposition 2.1.4, if $\tilde{C}$ is one of the matrices $\tilde{A}_j$ or $\tilde{B}$, $\tilde{C}$ defines a holomorphic function in the x_1-variable, for $|x_1| \leqslant \varepsilon_0$ (for some $\varepsilon_0 > 0$), with values in $L(X_s', X_s')$, $0 < t \leqslant t_0$, $s \leqslant s_0$, and the norm of this operator is bounded by some fixed constant C_0, not depending on x_1, t, s:

$$\|\tilde{C}(x_1)\|_{L(X_s', X_s')} \leqslant C_0.$$

Moreover it follows from the Cauchy inequalities that the operators D_j, $2 \leqslant j \leqslant p$, define continuous operators from X_s' to $X_{s'}'$, for $s' < s$, with norm:

$$\|D_j\|_{L(X_s', X_{s'}')} \leqslant \frac{C_1}{s-s'}$$

for a constant C_1 not depending on s, s', t.

Thus we may apply Theorem 3.2.1. Since $\vec{g}$ belongs to $\mathscr{S}_V^N(K(r))$, it defines an element in $X_r'(r+t)$, for some $t > 0$, and we find $\vec{f}$ in $X_r'(\delta_0(1-r-t))$.

Then replacing δ_0 by $\delta = C\delta_0$, $C \ll 1$, we have $\tilde{f} \in X^t_{\delta r}$ $(\delta r + t)$, for $t \leqslant \delta r$. Taking the inductive limit for $t > 0$, we get the result by Proposition 2.1.4.

c) We shall not repeat the proof, (of a purely algebraic nature), that the Cauchy problem (3.2.2) is solved by (3.3.1), and we refer to Trèves [1].

We just mention that we have to take $N = \#\{\alpha \in \mathbb{N}^p; |\alpha| < m\}$, and put $\tilde{f} = (D^\alpha f)_{|\alpha| < m}$. □

Let us give an example.

Example 3.3.1 Take $P = D_1^2 + D_2^2 + D_3^2 + a_0(x)$, on $\mathbb{C}^3$.

We set $\tilde{f} = (f_0, f_2, f_3, f_1)$. To a holomorphic function f we associate $(f, D_2 f, D_3 f, D_1 f)$.

Then the Cauchy problem:

$$Pf = g, \qquad f|_{x_1=0} = 0, \qquad (D_1 f)|_{x_1=0} = 0$$

is equivalent to the problem:

$$\begin{pmatrix} D_1 & 0 & 0 & -1 \\ 0 & D_1 & 0 & -D_2 \\ 0 & 0 & D_1 & -D_3 \\ a_0 & D_2 & D_3 & D_1 \end{pmatrix} (\vec{f}) = \begin{pmatrix} 0 \\ 0 \\ 0 \\ g \end{pmatrix}, \qquad (\vec{f})(0, x') = 0,$$

which may be written:

$$(I D_1 + \tilde{A}_2 D_2 + \tilde{A}_3 D_3 + \tilde{B}_0(x)) \vec{f} = \vec{g}, \qquad \vec{f}(0, x') = 0$$

for some matrices $\tilde{A}_2$, $\tilde{A}_3$, $\tilde{B}_0$.

Remark 3.3.2. We have in fact proved a generalization of Theorem 3.1.1 in which P is replaced by a square matrix.

Exercises to I.3

Ex. 3.1. Let (x_1, x') be the coordinates on $X = \mathbb{C}^n$, Y the hypersurface $\{x_1 = 0\}$. Let $\mathcal{O}_{Y|X}$ be the sheaf on Y of formal series with coefficients in $\mathcal{O}_Y$:

$$\mathcal{O}_{Y|X} \ni f(x_1, x') \Leftrightarrow f = \sum_j a_j(x') x_1^j, \qquad a_j \in \mathcal{O}_Y.$$

Prove that the Cauchy problem $(D_1 + D_2^2) f = g$, $f(0, x') = h(x')$ has a unique solution f in $\mathcal{O}_{Y|X}$ for a given (g, h) in $(\mathcal{O}_{Y|X}, \mathcal{O}_Y)$.

Ex. 3.2. Let $P = \sum_{j=1}^{n-1} D_j^2 + (x_n D_n)^2$. Solve the Cauchy problem $Pf = 0$, $f|_Y = 0$, $\left(\dfrac{\partial}{\partial x_1} f\right)\Big|_Y = 1/x_n$, (here $Y = \{x; x_1 = 0\}$), in the space of meromorphic functions with poles on $\{x_n = 0\}$.

Ex. 3.3. Let $a(x, D')$ be a microdifferential operator of order 0, defined in a neighborhood of dx_n, and not depending on D_1. Find an invertible operator $E(x, D')$ of order 0 such that $E^{-1} \circ (D_1 + a) \circ E = D_1$. (Hint: solve $\left(\dfrac{\partial}{\partial x_1} + a \right) E = 0$, $E|_{x_1 = 0} = I$, or refer to §.6.).

§ 4. Microdifferential Modules Associated to a Submanifold

4.1. The Sheaf $\mathscr{C}_{Z|X}$

Let X be an open set in E. We denote by $\Omega_X^{(p)}$ the sheaf of holomorphic p-forms on X and set $\Omega_X = \Omega_X^{\dim X}$. Recall that Ω_X is a locally free $\mathcal{O}_X$-module of rank one (here Ω_X is even free since X is affine). A volume element is by definition a generator of Ω_X. We also define:

$$(4.1.1) \qquad \Omega_X^{\otimes -1} = \mathscr{Hom}_{\mathcal{O}_X}(\Omega_X, \mathcal{O}_X).$$

If dx is a volume element on Ω_X then $dx^{\otimes -1}$ is the unique section of $\Omega_X^{\otimes -1}$ such that $dx^{\otimes -1} \otimes dx = 1$.

Let Z be a closed (complex analytic) submanifold of X, of codimension p. We denote by $\mathscr{I}_Z$ the subsheaf of $\mathcal{O}_X$ of sections which are zero on Z, and say that $\mathscr{I}_Z$ is the defining Ideal of Z. Locally on Z there exist p sections $f_1, \ldots, f_p$ of $\mathcal{O}_X$ which generate $\mathscr{I}_Z$ and such that:

$$df_1 \wedge \ldots \wedge df_p \neq 0 \text{ on } Z.$$

We shall say that such an $(f) = (f_1, \ldots, f_p)$ is a defining system for Z. Let $\mathscr{H}_{(f)}$ be the sheaf on Z of finite sums:

$$(4.1.2) \qquad \mathscr{H}_{(f)} = \left\{ \sum_{0 < |\alpha| \leqslant m} a_\alpha f^{-\alpha}; \ a_\alpha \in (\mathcal{O}_X)|_Z, \ \alpha \in \mathbb{N}^p \right\},$$

and let $\mathscr{L}_{(f)}$ be the subsheaf of $\mathscr{H}_{(f)}$ generated by sections which are sums of $a_\alpha f^{-\alpha}$, with $\alpha = (\alpha_1, \ldots, \alpha_{i-1}, 0, \alpha_{i+1}, \ldots, \alpha_p)$ for some i. We set:

$$(4.1.3) \qquad \mathscr{B}_{(f)} = \mathscr{H}_{(f)} / \mathscr{L}_{(f)}$$

and we denote by $\delta(f)$ the class of $1/f_1 \ldots f_p$ in $\mathscr{B}_{(f)}$.

One sees immediately that $\mathscr{B}_{(f)}$ is naturally endowed with the structure of a left $\mathscr{D}_X$-module, and that $\delta(f)$ generates this Module.

We denote by $\mathscr{I}_{(f)}$ the annihilator of $\delta(f)$ in $\mathscr{D}_X$. This is a sheaf of left ideals, defined by:

$$\mathscr{I}_{(f)} = \{P \in \mathscr{D}_X; \ P\delta(f) = 0\}$$

(The notation "$P \in \mathcal{D}_X$" is convenient, but quite incorrect, cf. Appendix C; it is a short way to say that P is a section of the sheaf $\mathcal{D}_X$ on some open set $U \subset X$).

When $f_i = x_i$ $(i \leqslant p)$ for a system of local coordinates, one sees easily that $\mathcal{I}_{(f)}$ coincides with the left Ideal generated by $x_1, \dots, x_p, D_{p+1}, \dots, D_n$.

Lemma 4.1.1. *The Ideal $\mathcal{I}_{(f)}$ depends only on Z and on $df_1 \wedge \dots \wedge df_p$. More precisely, if $(g) = (g_1, \dots, g_p)$ is another defining system for Z, then:*

$$dg_1 \wedge \dots \wedge dg_p = h\, df_1 \wedge \dots \wedge df_p \bmod \mathcal{I}_Z \Omega_X^p,$$

where h is a section of $\mathcal{O}_X$, $h \neq 0$ on Z, and we have:

$$(4.1.4) \qquad P \in \mathcal{I}_{(f)} \Leftrightarrow P \circ h \in \mathcal{I}_{(g)}.$$

Proof. We may assume that in a system of local coordinates $(x) = (x_1, \dots, x_n)$, $f_i = x_i$ $(i = 1, \dots, p)$. Then $(g_1, \dots, g_p, x_{p+1}, \dots, x_n)$ defines a system of local coordinates, that we denote $(y_1, \dots, y_n)$ (thus $y_i = g_i$ for $i \leqslant p$, $y_i = x_i$ for $i > p$). Let $A(x)$ be the $(n \times n)$-matrix $(a_{i,j}(x))$, such that:

$$y_j = \sum_i a_{j,i}(x) x_i.$$

Thus $a_{j,i} = \delta_j^i$ for $i, j > p$, and:

$$h(x) = \det(A(x)) \quad \bmod \left(\sum_{i=1}^p \mathcal{O}_X x_i \right),$$

$$dy_1 \wedge \dots \wedge dy_p = h(x)\, dx_1 \wedge \dots \wedge dx_p \quad \bmod \left(\sum_{i=1}^p x_i \Omega_X^{(p)} \right).$$

Let $\mathcal{I}$ be the left Ideal of $\mathcal{D}_X$ generated by $y_1, \dots, y_p, D_{y_{p+1}}, \dots, D_{y_n}$. It is clear that $\mathcal{I}$ contains $x_1, \dots, x_p$. Let us show that $\mathcal{I}$ contains $D_{x_j} \circ h$ for $j > p$.

Let $k = \det(A(x))$. It is enough to prove that $\mathcal{I}$ contains $D_{x_j} \circ k$ for $j > p$. We have:

$$D_{x_i} = \sum_{l,k} \left(\frac{\partial a_{l,k}}{\partial x_i} x_k + a_{l,k} \delta_k^i \right) D_{y_l},$$

thus for $j > p$:

$$D_{x_j} = D_{y_j} + \sum_{1 \leqslant l, k \leqslant p} \frac{\partial a_{l,k}}{\partial x_j} x_k D_{y_l}.$$

In matrix notation, writing $x = (x_1, \dots, x_n)$ as a row vector and $D_y = {}^t(D_{y_1}, \dots, D_{y_n})$ as a column vector:

$$D_{x_j} = D_{y_j} + \left\langle x, \frac{\partial}{\partial x_j}({}^t A), D_y \right\rangle \quad j > p.$$

Since $x = A^{-1} y$, we get for $j > p$:

$$D_{x_j} \circ k = k \circ D_{x_j} + \frac{\partial k}{\partial x_j}$$

$$= k \circ D_{y_j} + k \left\langle y, \, {}^t A^{-1} \frac{\partial}{\partial x_j} \, {}^t A, D_y \right\rangle + \frac{\partial k}{\partial x_j}$$

$$= k \circ D_{y_j} + k \left\langle {}^t \left({}^t A^{-1} \frac{\partial}{\partial x_j} \, {}^t A \right) D_y, \, {}^t y \right\rangle - k \, \mathrm{tr} \left({}^t A^{-1} \frac{\partial}{\partial x_j} \, {}^t A \right) + \frac{\partial k}{\partial x_j}.$$

Using the formula:

$$(4.1.5) \qquad\qquad \frac{\partial}{\partial x_j} \mathrm{Log} \det(B) = \mathrm{Tr} \left(B^{-1} \frac{\partial}{\partial x_j} (B) \right),$$

we get:

$$D_{x_j} \circ k = k \circ D_{y_j} + \sum_{j=1}^{p} B_j \circ y_j$$

for some vector fields B_j, which completes the proof of the lemma. □

We have exact sequences of left $\mathscr{D}_X$-modules:

$$0 \to \mathscr{I}_{(f)} \to \mathscr{D}_X \xrightarrow[\psi]{} \mathscr{B}_{(f)} \to 0,$$

where $\psi(P) = P\delta(f)$.

Since the two Modules $\mathscr{D}_X/\mathscr{I}_{(f)}$ and $\mathscr{B}_{(f)}$ are isomorphic we continue to denote by $\delta(f)$ the class of $1 \in \mathscr{D}_X \bmod \mathscr{I}_{(f)}$. Now if (g) is another basis of $\mathscr{I}_Z$, $\delta(g)$ the class $\bmod \mathscr{I}_{(g)}$ of $1 \in \mathscr{D}_X$, then the left Ideal which annihilates the section $h\delta(g)$ of $\mathscr{D}_X/\mathscr{I}_{(g)}$ is exactly $\mathscr{I}_{(f)}$. Thus the two Modules $\mathscr{D}_X/\mathscr{I}_{(g)}$ and $\mathscr{D}_X/\mathscr{I}_{(f)}$ are isomorphic by the correspondence $\delta(f) \leftrightarrow h\delta(g)$, and we can patch them to form a $\mathscr{D}_X$-module $\mathscr{B}_{Z|X}$ defined on X. Since $dg = h\,df$ (where $df = df_1 \wedge \dots \wedge df_p$), the section $\delta(f)\,df$ of $\mathscr{B}_{Z|X} \otimes_{\mathscr{O}_X} \Omega_X^{(p)}$ is canonically defined.

Definition 4.1.2. We set:

$$\delta_Z = \delta(f)\,df, \text{ section of } \mathscr{B}_{Z|X} \otimes_{\mathscr{O}_X} \Omega_X^{(p)},$$

$$\mathscr{C}_{Z|X} = \mathscr{E}_X \otimes_{\pi^{-1}\mathscr{D}_X} (\pi^{-1}\mathscr{D}_{Z|X}) = \mathscr{E}_X/\mathscr{E}_X \mathscr{I}_{(f)},$$

$$\mathscr{C}_{Z|X}(k) = \mathscr{E}_X(k)/\mathscr{E}_X \mathscr{I}_{(f)} \cap \mathscr{E}_X(k).$$

We say that $\mathscr{B}_{Z|X}$ (resp. $\mathscr{C}_{Z|X}$) is the sheaf of holomorphic hyperfunctions (resp. microfunctions) along Z. The section δ_Z is called the fundamental class of Z in X.

The left $\mathscr{E}_X(0)$-modules $\mathscr{C}_{Z|X}(k)$, $(k \in \mathbb{Z})$, define the "natural filtration" on $\mathscr{C}_{Z|X}$.

Remark that the sheaf $\mathscr{B}_{Z|X}$ is concentrated on Z. We shall see that $\mathscr{C}_{Z|X}$ is concentrated on $T_Z^* X$.

In the next section, we shall study these sheaves in detail when Z is a hypersurface.

When $Z = X$ we get $\mathscr{B}_{X|X} = \mathscr{O}_X$ and of course $\delta_X = 1$.

Remark 4.1.3. We shall give another construction for $\mathscr{B}_{Z|X}$ in Chapter II, § 3.

To a left Ideal $\mathscr{I}$ of $\mathscr{E}_X$, let us associate the symbol Ideal $\widetilde{\mathscr{I}}$ of $\mathscr{O}_{T^*X}$ generated by the principal symbols of sections of $\mathscr{I}$:

$$\begin{cases} f \in \widetilde{\mathscr{I}}_{(x,\xi)} \Leftrightarrow \exists P_1, \ldots, P_r \in \mathscr{I}_{(x,\xi)} \\ \exists a_1, \ldots, a_r \in \mathscr{O}_{T^*X, (x,\xi)}, \quad f = \sum_i a_i \sigma(P_i) \end{cases}$$

Remark that if f is homogeneous we may choose a_i homogeneous, and thus we may find $P \in \mathscr{I}_{(x,\xi)}$ with $f = \sigma(P)$, because $(\mathscr{O}_{T^*X})_{(x,\xi)}$ is faithfully flat over the graded Ring $\mathrm{gr}(\mathscr{E}_X)$ (bearing in mind, however the graded structure of $\mathrm{gr}(\mathscr{E}_X)$). We shall come back to these questions in Chapter II, § 2.

Proposition 4.1.4. *Let (f) be a defining system for Z. Then the sheaf ideal $\widetilde{\mathscr{I}}_{(f)}$ is exactly the sheaf ideal $\mathscr{I}_{T_Z^*X}$, that is the defining Ideal of the conormal bundle T_Z^*X of Z in X.*

Proof. Let $(x_1, \ldots, x_n)$ be a linear coordinate system on E. Since we still do not know how to perform a change of coordinates on $\mathscr{E}_X$, we may not assume that $f_i = x_i$ in these coordinates. Meanwhile we may find $(f_{p+1}, \ldots, f_n)$ such that $(f_1, \ldots, f_n)$ is a local system of coordinates. Then $f_1, \ldots, f_p$, $\dfrac{\partial}{\partial f_{p+1}}, \ldots, \dfrac{\partial}{\partial f_n}$ generate $\mathscr{I}_{(f)}$, and the assertion will follow from the Proposition 4.1.5 below. $\square$

But at this stage of the proof we may already notice that 1 is in the Ideal $\mathscr{E}_X \mathscr{I}_{(f)}$ outside of T_Z^*X by Proposition 1.3.5, and thus $\mathscr{E}_{Z|X}$ is supported by T_Z^*X.

Proposition 4.1.5. *Let $P_1, \ldots, P_d$ be microdifferential operators in a neighborhood of (x_0, ξ_0), which generate a left Ideal $\mathscr{I}$ of $\mathscr{E}_X$.*

Let m_j be the order of P_j, and assume:

i) $d\sigma(P_1) \wedge \ldots \wedge d\sigma(P_d) \neq 0$,

ii) *there exists $Q_{i,j,k}$ of order at most $m_i + m_j - m_k - 1$ such that*

$$[P_i, P_j] = \sum_k Q_{i,j,k} P_k.$$

Then the $\sigma(P_j)$'s generate the symbol Ideal $\widetilde{\mathscr{I}}$ of $\mathscr{I}$.

Proof. Let $(A_j)_j$ be sections of $\mathscr{E}_X$, $Q = \sum_j A_j P_j$.

We want to prove that $\sigma(Q)$ is generated by the $\sigma(P_j)$'s.

Let $m \in \mathbb{Z}$ such that $\mathrm{ord}\,(A_j) \leqslant m - m_j$. We argue by induction on $s = m - \mathrm{ord}\,(Q)$. If $s = 0$ the conclusion follows since:

$$\sigma_m(Q) = \sum_j \sigma_{m-m_j}(A_j)\sigma_{m_j}(P_j).$$

Assume $m > \mathrm{ord}\,(Q)$. Since $\sigma_m(Q) = 0$ there exist functions $b_{i,j}$, homogeneous of degree $m - m_i - m_j$, such that:

$$\begin{cases} \sigma_{m-m_j}(A_j) = \sum_i b_{i,j}\sigma_{m_j}(P_j), \\ b_{i,j} = -b_{j,i}. \end{cases}$$

Let us choose sections $B_{i,j} \in \mathscr{E}_X(m - m_i - m_j)$, such that:

$$\sigma(B_{i,j}) = b_{i,j}, \qquad B_{i,j} = -B_{j,i}.$$

Then

$$Q = \sum_j \left(A_j - \sum_i B_{i,j}P_i\right)P_j + \sum_{i<j} B_{i,j}[P_i, P_j]$$

$$= \sum_j C_j P_j + \sum_k R_k P_k$$

with $\mathrm{ord}\,(C_j) \leqslant m - m_j - 1$, and
$\mathrm{ord}\,(R_k) \leqslant m - m_i - m_j + (m_i + m_j - m_k) - 1$. $\square$

4.2. The Case when Z is a Hypersurface

Let Z be a hypersurface of X. Consider the left $\mathscr{D}_X$-ideal $\mathscr{I}_Z$ generated by the vector fields which are tangent to Z. We denote by Y_Z the class $1 \bmod \mathscr{I}_Z$ of the left $\mathscr{D}_X$-module $\mathscr{D}_X/\mathscr{I}_Z$.

If we choose a system of local coordinates $(x_1, \ldots, x_n)$ on X such that $Z = \{x \in X;\ x_1 = 0\}$, it is immediately verified that $\mathscr{I}_Z$ is generated by $x_1 D_1, D_2, \ldots, D_n$. We also write $Y(x_1)$ instead of Y_Z, although $Y(x_1)$ only depends on Z.

With the coordinates $(x_1, \ldots, x_n)$ we introduce the left $\mathscr{D}_X$-module with one generator u, and relations $D_1 x_1 D_1 u = D_2 u = \ldots = D_n u = 0$ and we write $\mathrm{Log}\,x_1$ instead of u. We have the exact sequence of left $\mathscr{D}_X$-modules:

$$(4.2.1) \qquad 0 \to \mathscr{O}_X \xrightarrow{\phi} \mathscr{D}_X(\mathrm{Log}\,x_1) \xrightarrow{\psi} \mathscr{D}_X Y(x_1) \to 0,$$

where $\phi(1) = x_1 D_1(\mathrm{Log}\,x_1)$, $\psi(\mathrm{Log}\,x_1) = Y(x_1) = \mathrm{Log}\,x_1 \bmod \mathscr{O}_X$, and we have the exact sequence (where $\mathscr{D}_X \delta(x_1) = \mathscr{B}_{Z|X}$):

$$(4.2.2) \qquad 0 \to \mathscr{D}_X \delta(x_1) \xrightarrow{\lambda} \mathscr{D}_X Y(x_1) \xrightarrow{\mu} \mathscr{O}_X \to 0,$$

where $\lambda(\delta(x_1)) = D_1 Y(x_1)$, $\mu(Y(x_1)) = 1 = Y(x_1) \bmod \mathscr{D}_X \delta(x_1)$.

Moreover we have an $\mathcal{O}_X$-linear (but not $\mathcal{D}_X$-linear) splitting:

(4.2.3) $$\mathcal{D}_X\, Y(x_1) \cong \mathcal{D}_X\, \delta(x_1) \oplus \mathcal{O}_X,$$

since any $P \in \mathcal{D}_X/\mathcal{D}_X(x_1 D_1, D_2, \ldots, D_n)$ may be written in a unique way:

$$P(x_1, x', D_1) = P_0(x) + P_1(x', D_1),$$

where $P_0(x) \in \mathcal{O}_X$, $P_1(x', D_1) \in \mathcal{B}_{Z|X} \cong \mathcal{D}_X/\mathcal{D}_X(x_1, D_2, \ldots, D_n)$.

Now we return to the sheaf $\mathcal{C}_{Z|X}$. As already mentioned, the unit 1 of $\mathcal{E}_X$ belongs to the Ideal $\mathcal{E}_X\, \mathcal{I}_{(x_1)}$ outside of $T_Z^* X$, and $\mathcal{C}_{Z|X}$ is therefore supported by $T_Z^* X$.

Let $\dot{T}^* X = T^* X - T_X^* X$, $\dot{\pi}$ the projection from $\dot{T}^* X$ to X. Since D_1 is invertible in a neighborhood of $\dot{T}_Z^* X$, the two left $\mathcal{E}_X$-modules $\mathcal{C}_{Z|X}$ and $\mathcal{E}_X \otimes_{\pi^{-1}\mathcal{D}_X} \pi^{-1}(\mathcal{D}_X\, Y_Z)$ are the same on $\dot{T}_Z^* X$. Thus $\mathcal{C}_{Z|X}$ has a canonical section Y_Z on $\dot{T}_Z^* X$ (as soon as Z is a hypersurface).

Proposition 4.2.1. *There exists a natural isomorphism* $\dot{\pi}_*\, \mathcal{C}_{Z|X} \cong \mathcal{D}_X\, Y_Z$. *In particular* $\mathcal{C}_{Z|X}$ *is generated by* Y_Z *over* $\mathcal{D}_X$ *on* $\dot{T}_Z^* X$.

Proof. We only prove this statement when Z is given by the equation $\{x_1 = 0\}$ in the coordinates $(x_1, \ldots, x_n)$. (We shall be allowed to change coordinates after Section 5, and Proposition 4.2.1 will not be used in this chapter).

Let $Q(x', D_1) = \sum\limits_{-\infty < j \leqslant m} a_j(x') D_1^j$ be a section of $\mathcal{C}_{Z|X}$. To Q we associate the differential operator:

$$P(x, D_1) = \sum_{j=0}^{m} a_j(x') D_1^{j+1} + b(x),$$

where $b(x) = \sum\limits_{j<0} a_j(x') \dfrac{x_1^{(-j-1)}}{(-j-1)!}$. Remark that $b(x)$ is a holomorphic function in a neighborhood of Z in view of Definition 1.3.1. Since $\dfrac{x_1^{k-1}}{(k-1)!} D_1^{-1} = D_1^{-k} \bmod (\mathcal{E}_X x_1)$, $(k>0)$, we get:

$$P(x, D_1) \circ D_1^{-1} = Q(x', D_1) \bmod (\mathcal{E}_X x_1)$$

Thus the morphism from $\mathcal{D}_X$ to $\mathcal{C}_{Z|X}|_{\dot{T}_X^* Z}$ which associates $P \circ D_1^{-1} \delta(x_1)$ to P is surjective, and its kernel is clearly generated by $x_1 D_1, D_2, \ldots, D_n$. $\quad\square$

We may formulate Proposition 4.2.1 differently by introducing the following sequence of holomorphic functions in one variable. For $n \in \mathbb{Z}$ set:

(4.2.4)
$$\begin{cases} \phi_n(t) = \dfrac{-1}{2i\pi}\, \dfrac{(-1)^n n!}{t^{n+1}} & \text{for } n \geqslant 0, \\[3mm] \phi_{-n}(t) = \dfrac{-1}{2i\pi}\, \dfrac{t^{n-1}}{(n-1)!} \left(\mathrm{Log}\, t - \sum\limits_{j=1}^{n-1} \dfrac{1}{j} + \gamma \right) & \text{for } n > 0, \end{cases}$$

where γ is the Euler constant. Then:

$$(4.2.5) \qquad\qquad D_t\phi_n = \phi_{n+1}, \quad n \in \mathbb{Z},$$

and to $P(x', D_1) = \sum\limits_{-\infty < j \leqslant m} a_j(x') D_1^j$ we associate $P(x', D_1)\phi_0 =$

$\sum\limits_{-\infty < j \leqslant m} a_j(x')\phi_j(x_1) = \sum\limits_{0 \leqslant j \leqslant m} a_j(x')\phi_j(x_1) + b(x) \operatorname{Log}(x_1) + c(x)$ where $c(x)$ is a

holomorphic function. The class of $P(x', D_1)\phi_0 \bmod \mathscr{O}_X$ defines an element of $\mathscr{B}_{Z|X} \oplus (\mathscr{O}_X)|_Z \operatorname{Log} x_1$. Thus we get an $\mathscr{O}_X$-linear isomorphism:

$$\dot{\pi}_* \mathscr{C}_{Z|X} \cong \mathscr{B}_{Z|X} \oplus (\mathscr{O}_X)|_Z$$

which is nothing but the composite of the isomorphism $\mathscr{D}_X Y_Z \cong \mathscr{B}_{Z|X} \oplus (\mathscr{O}_X)|_Z$, and the isomorphism: $\dot{\pi}_* \mathscr{C}_{Z|X} \cong \mathscr{D}_X Y_Z$.

4.3. The Sheaf $\mathscr{E}_{Y \to X}$

Let F be another finite dimensional complex vector space, and Y be an open set in F. Let f be a holomorphic map from Y to X. To f are naturally associated two maps from $Y \underset{X}{\times} T^* X$ to $T^* Y$ and to $T^* X$:

$$(4.3.1) \qquad\qquad T^* Y \xleftarrow{\rho} Y \underset{X}{\times} T^* X \xrightarrow{\bar{\omega}} T^* X.$$

Recall that $Y \underset{X}{\times} T^* X = \{(y, x, \xi) \in Y \times T^* X; f(y) = x\}$. Then $\bar{\omega}(y, x, \xi) = (x, \xi)$ and $\rho(y, x, \xi) = (y, {}^t f'(y)\xi)$.

We shall often identify Y with the graph of f in $Y \times X$. Then the projection from $T^*(Y \times X)$ to $Y \times T^* X$ induces an isomorphism:

$$(4.3.2) \qquad\qquad T_Y^*(Y \times X) \cong Y \underset{X}{\times} T^* X,$$

(where the left hand side denotes the conormal bundle to the graph of f in $Y \times X$).

We shall consider the sheaf $\mathscr{B}_{Y|Y \times X}$ as a sheaf over Y, and the sheaf $\mathscr{C}_{Y|Y \times X}$ as a sheaf over $Y \underset{X}{\times} T^* X$. The projection $T_Y^*(Y \times X) \xrightarrow{\sim} Y \underset{X}{\times} T^* X$ gives an isomorphism between $\Omega_{Y \times X} \otimes_{\mathscr{O}_Y} \Omega_Y^{\otimes -1}$ and $\mathscr{O}_{Y \times X} \otimes_{\mathscr{O}_X} \Omega_X$. Since the fundamental class δ_Y of Y in $Y \times X$ gives a section of $\mathscr{B}_{Y|Y \times X} \otimes_{\mathscr{O}_{Y \times X}} \Omega_{Y \times X} \otimes_{\mathscr{O}_Y} \Omega_Y^{\otimes -1}$ we may consider it as a section of $\mathscr{B}_{Y|Y \times X} \otimes_{\mathscr{O}_X} \Omega_X$.

If $(x_1, \ldots, x_n)$ is a system of local coordinates on X, and if $f(y) = (f_1(y), \ldots, f_n(y))$ in those coordinates, the projection $T_Y^*(Y \times X) \to Y \underset{X}{\times} T^* X$ associates the section dx_j of $Y \underset{X}{\times} T^* X$ to the section $d(x_j - f_j(y))$ of $T_Y^*(Y \times X)$. Hence this projection associates $\delta(x - f(y)) dx_1 \wedge \ldots \wedge dx_n$ to $\delta(x - f(y)) d(x_1 - f_1(y)) \wedge \ldots \wedge d(x_n - f_n(y))$.

Definition 4.3.1. We denote by $\mathscr{E}_{Y\to X}$ the sheaf $\mathscr{C}_{Y|Y\times X}\otimes_{\mathscr{O}_X}\Omega_X$ on $Y\underset{X}{\times}T^*X$
and by $\mathscr{D}_{Y\to X}$ the sheaf $\mathscr{B}_{Y|Y\times X}\otimes_{\mathscr{O}_X}\Omega_X$ on Y. We denote by $1_{Y\to X}$ its canonical section.

  Remark that $\mathscr{D}_{Y\to X}=\mathscr{E}_{Y\to X}|_{Y\times T^*_X X}$. Moreover $\mathscr{E}_{Y\to X}$ is naturally a left
$\rho^{-1}(\mathscr{E}_Y)$-module (thus $\mathscr{D}_{Y\to X}$ a left $\mathscr{D}_Y$-module). We shall see in the next
section that it is also naturally a right $\overline{\omega}^{-1}\mathscr{E}_X$-module.
  We set:
(4.3.3) $\qquad\qquad\qquad \mathscr{E}_{Y\to X}(k)=\mathscr{C}_{Y|Y\times X}(k)\otimes_{\mathscr{O}_X}\Omega_X$
thus $1_{Y\to X}$ belongs to $\mathscr{E}_{Y\to X}(0)$.

Remark 4.3.2. The notation $\mathscr{C}_{Y|Y\times X}\otimes_{\mathscr{O}_X}\Omega_X$, for example, is not correct, and
should be replaced by $\mathscr{C}_{Y|Y\times X}\otimes_{p^{-1}\mathscr{O}_X}p^{-1}\Omega_X$, where p denotes the projection
from $T^*(Y\times X)$ to X, but we shall often forget to write symbols such as p^{-1}
in these situations.

Exercises to I.4

Ex. 4.1. Let $P=\sum\limits_{j=1}^{n} D_j^2$ on $\mathbb{C}^n$, $Y=\{x\in\mathbb{C}^n; x_1=0\}$, $Z=\{x\in\mathbb{C}^n; x_1+ix_2=0\}$.
Prove that the Cauchy problem:

$$Pf=g, \quad f|_Y=h$$

has a unique solution $f\in\mathscr{C}_{Z|X}|_{\dot{T}^*_Z X}$ (in the neighborhood of Y) for a given
$g\in\mathscr{C}_{Z|X}|_{\dot{T}^*_Z X}$, $h\in\mathscr{C}_{Z\cap Y|Y}|_{\dot{T}^*_{Z\cap Y} Y}$ (cf. § 3).

Ex. 4.2. Let $\mathscr{I}$ be the left Ideal of $\mathscr{D}_{\mathbb{C}^n}$ generated by D_1^2 and $x_1 D_1+D_2$.
Calculate $\tilde{\mathscr{I}}$ the symbol Ideal of $\mathscr{I}$ in $\mathscr{O}_{T^*X}$.

Ex. 4.3. Let Z be the hypersurface $\left\{x\in\mathbb{C}^n; f(x)=x_1+\sum\limits_{j=2}^{n}x_j^2=0\right\}$. Calculate $T^*_Z X$. Find $(n-1)$ vector fields $X_2,\ldots,X_n$ such that $\mathscr{I}_{(f)}$ is generated by
f and the X_j's.

Ex. 4.4. Let $\overline{\mathbb{C}}^n$ be the anti-holomorphic manifold associated to $\mathbb{C}^n$ (the holomorphic functions on $\overline{\mathbb{C}}^n$ are the anti-holomorphic functions on $\mathbb{C}^n$), and
let $X=\mathbb{C}^n\times\overline{\mathbb{C}}^n$. Let P be a section of $\mathscr{D}_{\mathbb{C}^n}$, and let $\mathscr{I}_{\overline{\partial}}$ be the left Ideal of $\mathscr{D}_X$
generated by the holomorphic vector fields of $\overline{\mathbb{C}}^n$ ($\mathscr{I}_{\overline{\partial}}$ is thus the Ideal generated by the Cauchy-Riemann operators). Calculate the symbol Ideal of
the Ideal of $\mathscr{D}_X$ generated by P and $\mathscr{I}_{\overline{\partial}}$.

Ex. 4.5. State and prove the analogous to Proposition 4.2.1 with $\mathscr{C}_{Z|X}$ replaced by $\hat{\mathscr{C}}_{Z|X}=\hat{\mathscr{E}}_X\otimes_{\mathscr{E}_X}\mathscr{C}_{Z|X}$.

§ 5. Quantized Contact Transformations

5.1. Division by an Ideal

As in the preceding section, E and F are finite dimensional complex vector spaces, X is open in E, Y is open in F.

Assume given: an open set U in $T^*(X \times Y) = T^*X \times T^*Y$, a closed conic (smooth) manifold V in U, a left Ideal $\mathscr{I}$ of $\mathscr{E}_{X \times Y}$ on U, with symbol Ideal $\overline{\mathscr{I}}$ in $\mathscr{O}_{T^*(X \times Y)}$. We denote by p_1 and p_2 the natural projections from $T^*(X \times Y)$ to T^*X and to T^*Y.

Proposition 5.1.1. *Assume:*

i) *p_1 is an immersion (i.e.: its differential is injective) from V into T^*X.*

ii) *The defining Ideal $\mathscr{I}_V$ of V in $T^*(X \times Y)$ is contained in $\overline{\mathscr{I}}$. Then we have on V:*

$$(5.1.1) \qquad\qquad p_1^{-1}(\mathscr{E}_X) + \mathscr{I} = \mathscr{E}_{X \times Y}.$$

In other words, under hypothesis i) and ii), any section P of $\mathscr{E}_{X \times Y}$ may locally be decomposed as $P' + R$, P' a section of $\mathscr{E}_X$, R a section of $\mathscr{I}$.

Proof. We may assume $Y = \mathbb{C}^p$. By writing $X \times Y = (X \times \mathbb{C}^{p-1}) \times \mathbb{C}$ we may proceed by induction on dim Y, and it is enough to prove the theorem when $p = 1$. Let (x, ξ) and (y, η) be the coordinates on T^*X and T^*Y. Since p_1 is an immersion, we may locally find functions on V, $f(x, \xi)$ and $g(x, \xi)$ homogeneous with respect to ξ of degree 0 and 1 respectively, such that:

$$(5.1.2) \qquad\qquad y - f \in \mathscr{I}_V, \qquad \eta - g \in \mathscr{I}_V.$$

Since we assumed $\mathscr{I}_V \subset \overline{\mathscr{I}}$, there (locally) exist microdifferential operators P_0 and Q_1, of respective orders 0 and 1 and principal symbols $y - f(x, \xi)$ and $\eta - g(x, \xi)$, such that P_0 and Q_1 belong to $\mathscr{I}$.

We apply Theorem 2.2.1 and divide P_0 by Q_1. We find:

$$P_0 = S \circ Q_1 + R,$$

where $[y, R] = 0$, $\text{ord}(R) \leqslant 0$. Since $\dfrac{\partial}{\partial \eta} \sigma_0(P_0) \equiv 0$, $\dfrac{\partial}{\partial \eta} \sigma_1(Q_0) \equiv 1$, we have $\sigma_0(P_0) = \sigma_0(R)$. Thus replacing P_0 by $P_0 - S \circ Q_1$ which is a section of the Ideal $\mathscr{I}$, we may assume:

$$\sigma_0(P_0) = y - f(x, \xi), \qquad [y, P_0] = 0.$$

Now let P be a section of $\mathscr{E}_{X \times Y}$. By dividing P by Q_1 and the remainder by P_0 we get:

$$P = S_1 \circ Q_1 + R_1, \qquad [y, R_1] = 0,$$

$$R_1 = S_2 \circ P_0 + R_2, \qquad [y, R_2] = [D_y, R_2] = 0,$$

(the identity $[y, R_2] \equiv 0$ follows from the same identity with R_2 replaced by R_1 or P_0, cf. the remarks after Theorem 2.2.1).

We have found $R_2 \in p_1^{-1}(\mathscr{E}_X)$ such that $P - R_2 \in \mathscr{I}$.   □

Proposition 5.1.2. *Assume now:*

i) p_1 *is a submersion (i.e.: its differential is surjective) from V to T^*X.*

ii) *The symbol Ideal $\mathscr{I}$ is contained in $\mathscr{I}_V$.*
Then we have:

(5.1.3) $p_1^{-1}(\mathscr{E}_X) \cap \mathscr{I} = \{0\}$ on V

Proof. Let P be a section of $p_1^{-1}(\mathscr{E}_X)$ and assume that P belongs to $\mathscr{I}$. Then $\sigma(P) \circ p_1$ is zero on V, and p_1 being open on V, $\sigma(P) \equiv 0$. Thus $P = 0$ in a neighborhood of V.   □

Theorem 5.1.3. *Let V be a closed conic manifold in the open set U of $T^*(X \times Y)$ and let $\mathscr{I}$ be a left Ideal of $\mathscr{E}_{X \times Y}$ on U with symbol Ideal $\mathscr{I}$. Assume:*

i) *The projections p_1 and p_2 define (complex) diffeomorphisms from V onto open sets U_X and U_Y of T^*X and T^*Y respectively.*

ii) *The symbol Ideal $\mathscr{I}$ coincides with the defining Ideal $\mathscr{I}_V$ of V in $T^*(X \times Y)$.*

Then: a) *For any section P of $\mathscr{E}_X$ on U_X there exists a unique section Q of $\mathscr{E}_Y$ on U_Y such that $P - Q$ belong to $\mathscr{I}$ on V (we consider P or Q as sections of $\mathscr{E}_{X \times Y}$).*

b) *Let $\hat{\phi}$ be the morphism from $p_{2*}(p_1^{-1}(\mathscr{E}_X|_{U_X})|_V)$ to $\mathscr{E}_Y|_{U_Y}$ so defined. Then $\hat{\phi}$ is an anti-isomorphism of Algebras:*

(5.1.4) $\hat{\phi}(P_1 \circ P_2) = \hat{\phi}(P_2) \circ \hat{\phi}(P_1)$

c) *The morphism $\hat{\phi}$ preserves order, and if we denote by ϕ the map from U_X to U_Y whose graph is V, then for any section P of $\mathscr{E}_X|_{U_X}$, $\sigma(\hat{\phi}(P)) = \sigma(P) \circ \phi^{-1}$.*

Proof. a) By Proposition 5.1.1 and 5.1.2 we have the sheaf isomorphisms on V:

$$(p_1^{-1}\mathscr{E}_X|_{U_X}) \oplus \mathscr{I} \cong \mathscr{E}_{X \times Y}|_V \cong (p_2^{-1}\mathscr{E}_Y|_{U_Y}) \oplus \mathscr{I}.$$

b) The two microdifferential operators P_1 and $\hat{\phi}(P_2)$ commute. Thus:

$$P_1 \circ P_2 - \hat{\phi}(P_2) \circ \hat{\phi}(P_1) = (P_1 \circ P_2 - P_1 \circ \hat{\phi}(P_2)) - (\hat{\phi}(P_2) \circ \hat{\phi}(P_1) - \hat{\phi}(P_2) \circ P_1),$$

and this operator belongs to $\mathscr{I}$.

c) The two operators P and $\hat{\phi}(P)$ have the same order, otherwise $\sigma(P)$ or $\sigma(\hat{\phi}(P))$ would be zero on V, thus P or $\hat{\phi}(P)$ would be zero. Then

$$\sigma(P - \hat{\phi}(P)) = \sigma(P) - \sigma(\hat{\phi}(P)),$$

and this function is zero on V, since $P - \hat{\phi}(P)$ belongs to $\mathscr{I}$.   □

5.2. Adjoint

Let X' be another copy of X, and let p_1 and p_2 be the projections from $T^*(X \times X')$ to T^*X and T^*X' respectively. We sometimes identify X with the diagonal Δ of $X \times X'$, by the first projection, and T^*X with $T^*_\Delta(X \times X')$, the conormal bundle to Δ in $X \times X'$.

We denote by "a" the antipodal map on T^*X. Hence $a((x;\xi)) = (x; -\xi)$. We write p_2^a instead of $p_2 \circ a$, and we recall that the sheaf $a^{-1}(\mathcal{E}_X)$ on T^*X satisfies $(a^{-1}(\mathcal{E}_X))_{(x;\xi)} = (\mathcal{E}_X)_{(x; -\xi)}$.

Let dx be a volume element on X, that is a nowhere vanishing section of Ω_X. Since $\mathcal{B}_{\Delta|X \times X'} = \mathcal{D}_{X \to X'} \otimes_{\mathcal{O}_X} \Omega_X^{\otimes -1}$, the section $1_{X \to X'} \otimes dx^{\otimes -1}$ is well defined in $\mathcal{B}_{\Delta|X \times X'}$, and generates this left $\mathcal{D}_{X \times X'}$-module. We call this section $\delta(x-x')$, and denote $\mathcal{I}_\Delta$ its annihilator in $\mathcal{E}_{X \times X'}$. Let $(x_1, \ldots, x_n)$ be a system of local coordinates on X with $dx = dx_1 \wedge \ldots \wedge dx_n$, and let $(x_1', \ldots, x_n')$ be the same coordinates on X'. Then the left Ideal $\mathcal{I}_\Delta$ is generated by the operators $(x_j - x_j', D_{x_j} + D_{x_j'}; 1 \leqslant j \leqslant n)$. The symbol Ideal $\mathcal{I}_\Delta$ coincide with the defining Ideal of $T^*_\Delta(X \times X')$, that is, with the defining Ideal of the graph of the antipodal map. Applying Theorem 5.1.3 we get, by identifying X and X':

Corollary 5.2.1. *Let dx be a volume element on X. Then there exists a well defined morphism from $a^{-1}(\mathcal{E}_X)$ to $\mathcal{E}_X$, which associates to a section P of $a^{-1}(\mathcal{E}_X)$ the unique section P^* of $\mathcal{E}_X$ such that $P - P^*$ belongs to the annihilator Ideal of $1_{X \to X} \otimes dx^{\otimes -1}$. Moreover we have:*

$$(5.2.1) \quad P^{**} = P, \quad (P \circ Q)^* = Q^* \circ P^*, \quad \sigma(P^*)(x, \xi) = \sigma(P)(x, -\xi)$$

We shall say that P^* is the adjoint of P (with respect to dx). Let $\sum\limits_{-\infty < j \leqslant m} p_j(x, \xi)$ be the total symbol of P. By applying the Leibniz formula, and starting with:

$$(x_j)^* = x_j, \quad (D_j)^* = -D_j, \quad 1 \leqslant j \leqslant n,$$

we get for the total symbol of P^*:

$$(5.2.2) \quad P^* = \sum_{j \leqslant m} q_j(x, \xi), \quad q_j(x, \xi) = \sum_{j = k - |\alpha|} \frac{(-1)^{|\alpha|}}{\alpha!} \partial_\xi^\alpha \partial_x^\alpha p_k(x, -\xi).$$

Now let dx and $d\tilde{x}$ be two volume elements on X, and let us denote respectively by P^* and $P^\sim$ the adjoint of P with respect to each of these volume elements. The sections dx and $d\tilde{x}$ are related by $dx = h \, d\tilde{x}$, for some nowhere vanishing holomorphic function h.

Set $u = 1_{X \to X'} \otimes dx^{\otimes -1}$, $v = 1_{X \to X'} \otimes d\tilde{x}^{\otimes -1}$, and let $\mathcal{I}_u$ and $\mathcal{I}_v$ be the annihilator Ideal of u and v respectively. Let h' be the image of h on X'. We have $v = h u$. Since $(h - h')u = 0$ we also have $v = h'u$. Moreover:

$$P - P^* \in \mathcal{I}_u, \quad P - P^\sim \in \mathcal{I}_v$$

Thus $h' \circ (P - P^*) \in \mathscr{I}_u$, $(P - P^\sim) \circ h' \in \mathscr{I}_u$, and since $h' \circ P = P \circ h'$, we get:

$$h' \circ P^* - P^\sim \circ h' \in \mathscr{I}_u.$$

Since this operator belongs to $p_2^{-1}(\mathscr{E}_{X'})$, this implies $h' \circ P^* = P^\sim \circ h'$. If we identify X and X' we have obtained:

$$P^\sim = h \circ P^* \circ h^{-1}$$

Proposition 5.2.2. *Let $\mathscr{M}$ be a left $\mathscr{E}_X$-module. Then $\mathscr{M} \otimes_{\mathscr{O}_X} \Omega_X$ is naturally endowed with the structure of a right $\mathscr{E}_X$-module by*

$$(u \otimes dx) P = (P^* u) \otimes dx,$$

where dx is a volume element and P^* is the adjoint of P with respect to dx. Similarly if $\mathscr{N}$ is a right $\mathscr{E}_X$-module then $\mathscr{N} \otimes_{\mathscr{O}_X} \Omega_X^{\otimes -1}$ is naturally endowed with the structure of a left $\mathscr{E}_X$-module.

Proof. We have to check that the action of P on $\mathscr{M} \otimes_{\mathscr{O}_X} \Omega_X$ does not depend on the choice of the volume element. Let $d\tilde{x} = h^{-1} dx$. Then:
$(u \otimes dx) P = (hu \otimes d\tilde{x}) P = \tilde{P}(hu) \otimes d\tilde{x} = h(P^* u) \otimes d\tilde{x} = P^* u \otimes dx$.

The case of a right $\mathscr{E}_X$-module is similar. We set $P(dx^{\otimes -1} \otimes v)$ $= dx^{\otimes -1} \otimes P^* v$. □

Corollary 5.2.3. *The sheaf $\mathscr{E}_{Y \to X}$ (cf. § 4) is naturally endowed with a structure of $(\rho^{-1} \mathscr{E}_Y, \overline{\omega}^{-1} \mathscr{E}_X)$-bimodule.*

(that is, $\mathscr{E}_{Y \to X}$ is a right $\overline{\omega}^{-1} \mathscr{E}_X$-module and a left $\rho^{-1} \mathscr{E}_Y$-module).

5.3. Quantized Contact Transformations

We refer to Appendix A for the geometrical notions used here. As mentioned in this Appendix, "homogeneous symplectic transformations" should be a more appropriate formulation, but we use "contact transformations" for short.

Let X and Y be two open sets of the vector spaces E and F, as in § 5.1. and let U and V be two open sets in $T^* X$ and $T^* Y$ respectively. Assume given a contact transformation ϕ from U to V, and let Λ_ϕ be the graph of ϕ in $U \times V$. Let V^a be antipodal to V, and let Λ_ϕ^a be the image in $U \times V^a$ of Λ_ϕ by the antipodal map of $T^* Y$:

(5.3.1) $\Lambda_\phi^a = \{(x, y; \xi, \eta) \in U \times V^a; \quad (y; -\eta) = \phi(x, \xi)\}$.

The manifold Λ_ϕ^a is conic and Lagrangean.

Theorem 5.3.1. *Assume to be given:*

i) *An Ideal $\mathcal{I}$ of $\mathcal{E}_{X \times Y}$ on $U \times V^a$ such that $\tilde{\mathcal{I}}$, the symbol Ideal, coincides with $\mathcal{I}_{\Lambda_\phi^a}$ the defining Ideal of Λ_ϕ^a in $\mathcal{O}_{T^*(X \times Y)}$.*

ii) *A volume element dy on Y.*

Then there exists an Algebra isomorphism $\hat{\phi}$, from $\phi_*(\mathcal{E}_X|_U)$ onto $\mathcal{E}_Y|_V$, satisfying for any section P of $\mathcal{E}_X$:

$$(5.3.2) \qquad\qquad \sigma(\hat{\phi}(P)) = \sigma(P) \circ \phi^{-1}.$$

Proof. The manifold Λ_ϕ^a is the graph of the composite $a \circ \phi$. We may apply Theorem 5.1.3 and find an anti-isomorphism from $(a \circ \phi)_*(\mathcal{E}_X|_U)$ onto $(\mathcal{E}_Y|_{V^a})$. If we compose it with the anti-isomorphism defined by taking the adjoint with respect to dy (Corollary 5.2.1) we get the theorem.   □

Definition 5.3.2. One says that $\hat{\phi}$ is a quantized contact transformation above ϕ.

Let (x) and (y) be coordinates on X and Y, (x, ξ) and (y, η) the associated coordinates on T^*X and T^*Y respectively. Then ϕ is defined by the relations:

$$(5.3.3) \qquad y_j = f_j(x, \xi), \quad \eta_j = g_j(x, \xi), \quad j = 1, \ldots, n,$$

where f_j (resp. g_j) is homogeneous of degree 0 (resp. 1) with respect to ξ, and the f_j's and g_j's satisfy the relations:

$$(5.3.4) \qquad \{f_j, f_k\} = 0, \quad \{g_j, g_k\} = 0, \quad \{f_j, g_k\} = -\delta_k^j.$$

To quantize ϕ is equivalent to finding microdifferential operators P_j and Q_k on U with principal symbols f_j and g_k ($1 \leq j, k \leq n$) satisfying the relations:

$$[P_j, P_k] = 0, \quad [Q_j, Q_k] = 0, \quad [P_j, Q_k] = -\delta_k^j.$$

In fact if $\hat{\phi}$ is a quantized contact transformation above ϕ, then the operators P_j and Q_k on U are defined by:

$$P_j = \hat{\phi}^{-1}(y_j), \quad Q_k = \hat{\phi}^{-1}(D_{y_k}),$$

and the relations follow from the corresponding relations for the y_j's and the D_{y_k}'s. Conversely assume the operators P_j, Q_k to be given on U and consider the left Ideal $\mathcal{I}$ of $\mathcal{E}_{X \times Y}$ on $U \times V$ generated by the operators

$$y_j - P_j(x, D_x), \quad D_{y_k} + Q_k(x, D_x), \quad (1 \leq j, k \leq n).$$

Since all these operators commute, the symbol Ideal $\tilde{\mathcal{I}}$ is just $\mathcal{I}_{\Lambda_\phi^a}$ by Proposition 4.1.5 and we may apply Theorem 5.1.3 (with the volume element $dy = dy_1 \wedge \ldots \wedge dy_n$).

As an immediate application we get the following.

Let Z be another open set in some vector space, W an open set in T^*Z, and ψ a contact transformation from V to W.

Proposition 5.3.3. *Let $\hat{\psi}$ be a quantized contact transformation above ψ. Then $\hat{\psi} \circ \hat{\phi}$ is a quantized contact transformation above $\psi \circ \phi$.*

Any contact transformation ϕ is locally the composite $\phi_2 \circ \phi_1$ of two contact transformations ϕ_1 and ϕ_2 such that each graph $\Lambda^a_{\phi_i}$ ($i = 1, 2$) is a conormal bundle $T^*_{Z_i}(X \times Y)$ for some submanifold Z_i of $X \times Y$, (Theorem A.4.1).

Since we may locally quantize ϕ_1 and ϕ_2 by Proposition 4.1.4, we get:

Theorem 5.3.4. *Let ϕ be a contact transformation from $U \subset T^* X$ to $V \subset T^* Y$. Then locally on U there exists a quantized contact transformation $\hat{\phi}$ above ϕ.*

5.4. Examples

Example 5.4.1. Legendre transformation (cf. Sato-Kashiwara-Kawaï [1, p. 400]). On $\mathbb{C}^n \times \mathbb{C}^n$ consider the function $f(x, y) = x_n - y_n + \sum_{j=1}^{n-1} x_j y_j$ and the hypersurface $Z = \{(x, y); \, f(x, y) = 0\}$. The manifold $\dot{T}^*_Z(\mathbb{C}^n \times \mathbb{C}^n)$ is a graph Λ^a_ϕ, where ϕ is a contact transformation from $U = \{(x, \xi) \in T^* \mathbb{C}^n; \, \xi_n \neq 0\}$ on $V = \{(y, \eta) \in T^* \mathbb{C}^n; \, \eta_n \neq 0\}$. In fact:

$$(5.4.1) \qquad (x, y; \xi, \eta) \in \dot{T}^*_Z(\mathbb{C}^n \times \mathbb{C}^n)^a \Leftrightarrow \begin{cases} f(x, y) = 0, \\ \xi_j = \lambda y_j, \quad \eta_j = -\lambda x_j \, (j < n), \\ \xi_n = \lambda, \quad \eta_n = \lambda, \quad \lambda \in \mathbb{C}^\times, \end{cases}$$

which is equivalent to:

$$(5.4.2) \qquad \begin{cases} y_j = \xi_j \xi_n^{-1} \, (j < n), \quad y_n = \langle x, \xi \rangle \xi_n^{-1}, \\ \eta_j = -x_j \xi_n \, (j < n), \quad \eta_n = \xi_n. \end{cases}$$

We may quantize ϕ by:

$$(5.4.3) \qquad \begin{cases} y_j = D_{x_j} D_{x_n}^{-1} \, (j < n), \quad y_n = D_{x_n}^{-1} \circ \langle x, D_x \rangle, \\ D_{y_j} = -x_j D_{x_n} \, (j < n), \quad D_{y_n} = D_{x_n} \end{cases}$$

It is sometimes useful to consider a "partial Legendre transformation", by taking the conormal bundle to the submanifold:

$$S = \left\{ (x, y) \in \mathbb{C}^n \times \mathbb{C}^n; \, x_j - y_j = 0 \, (1 \leqslant j \leqslant p < n), \, x_n - y_n + \sum_{j=p+1}^{n} x_j y_j = 0 \right\}.$$

Let L be the submanifold $\{x \in \mathbb{C}^n; \, x_{p+1} = \ldots = x_n = 0\}$. The partial Legendre transformation whose graph is $\dot{T}^*_S(\mathbb{C}^n \times \mathbb{C}^n)^a$ interchanges $\dot{T}^*_L \mathbb{C}^n$ and $\dot{T}^*_K \mathbb{C}^n$, where $K = \{y \in \mathbb{C}^n; \, y_n = 0\}$. Thus, "microlocally" (i.e. locally on $\dot{T}^* X$), submanifolds are equivalent to hypersurfaces.

Example 5.4.2. Change of coordinates. Let χ be a bi-holomorphic map between two open sets X and X' of E. We identify X with the graph of χ. Then $T^*_X(X \times X')^a$ is given in the coordinates $(x, y; \xi, \eta)$ on $T^*(X \times X')$, by:

$$(5.4.4) \quad \begin{cases} y - \chi(x) = 0, \\ \xi = \lambda \partial_x(y - \chi(x)), \quad \eta = -\lambda \partial_y(y - \chi(x)), \quad \lambda \in \mathbb{C}^n, \end{cases}$$

thus $T^*_X(X \times X')^a$ is the graph of the map $\phi = (\chi, {}^t\chi'(x)^{-1})$ from T^*X to T^*X'. We may quantize ϕ in a unique way so that $\hat{\phi}$ sends vector fields onto vector fields, by setting:

$$(5.4.5) \quad y = \chi(x), D_y = {}^t(\chi'(x))^{-1} D_x .$$

The existence follows from the preceding considerations, and the uniqueness from the fact that the D_{y_j}'s should be first order differential operators (since ϕ is defined in the neighborhood of the zero section), with prescribed principal symbols.

Theorem 5.4.3. *Let χ be a bi-holomorphic map between two open sets X and X' of E, $\phi = (\chi, {}^t\chi'^{-1})$ the associated contact transformation from T^*X to T^*X'. Then there exists a unique isomorphism $\hat{\phi}$ from $\phi_* \mathscr{E}_X$ to $\mathscr{E}_{X'}$ which extends the natural isomorphism from $\chi_*(\mathscr{D}_X)$ to $\mathscr{D}_{X'}$.*

In particular if we have choosen a basis on E, we may write $x = (x_1, \ldots, x_n), \chi(x) = (\chi_1(x), \ldots, \chi_n(x)) = (y_1, \ldots, y_n)$, and one says that χ is a holomorphic change of coordinates.

Since we know how to carry out a change of coordinates on the Ring $\mathscr{E}_X$, we may now define this sheaf on the cotangent bundle to a complex manifold: we define it on T^*U_j, for local charts U_j, and glue those different sheaves by applying Theorem 5.4.3.

Example 5.4.4. Let M be a real analytic manifold, X a complexification of M. Sato's microfunctions (Sato [1]) as well as wave front sets of distributions (Hörmander [3]) are objects which naturally "live" on $T^*_M X$, the conormal bundle to M in X.

Let us show that this bundle is locally isomorphic, outside of the zero section, (by a contact transformation), to the conormal bundle of a strictly pseudo-convex set (cf. Kashiwara-Kawaï [2], Schapira [2], Lebeau [1]). Let (z, ζ) and (w, θ) be the coordinates of two copies of $T^*\mathbb{C}^n$, with $z = x + iy$, $\zeta = \xi + i\eta$, $(x, y, \xi, \eta$ being real). Then $T^*_{\mathbb{R}^n}\mathbb{C}^n = \{(z, \zeta); y = \zeta = 0\}$. Let Z be the hypersurface:

$$Z = \left\{ (z, w) \in \mathbb{C}^n \times \mathbb{C}^n; \quad f(z, w) = \sum_{j=1}^n (z_j - w_j)^2 + 1 = 0 \right\}.$$

We have:

$$(z, w;\ \zeta, \theta) \in T^*_{\dot{Z}}(\mathbb{C}^n \times \mathbb{C}^n)^a \Leftrightarrow \begin{cases} f(z, w) = 0, \\ \zeta = \lambda(z - w), \quad \theta = \lambda(z - w), \quad \lambda \in \mathbb{C}. \end{cases}$$

Thus $\zeta = \theta$, $w_j = z_j \pm \sqrt{-1}\ \dfrac{\zeta_j}{\sqrt{\sum\limits_j \zeta_j^2}}$ and $\dot{T}^*_{\dot{Z}}(\mathbb{C}^n \times \mathbb{C}^n)^a$ is the union of the graphs

of two contact transformations $\phi^\pm$ from $U = \{(z, \zeta) \in T^*\mathbb{C}^n;\ \sum\limits_j \zeta_j^2 \notin \mathbb{R}_-\}$
onto the same open set V of $T^*\mathbb{C}^n$ with coordinates (w, θ). We may also
obtain the contact transformations $\phi^\pm$ by means of the "generating functions" $g^\pm$:

$$g^\pm(z, \theta) = \langle z, \theta \rangle \pm \sqrt{-1}\sqrt{\sum_j \theta_j^2}\ ,$$

since $\zeta = g'_z(z, \theta)$, $w = g'_\theta(z, \theta)$.

If we restrict $\phi^\pm$ to $\dot{T}^*_{\mathbb{R}^n}\mathbb{C}^n$, then $\phi^\pm$ defines an isomorphism from this
bundle onto $(T^*_{\partial B}\mathbb{C}^n)^\pm$, the "exterior" or "interior" conormal bundle to the
boundary ∂B of the tube:

$$B = \{w \in \mathbb{C}^n;\ |\operatorname{Im} w| < 1\}.$$

We may quantize ϕ^+ and ϕ^- by setting $D_{w_j} = D_{z_j}$, $w_j = z_j \pm \sqrt{-1}\ D_{z_j}/\sqrt{\sum\limits_j D_{z_j}^2}$,

$j = 1, \ldots, n$. Remark that $\hat{\phi}^+$ and $\hat{\phi}^-$ keep microdifferential operators with constant coefficients and are invariant by translation and rotation.

Exercises to I.5

Ex. 5.1. Find an analytical change of coordinates on $\mathbb{C}^n$ which interchanges D_1
and $D_1 + ix_1^3 D_2$.

Ex. 5.2. Find all quantized canonical transformations above the identity globally
defined all over $T^*\mathbb{C}^n$.

Ex. 5.3 Let $W_n = \mathbb{C}[x, D_x]$ be the Weyl algebra on $\mathbb{C}^n$, that ist the ring of differential operators on $\mathbb{C}^n$ with polynomial coefficients. Find an isomorphism of W_n which is not induced by a quantized contact transformation.

Ex. 5.4. Let $Y = \{x \in \mathbb{C}^n;\ x_1 = 0\}$. Prove that the left $\mathscr{E}_X$-modules $\mathscr{C}_{Y|\mathbb{C}^n}$ and $\mathscr{C}_{\{0\}|\mathbb{C}^n}$
may be interchanged by a quantized contact transformation near
$(0;\ 1, 0, \ldots, 0)$.

Ex. 5.5. State and prove an analogous to Theorem 3.1.1, where V is replaced by
$W = \{(x, \xi) \in T^*\mathbb{C}^n;\ \xi_1 = \ldots = \xi_q = x_{q+1} = \ldots x_p = 0\}$ $\quad(1 \leqslant q \leqslant p < n)$ and
$P = \sum\limits_{0 \leqslant |\alpha| \leqslant m} A_\alpha Q^\alpha$, where $Q = (D_1, \ldots, D_q, x_{q+1}D_{q+1}, \ldots, x_p D_p)$ in a neigh-

borhood of $(0; dx_n)$ (use a quantized partial Legendre transformation to reduce the problem to Theorem 3.1.1).

Ex. 5.6 Extend the results of this section to the sheaf $\hat{\mathscr{E}}_X$ (use Ex.2.2, or use the fact that $\hat{\mathscr{E}}_X(m) = \varprojlim_{j \geq 0} \mathscr{E}_X(m)/\mathscr{E}_X(m-j)$).

Ex. 5.7. Let $\phi(x, \eta)$ be a holomorphic function on $\mathbb{C}^n_x \times \dot{\mathbb{C}}^n_\eta$, in a neighborhood of some point (x^0, η^0), homogeneous of degree 1 with respect to η. Set:

$$\xi = \frac{\partial \phi}{\partial x}, \quad y = \frac{\partial \phi}{\partial \eta},$$

(i.e.: $\xi_j = \dfrac{\partial \phi}{\partial x_j}, y_k = \dfrac{\partial \phi}{\partial \eta_k}$). Prove that if the determinant of the matrix $\left(\dfrac{\partial^2 \phi}{\partial x_i \partial \eta_j} \right)$ is not zero at (x^0, η^0), we may calculate η as a function of (x, ξ) and prove that the map $(x, \xi) \rightarrow (y, \eta)$ so defined is a contact transformation (one will verify that $d(\xi \, dx - \eta \, dy) = 0$).

Ex. 5.8. Let $\psi(\xi)$ be a holomorphic function, homogeneous of degree 1 in a neighborhood of ξ^0. Prove that the map:

$$(x, \xi) \rightarrow (x + \psi'(\xi), \xi)$$

is a contact transformation.

Ex. 5.9. Calculate the closure of the conormal bundle to the manifold

$$S = \left\{ x \in \mathbb{C}^n; \quad \sum_j x_j^2 = 0, \quad x \neq 0 \right\}.$$

Ex. 5.10. Calculate the bicharacteristic curves of the operator $D_1^2 - \sum_{j=2}^n D_j^2$ on $T^* \mathbb{C}^n$. (For the definition of bicharacteristic leaves, cf. Appendix A).

Ex. 5.11. Calculate the bicharacteristic curve of the operator $D_1^2 - x_1 D_2^2$ issued from the point $(0, 0; 0, 1)$ of $T^* \mathbb{C}^n$.

Ex. 5.12. Let $X = \mathbb{C}^n$, $Y = \{x \in X; x_1 = 0\}$, ρ the natural projection $Y \times_X T^* X \rightarrow T^* Y$. Let Z be the submanifold $\{x' \in Y; x_2 = 0\}$ of Y, and finally let P be the operator $\sum_{j=1}^n D_j^2$.

a) Calculate the bicharacteristic curves issued from

$$\rho^{-1}(T_Z^* Y) \cap \{(x, \xi); \sigma(P)(x, \xi) = 0\}$$

b) Same question with P replaced by the operator $Q = D_1^2 - (x_1 - x_2) D_3^2$.

§ 6. Systems with Simple Characteristics

6.1. Equivalence of Operators

Let $(x_1, \ldots, x_n)$ be a system of local coordinates on X, (x, ξ) the associated coordinates on T^*X. We set $x = (x_1, x')$, $D_x = (D_1, D')$. We write $A(x, D')$ to mention that the microdifferential operator A does not depend on D_1 (i.e.: $[x_1, A] = 0$).

Theorem 6.1.1. *Let $A(x, D')$ be a $(m \times m)$-matrix whose entries are zero order microdifferential operators not depending on D_1, defined at the neighborhood of some point $(x^0; 0, \xi'^0)$. Let I be the $(m \times m)$-unit matrix. Then there exists an invertible $(m \times m)$-matrix $E(x, D')$, whose entries are zero order microdifferential operators not depending on D_1, such that:*

$$E^{-1} \circ (D_1 I + A) \circ E = D_1 I.$$

Proof. We may assume $x^0 = (0, x'^0)$. Applying Theorem 3.1.1 (and Remark 3.3.2) we find $E(x, D')$, of order zero, which solves the Cauchy problem:

$$(6.1.1) \qquad \begin{cases} \dfrac{\partial E}{\partial x_1} + A \circ E = 0, \\ E(0, x', D') = I. \end{cases}$$

Since the determinant of the matrix of the zero order symbols of the entries of E does not vanish at $(x^0; 0, \xi'^0)$, the matrix E is invertible in a neighborhood of that point (Proposition 1.3.5 and Ex. 1.4). Moreover:

$$(D_1 I + A) \circ E = \frac{\partial E}{\partial x_1} + A \circ E + E \circ D_1 I$$

$$= E \circ D_1 I. \quad \square$$

Corollary 6.1.2. *Let P be an operator of order m which is of the type $P = \sum\limits_{0 \leqslant j \leqslant m} A_j D_1^j$, where the A_j's are zero order microdifferential operators, and $A_m \equiv 1$. Then the two left $\mathscr{E}_X$-modules $\mathscr{E}_X/\mathscr{E}_X P$ and $\mathscr{E}_X/\mathscr{E}_X D_1^m$ are locally isomorphic.*

Proof. The two Modules have their supports on the manifold $V = \{(x, \xi); \xi_1 = 0\}$, since P and ξ_1^m are invertible outside of V. Thus it is enough to prove the result in a neighborhood of $(x^0; 0, \xi'^0)$. By applying the Weierstrass preparation theorem (Corollary 2.2.3) we find an invertible operator W, and operators $B_j(x, D')$ such that:

$$W \circ P = \sum_{0 \leqslant j \leqslant m} B_j(x, D') D_1^j, \quad \operatorname{ord}(B_j) \leqslant m - j, \quad [x_1, B_j] = 0.$$

If we remark that the homogeneous part of order $k \geqslant 0$ of the total symbol of P (thus of $W \circ P$) vanishes to an order $\geqslant k$ on the manifold V we find that the order of the B_j's are at most zero.

Replacing P by $W \circ P$, we may assume from the beginning that P is a Weierstrass polynomial.

Let B be the $(m \times m)$-matrix:

$$B = \begin{pmatrix} B_{m-1} & -1 & 0 & \cdot & 0 \\ \cdot & 0 & -1 & \cdot & \cdot \\ \cdot & \cdot & 0 & \cdot & 0 \\ \cdot & \cdot & \cdot & \cdot & -1 \\ B_0 & 0 & 0 & \cdot & 0 \end{pmatrix}.$$

We define the morphism from $\mathscr{E}_X/\mathscr{E}_X P$ to $\mathscr{E}_X^m/\mathscr{E}_X^m (D_1 I + B)$ by sending v to the class of $(1, 0, \ldots, 0) \bmod \mathscr{E}_X^m (D_1 I + B)$. This map is well defined since $(P, 0, \ldots, 0) = (D_1^{m-1}, \ldots, 1)(D_1 I + B)$, and it is left to the reader to check that it is an isomorphism. $\square$

Remark 6.1.3. If P is of order 1, with $\sigma(P) = \xi_1$, we have proved that there exist two invertible microdifferential operators of order zero, A and B such that $A \circ P \circ B^{-1} = D_1$. In fact we may choose A and B such that $A = B$ (cf. Sato-Kashiwara-Kawaï [1]).

6.2. The Regular Involutive Case

Theorem 6.2.1. *Let $\mathscr{I}$ be a left Ideal of $\mathscr{E}_X$ defined on an open set U of $T^* X$. Let $\mathscr{M}$ be the left $\mathscr{E}_X$-module $\mathscr{E}_X/\mathscr{I}$, and let $\mathscr{J}$ be the symbol Ideal of $\mathscr{I}$. Assume $\mathscr{J}$ is generated by $\xi_1, \ldots, \xi_p$. Then locally on U, $\mathscr{M}$ is isomorphic to the Module $\mathscr{E}_X/\mathscr{E}_X (D_1, \ldots, D_p)$.*

Proof. We shall follow the proof of Sato-Kashiwara-Kawaï [1].

a) First we remark that if $D_1, \ldots, D_p$ belong to an Ideal $\mathscr{I}$ such that $\mathscr{J}$ is generated by $\xi_1, \ldots, \xi_p$, then $D_1, \ldots, D_p$ generate $\mathscr{I}$. In fact if P is a section of $\mathscr{I}$ we may write, after successive applications of the division theorem:

$$P = \sum_{1 \leqslant j \leqslant p} A_j D_j + R(x, D''),$$

where $D'' = (D_{p+1}, \ldots, D_n)$. Since R will be a section of $\mathscr{I}$, $\sigma(R)$ will vanish for $\xi_1 = \ldots = \xi_p = 0$, thus $R = 0$.

b) By hypothesis there exist operators $P_i = D_i + R_i(x, D)$, $i = 1, \ldots, p$, which belong to $\mathscr{I}$, where the R_i are of order zero. Let us proceed by induction and assume $P_j = D_j$ for $j = 1, \ldots, k$.

c) We may assume that for all i, $1 \leqslant i \leqslant p$, R_i does not depend on $D_i, D_{i+1}, \ldots, D_p$. In fact assume this is true for $i > l$ and consider $R_l(x, D)$. We first divide R_l by P_p which allows us to assume that the R_l are independent of D_p. Then we divide it by P_{p-1}, which is itself independent of D_p, and the remainder will depend neither on D_p nor on D_{p-1}. We continue this process up to order l.

d) In particular we have obtained:

$$P_{k+1} = D_{k+1} + R_{k+1}(x; D_1, \ldots, D_k, D_{p+1}, \ldots, D_n).$$

Since $D_1, \ldots, D_k$ belong to $\mathscr{I}$ by the induction assumption, we may divide R_{k+1} by $D_1, \ldots, D_k$ and assume:

$$P_{k+1} = D_{k+1} + R_{k+1}(x; D_{p+1}, \ldots, D_n)$$

e) The brackets $[D_j, P_{k+1}], j = 1, \ldots, k$ belong to $\mathscr{I}$, but do not depend on $D_1, \ldots, D_p$. Since their principal symbols should vanish on the manifold $\{(x, \xi); \xi_1 = \ldots = \xi_p = 0\}$, they are identically zero, which means that R_{k+1} is independent of $x_1, \ldots, x_k$. Set $\tilde{x} = (x_{k+1}, \ldots, x_n)$, $\tilde{D} = (D_{k+1}, \ldots, D_n)$. We may find by Theorem 6.1.1 an invertible operator $A(\tilde{x}, \tilde{D})$ of order zero such that $A \circ P_{k+1} \circ A^{-1} = D_{k+1}$.

f) Let $\mathscr{I}_A$ be the left Ideal in $\mathscr{E}_X$ defined by:

$$Q \in \mathscr{I}_A \Leftrightarrow Q \circ A \in \mathscr{I}.$$

The two Modules $\mathscr{E}_X/\mathscr{I}$ and $\mathscr{E}_X/\mathscr{I}_A$ are clearly isomorphic and D_{k+1} belongs to $\mathscr{I}_A$. Moreover $D_1, \ldots, D_k$ belong to $\mathscr{I}_A$ since these operators commute with A, which completes the proof.   □

Definition 6.2.2. Let V be a closed conic involutive manifold of $U \subset T^*X$, $\mathscr{I}_V$ the defining Ideal of V in $\mathscr{O}_{T^*X}$. A Module with simple characteristics along V is an $\mathscr{E}_X$-module $\mathscr{M}$ which, locally on U, admits one generator u such that if $\mathscr{I}$ is the annihilator Ideal of u in $\mathscr{E}_X$, $\mathscr{I}$ the symbol Ideal of $\mathscr{I}$, then $\mathscr{I} = \mathscr{I}_V$.

We shall say that such a generator is a simple generator of $\mathscr{M}$.

Remark that if $\mathscr{M}$ is a Module with simple characteristics along V, then it is supported by V.

Corollary 6.2.3. Let $\mathscr{M}$ and $\mathscr{N}$ be two left $\mathscr{E}_X$-modules with simple characteristics along the regular involutive manifold V (recall that V regular means that $\omega|_V \neq 0$). Then $\mathscr{M}$ and $\mathscr{N}$ are locally isomorphic.

Proof. We may perform a quantized contact transformation, and assume V defined by the equation $\xi_1 = \ldots = \xi_p = 0$ (Corollary A.4.5). Then we apply Theorem 6.2.1.   □

Remark 6.2.4. Take $X = \mathbb{C}$, $\mathscr{M} = \mathscr{D}_\mathbb{C}/\mathscr{D}_\mathbb{C}D$. Let $u = 1 \mod(\mathscr{D}_\mathbb{C}D)$. Then u is a simple generator of $\mathscr{M}$. Now consider $v = xu$. Then v also generates $\mathscr{M}$

since $Dv=u$, but v is not a simple generator of $\mathscr{M}$. In fact $\mathscr{I}_v$, the annihilator of v, is generated by the two operators $xD-1$ and D^2, and $\mathscr{I}_v$, the symbol Ideal by the two functions $x\xi$ and ξ^2, and $\mathscr{I}_v$ is not a reduced Ideal of $\mathscr{O}_{T^*\mathbb{C}}$.

6.3. Holonomic Systems with Simple Characteristics

Definition 6.3.1. A holonomic system with simple characteristics is an $\mathscr{E}_X$-module with simple characteristics along a Lagrangean manifold.

Let Λ be the manifold $\{(x,\xi)\in\dot{T}^*\mathbb{C}^n; x_1=\xi_2=\ldots=\xi_n=0\}$.

Theorem 6.3.2. *Let $\mathscr{M}$ be a holonomic system with simple characteristics along Λ, on an open set $U\subset\dot{T}^*\mathbb{C}^n$. Then locally on U there exists $\alpha\in\mathbb{C}$ such that $\mathscr{M}$ is isomorphic to the Module $\mathscr{E}_{\mathbb{C}^n}/\mathscr{I}_\alpha$ where $\mathscr{I}_\alpha$ is the left Ideal generated by $x_1D_1-\alpha, D_2, \ldots, D_n$. Moreover if $U\cap\Lambda$ is connected and non empty then α is unique modulo $\mathbb{Z}$.*

Proof. As in the proof of Theorem 6.2.1 we may locally find an Ideal $\mathscr{I}$ which contains $D_2, \ldots, D_n$ such that $\mathscr{M}\cong\mathscr{E}_X/\mathscr{I}$. Since $x_1\xi_1$ belongs to $\mathscr{I}$ there exists an operator R of order zero such that x_1D+R belong to $\mathscr{I}$. By dividing R by $D_2, \ldots, D_n$ and by applying Corollary 2.2.3, we may assume $R(x', D_1)$ not to depend on $x_1, D_2, \ldots, D_n$. Since the brackets $[D_j, R(x', D_1)]$ $(2\leqslant j\leqslant n)$ belong to $\mathscr{I}$, the principal symbols of $\dfrac{\partial}{\partial x_j}R(x', D_1)$ should vanish on Λ. Since this operator does not depend on $x_1, D_2, \ldots, D_n$, is must be identically zero. Thus we have found operators:

$$x_1D_1+R(D_1), \qquad D_2, \ldots, D_n,$$

belonging to $\mathscr{I}$, with $R(D_1)\in\mathscr{E}_\mathbb{C}(0)$, $R(D_1)$ independent of x_1.

Lemma 6.3.3. *Let (t,τ) be the coordinates on $T^*\mathbb{C}$. Let $A(D_t)$ be a microdifferential operator of order strictly inferior to -1, not depending on t (thus defined for $\tau\neq0$). Then there exists a zero order microdifferential operator $E(D_t)$, not depending on t, with principal symbol equal to 1, such that:*

$$(t+A(D_t))\circ E=E\circ t.$$

Proof of Lemma 6.3.3. Let $A(\tau)$ be the total symbol of A. One solves:

$$\begin{cases} \dfrac{\partial E}{\partial\tau} - A(\tau)E(\tau)=0, \\ E(0)=1. \end{cases}$$

Thanks to the hypothesis $\operatorname{ord}(A) < -1$ one checks immediately that the formal serie (with respect to τ^{-1}) $E(\tau)$ is the total symbol of a microdifferential operator E.   □

End of the proof of Theorem 6.3.2. Let $\alpha = -\sigma_0(R(D_1))$, $S(D_1) = R(D_1) + \alpha$. Applying the lemma to the operator $(x_1 D_1 + S) \circ D_1^{-1}$ we get:

$$E^{-1}(D_1) \circ (x_1 D_1 - \alpha + S(D_1)) \circ E(D_1) = x_1 D_1 - \alpha,$$

which completes the proof of the existence of the Ideal $\mathscr{I}_\alpha$. Uniqueness follows from:

Lemma 6.3.4. *Let α and β be two complex numbers. Then in some neighborhood of $(0; 1) \in T^* \mathbb{C}$ we have:*

$$\mathscr{H}om_{\mathscr{E}_{\mathbb{C}}}(\mathscr{E}_{\mathbb{C}}/\mathscr{I}_\alpha, \mathscr{E}_{\mathbb{C}}/\mathscr{I}_\beta) = \begin{cases} 0 & \text{for } \alpha - \beta \notin \mathbb{Z}, \\ \mathbb{C} & \text{for } \alpha - \beta \in \mathbb{Z}, \end{cases}$$

where $\mathscr{I}_\gamma$ denotes the Ideal $\mathscr{E}_{\mathbb{C}}(t D_t - \gamma)$, $(\gamma = \alpha, \beta)$.

Proof. Any section P of $\mathscr{E}_{\mathbb{C}}$ at the neighborhood of $(0; 1)$ may be decomposed as (Theorem 2.2.1):

$$P(t, D_t) = \sum_{-\infty < j \leqslant m} a_j D_t^j + Q \circ (t D_t - \beta)$$

where $a_j \in \mathbb{C}$.

Let v be the class of $1 \in \mathscr{E}_{\mathbb{C}} \bmod \mathscr{E}_{\mathbb{C}}(t D_t - \beta)$ and let u be a section of $\mathscr{E}_{\mathbb{C}} v = \mathscr{E}_{\mathbb{C}}/\mathscr{E}_{\mathbb{C}}(t D_t - \beta)$. Then $u = Pv$, for some $P = \sum_j a_j D_t^j$. Assume u belongs to $\mathscr{H}om_{\mathscr{E}_{\mathbb{C}}}(\mathscr{E}_{\mathbb{C}}/\mathscr{I}_\alpha, \mathscr{E}_{\mathbb{C}}/\mathscr{I}_\beta)$, that is assume $(t D_t - \alpha) u = 0$. We get:

$$(t D_t - \alpha)\left(\sum_j a_j D_t^j v\right) = 0, \quad (t D_t - \beta) v = 0,$$

thus:

$$\sum_{-\infty < j \leqslant m} a_j(\beta - \alpha - j) D_t^j v = 0,$$

in other words $\sum_{-\infty < j \leqslant m} a_j(\beta - \alpha - j) D_t^j$ should be in the left Ideal $\mathscr{E}_{\mathbb{C}}(t D_t - \beta)$. But the uniqueness of the division by $(t D_t - \beta)$ implies $\sum_j a_j(\beta - \alpha - j) D_t^j = 0$, thus $a_j(\beta - \alpha - j) = 0 \ \forall j$.

If $\alpha - \beta \notin \mathbb{Z}$, $a_j = 0$ for all j, and $u = 0$. If $\alpha - \beta$ is an integer then the space of solutions is one dimensional.   □

Exercises to I.6

Ex. 6.1. Calculate explicitly the operator E given by Lemma 6.3.3 such that $E \circ (t D_t + D_t^{-1}) \circ E^{-1} = t D_t$.

Ex. 6.2. Extend Lemma 6.3.3 to square $(m \times m)$-matrices $tI + A(D_t)$.

Ex. 6.3. Let $\mathscr{D}_{\mathbb{C}} u^{\alpha}$ be the left $\mathscr{D}_{\mathbb{C}}$-module $\mathscr{D}_{\mathbb{C}} / \mathscr{D}_{\mathbb{C}} (tD_t - \alpha)$. For which couples $(\alpha, \beta) \in \mathbb{C} \times \mathbb{C}$ are the two Modules $\mathscr{D}_{\mathbb{C}} u^{\alpha}$ and $\mathscr{D}_{\mathbb{C}} u^{\beta}$ isomorphic?

(Answer: i) $\alpha \notin \mathbb{Z}$, $\alpha - \beta \in \mathbb{Z}$, ii) $\alpha \in \mathbb{N}$, $\beta = 0$, iii) $-\alpha \in \mathbb{N}$, $\alpha \neq 0$, $\beta = -1$). (cf. Malgrange [2]).

Ex. 6.4. Let $b(\cdot)$ be a polynomial on $\mathbb{C}$. Calculate the kernel and the cokernel of the differential operator $b(tD_t)$ acting:

a) on $\mathscr{B}_{\{0\} | \mathbb{C}}$,   b) on $\mathscr{C}_{\{0\} | \mathbb{C}} |_{\dot{T}^* \mathbb{C}}$.

Ex. 6.5. Let $\Lambda = T_Z^* X$. Prove that the only $\mathscr{D}_X$-module with simple characteristics on Λ is the sheaf $\mathscr{B}_{Z|X}$.

Notes

Analytic microdifferential operators first appeared in the work of L. Boutet de Monvel and P. Kree [1] in a more restrictive situation (they were only defined all over $\dot{T}^* \Omega$ for Ω an open subset of $\mathbb{R}^n$). Later the sheaf $\mathscr{E}_X$ was introduced by M. Sato, M. Kashiwara, T. Kawaï [1] as a sub-Ring of a Ring $\mathscr{E}_X^{\mathbb{R}}$, this last sheaf being defined cohomologically, using Sato's microlocalization.

The results in § 1.3 are due to Boutet de Monvel-Kree [1], and the division theorems of § 2 are due to Sato-Kashiwara-Kawaï [1]. Our proof of Theorem 2.1.1 is slightly different from Sato-Kashiwara-Kawaï's, since, following a suggestion of C. Houzel, we begin by an abstract division theorem in Banach algebras.

The microdifferential Cauchy-Kowalewski theorem (Theorem 3.1.1) is due to the author, but is closely related to a theorem of J.M. Bony and P. Schapira [2, Theorem 3.1.2] (cf. also Bony [1]). The abstract Cauchy-Kowalewski theorem (Theorem 3.2.1) has a long history starting with Nagumo [1] (cf. Nirenberg [1]).

The sheaves $\mathscr{B}_{Z|X}$, $\mathscr{C}_{Z|X}$, $\mathscr{E}_{Y \to X}$ are constructed in Sato-Kashiwara-Kawaï [1], with the same method as for $\mathscr{E}_X$, that is, with cohomological tools. Our construction is not so elegant, but might be more direct and accessible. Proposition 4.1.5 was announced in Sato-Kashiwara-Kimura-Oshima [1]. The sequence of functions $\phi_n(t)$ (formula (4.2.4)) has been used by many authors (e.g. Hamada [1], Sato-Kashiwara-Kawaï [1]).

Quantized contact transformations were initiated in the work of V.P. Maslov [1], but in the analytic framework they are due to Sato-Kashiwara-Kawaï [1]. Here we follow M. Kashiwara [5] for the proof of Theorem 5.3.1.

In Sato-Kashiwara-Kawaï [1], Theorem 6.1.1 is proved directly, but here we obtain it as a corollary of the microdifferential Cauchy-Kowalewski theorem.

The proofs of Theorem 6.2.1 and 6.3.2 follow Sato-Kashiwara-Kawaï [1].

Finally let us mention an extension of the division theorem by L. Boutet de Monvel [1] which provides another proof of Theorem 5.3.1. Let us also mention the different approach of J. Sjöstrand [1] of analytic microdifferential operators.

Chapter II. $\mathscr{E}_X$-modules

Summary

This chapter is devoted to the study of the algebraic structure of the sheaf of rings $\mathscr{E}_X$ and of the usual operations on Modules over $\mathscr{E}_X$. The Ring $\mathscr{E}_X$ has a $\mathbb{Z}$-filtration induced by the order, and many properties of a $\mathbb{Z}$-filtered Module $\mathscr{M}$ over a $\mathbb{Z}$-filtered Ring $\mathscr{A}$ can be obtained from the corresponding ones on $\mathrm{gr}(\mathscr{M})$, the associated graded Module over $\mathrm{gr}(\mathscr{A})$, if we make the assumption that the filtration is what we call a "zariskian filtration". We study the theory of filtered modules in the abstract in § 1, with special attention to characteristic ideals, flatness, homological dimension, coherency. The important theorem of O. Gabber [1] on the involutivity of the characteristic ideal is stated, but it seemed irrelevant to us to duplicate his very clear proof.

Next we have to show that the filtration over $\mathscr{E}_X$ is in fact a zariskian one. The proof follows that of Sato-Kashiwara-Kawaï [1] and is classical in its approach: induction on the dimension with the help of the division theorems. The main properties of $\mathscr{E}_X$ then follow immediately: the sheaf $\mathscr{E}_X$ is coherent and noetherian, and the characteristic cycle of a coherent $\mathscr{E}_X$-module is a conic analytic involutive cycle. As a corollary of the involutivity we get that the homological dimension of $\mathscr{E}_X$ is equal to n the (complex) dimension of X, and even that any coherent Module locally admits a free resolution of length at most n.

It is beyond the scope of this book to make a detailed study of holonomic Modules (i.e.: coherent Modules whose characteristic variety is Lagrangean). We just give the first properties of the category of holonomic Modules, and study the "adjoint" functor on this category.

We end § 2 by introducing the "dummy variable" trick which enables to work outside of the zero section of T^*X, and by finally studying the relations between coherent $\mathscr{D}_X$-modules and globally defined coherent $\mathscr{E}_X$-modules.

Next we explain how one manipulates coherent $\mathscr{E}_X$-modules with the same ease as is possible for functions: tensor products for two Modules over two different manifolds, inverse images for left $\mathscr{E}_X$-modules (analogous to restriction for functions), direct images of right Modules (in line with integration of differential forms). All these constructions have been done in

Sato-Kashiwara-Kawaï [1] using a deep cohomological method, but we proceed here in an elementary way, beginning with the study of these operations for Modules like $\mathscr{C}_{Z|X}$, then passing to the general case. Non-characteristic restrictions of $\mathscr{E}_X$-modules are studied in detail, and we calculate exactly, following M. Kashiwara [5], the characteristic cycle of the induced system in this case.

On the other hand direct images for proper maps are not studied here, and reference is made to Kashiwara [3] and Pham [1] for further developments on this subject (cf. also Chapter III, § 4).

Finally we give a general formulation of the Cauchy problem for general systems, as preparation for the next chapter.

§ 1. Filtered Rings and Modules

1.1. Noetherian and Zariskian Filtrations

Throughout § 1 we shall consider unitary rings, in general non commutative. Unless otherwise specified we consider only left modules. Thus "module" will mean left module, (cf. Appendix B).

A filtered ring FA (over $\mathbb{Z}$), is a ring A endowed with a family of subgroups A_k, $k \in \mathbb{Z}$, such that:

$$(1.1.1) \quad \begin{cases} A_k \subset A_{k+1}, \quad \bigcup_k A_k = A, \\ 1 \in A_0, \quad A_k \cdot A_l \subset A_{k+l} \quad (k, l \in \mathbb{Z}). \end{cases}$$

An FA-module, or a filtered A-module, FM, is an A-module M endowed with a family of subgroups M_k such that:

$$(1.1.2) \quad \begin{cases} M_k \subset M_{k+1}, \quad \bigcup_k M_k = M, \\ A_l \cdot M_k \subset M_{k+l} \quad (k, l \in \mathbb{Z}). \end{cases}$$

We sometimes write $F_k M$ instead of M_k to specify the filtration FM we consider on M.

The order of an element u of FM is the smallest $k \in \mathbb{Z} \cup \{-\infty\}$ such that $u \in M_k \setminus M_{k-1}$.

Two filtrations FM and $F'M$ on M are said to be equivalent if there exists an $r \in \mathbb{N}$ such that:

$$(1.1.3) \quad F_{k-r} M \subset F'_k M \subset F_{k+r} M \quad \forall k \in \mathbb{Z}.$$

We shall have to consider the A-module A endowed with the "r-shifted filtration":

(1.1.4) $(FA_{[r]})_k = A_{k-r}$.

Let FM and FN be two FA-modules.

An FA-linear map from FM to FN is an A-linear map ψ from M to N such that:

$$\psi(M_k) \subset N_k \qquad \forall k \in \mathbb{Z}.$$

In general we keep the same notation for the A-linear and the FA-linear maps. We endow Ker ψ with the induced filtration:

(1.1.5) $(\text{Ker }\psi)_k = \text{Ker }\psi \cap M_k$

and Im ψ with the image filtration:

(1.1.6) $(\text{Im }\psi)_k = \psi(M_k)$.

A complex of FA-modules:

(1.1.7) $\ldots \to FM_j \xrightarrow[\psi_j]{} FM_{j+1} \to \ldots$

is a complex of A-modules such that each ψ_j is FA-linear. Such a complex is called an exact sequence of FA-modules if moreover:

(1.1.8) $\text{Im }\psi_j = \text{Ker }\psi_{j+1} \qquad \forall j$, as FA-modules.

Thus one should take care that a complex may be exact as a complex of A-modules, but not as a complex of FA-modules. (The category of FA-modules and FA-linear maps is not abelian). For example the identity $FA \to FA_{[-1]}$ is FA-linear, but the inverse is not FA-linear, hence this map is not an isomorphism of FA-modules.

The filtration on a direct sum of two FA-modules FN and FM will be defined by:

(1.1.9) $(FN \oplus FM)_k = N_k \oplus M_k$.

As is well known (cf. Bourbaki [1]) it is possible to associate a topology on M to the filtration FM: the M_k's define a fundamental system of neighborhoods of 0. In particular M is separated iff $\bigcap_k M_k = \{0\}$, and a subgroup $G \subset M$ is closed iff:

(1.1.10) $G = \bigcap_l (G + M_l)$.

Remark that M is separated iff the only element of order $-\infty$ is 0. We denote[1] by GA the graded ring of FA:

$$GA = \bigoplus_k A_k/A_{k-1},$$

---

[1] We shall also use in this book the notation gr(A) or gr(M) instead of GA or GM, especially when $A = \mathscr{D}_X$ or $A = \mathscr{E}_X$.

and to an FA-module FM we associate the graded GA-module

$$GM = \bigoplus_k M_k/M_{k-1}.$$

We shall only consider the structure of graded ring on GA, not to be confused with the underlying ring structure. Thus "GA-module" means "graded GA-module".

The natural map from FM to GM is called the "principal symbol" map, and denoted by $\sigma(\cdot)$. When there is no risk of confusion we also call σ the symbol map.

We also put:

$$GA_k = A_k/A_{k-1}$$

and define the "symbol of order k", σ_k, as the map:

(1.1.11) $$\sigma_k: A_k \to GA_k$$

An FA-linear map ψ naturally defines a GA-linear map from GM to GN, denoted by $\sigma(\psi)$, or $\overline{\psi}$.

Definition 1.1.1. i) A finite free FA-module FL is an FA-module isomorphic to a finite direct sum of modules $FA_{[i]}$, $i \in \mathbb{Z}$.

ii) A finite free s-presentation of an FA-module FM is an exact sequence of FA-modules:

(1.1.12) $$FL_{-s} \to \dots \to FL_0 \to FM \to 0,$$

where the FL_j's are finite free.

iii) A finite type FA-module FM is an FA-module with admits a finite free 0-presentation. In that case one says that the filtration FM on M is "a good filtration".

It is clear that a finite type FA-module is a finite type A-module, and conversely any finite type A-module admits good filtrations. Moreover all good filtrations on a finite type A-module are equivalent.

Definition 1.1.2. i) We shall say that the filtered ring FA is (left) noetherian (or also if the filtration is unambiguously defined, that the ring A is filtered noetherian), if any finite free 0-presentation of an FA-module extends as a finite free 1-presentation.

ii) We shall say that FA is zariskian if FA is noetherian and moreover for any $a \in A_{-1}$, $1 - a$ is invertible in A_0.

If FA is noetherian, if FM is an FA-module of finite type, and if FN is a sub-FA-module of FM (for the induced filtration), then FN is of finite type. In fact we take a finite free 0-presentation of FM

$$\psi: FL \to FM \to 0$$

and denote by F I the inverse image by ψ of F N. It is enough to prove that F I (for the induced filtration by F L) is of finite type.

Consider the 0-presentation of F L/F I:

$$FL \to FL/FI \to 0.$$

The ring F A being noetherian, this 0-presentation extends as a 1-presentation, which proves that F I is of finite type.

Of course if F A is noetherian, then A is noetherian. We shall give examples in Section 1.5.

Proposition 1.1.3. *Assume* F A *zariskian. Then:*

a) *If* F M *is an* F A-*module of finite type, then it is separated.*

b) *Let* F M *be an* F A-*module of finite type,* F N *a submodule endowed with the induced filtration. Then* N *and all the* N_k*'s are closed in* F M.

c) *Let* F M *be an* F A-*module of finite type. Then* G M *is of finite type over* G A, *and if* $u_1, \ldots, u_r$ *are elements of* F M *of order* $k_1, \ldots, k_r$ *whose principal symbols generate* G M, *then the* u_j*'s generate* M *and:*

$$(1.1.13) \qquad\qquad M_k = \sum_{j=1}^{r} A_{k-k_j} u_j \qquad \forall k.$$

d) *Consider a complex of* F A-*modules*

$$(1.1.14) \qquad\qquad FL \xrightarrow{\phi} FM \xrightarrow{\psi} FN$$

and the associated complex of G A-*modules:*

$$(1.1.15) \qquad\qquad GL \xrightarrow{\overline{\phi}} GM \xrightarrow{\overline{\psi}} GN.$$

Assume F M *of finite type. Then if (1.1.15) is exact, then (1.1.14) is exact.*

e) *Let* F M *be an* F A-*module of finite type. If* G M *is free, then* F M *is finite free.*

Proof. a) Let $u \in \bigcap_k M_k$, and set $L = Au$. The two filtrations $L_k = A_k u$ and $F_k L = L \cap M_k$ on L are equivalent since both are of finite type. Thus $A_{-1} u$ contains $L \cap M_k$ for some k, and since $L \cap M_k = L$, u belongs to $A_{-1} u$. Thus $u = au$ for some $a \in A_{-1}$, and $1-a$ being invertible, $u = 0$.

b) since F M/F N is of finite type, it is separated and N is closed. Thus:

$$\bigcap_l (N_k + M_l) \subset N \cap (N_k + M_l) \subset N_k + N_l \subset N_k \quad \text{for} \quad l \leqslant k.$$

c) There exists a finite free 0-presentation of F M, $FL \to FM \to 0$. This induces a finite free 0-presentation $GL \to GM \to 0$ of G M, since G L is finite free over G A. Thus G M is of finite type. Set $M'_k = \sum_j A_{k-k_j} u_j$. Then $M_k = \bigcap_l (M'_k + M_l)$, and by b) we get $M_k = M'_k$.

d) Let $u \in M_k$, such that $\psi(u) = 0$. Then $\sigma(u) = \overline{\psi}(\overline{v})$ for some $\overline{v} \in L_k/L_{k-1}$. Choose $v \in L_k$ with $\sigma(v) = \overline{v}$. We get:

$$u = \phi(v) + w, \quad w \in M_{k-1}.$$

Since $\psi(w) = 0$, we may recommence this procedure with w. Thus for all l we get:

$$u \in \phi(L_k) + M_l.$$

Since $\phi(L_k)$ is closed, $u \in \phi(L_k)$.

e) Let $u_1, \ldots, u_r$ be elements of FM of order $k_1, \ldots, k_r$ such that $\sigma(u_1), \ldots, \sigma(u_r)$ is a basis of GM. Set $FL = \bigoplus_{j=1}^{r} A_{[k_j]}$ and let ψ be the FA-linear map from FL to FM which associates u_j to the element 1 of order k_j in $A_{[k_j]}$.

Endow $\mathrm{Ker}\,\psi$ with the induced filtration. By c) we get an exact sequence of FA-modules:

$$0 \to \mathrm{Ker}\,\psi \to FL \xrightarrow{\psi} FM \to 0 .$$

Since the associated graded morphism is an isomorphism, $G(\mathrm{Ker}\,\psi) = 0$, and the filtrations being separated, $\mathrm{Ker}\,\psi = 0$. □

Proposition 1.1.4. *Let FM and FN be two FA-modules of finite type, ψ an FA-linear map from FM to FN, $\overline{\psi}$ the graded linear map from GM to GN associated to ψ. We endow $\mathrm{Ker}\,\psi$ (resp. $\mathrm{Im}\,\psi$) with the induced (resp. image) filtration, and we denote by $\mathrm{Im}'\psi$ the module $\mathrm{Im}\,\psi$ endowed with the induced filtration by FN. Then:*
  a) $G(\mathrm{Ker}\,\psi) \subset \mathrm{Ker}\,\overline{\psi}$.
  b) $\mathrm{Im}(\overline{\psi}) \subset G(\mathrm{Im}'\psi)$.
  c) $G(\mathrm{Ker}\,\psi) = \mathrm{Ker}\,\overline{\psi} \Rightarrow \mathrm{Im}\,\overline{\psi} = G(\mathrm{Im}'\psi)$.
Assume FA zariskian. Then:
  d) $\mathrm{Im}\,\overline{\psi} = G(\mathrm{Im}'\psi) \Rightarrow G(\mathrm{Ker}\,\psi) = \mathrm{Ker}\,\overline{\psi}$.

Proof. We have:

$$G(\mathrm{Ker}\,\psi) = \bigoplus_k (\psi^{-1}(0) \cap M_k)/(\psi^{-1}(0) \cap M_{k-1}),$$
$$\mathrm{Ker}\,\overline{\psi} = \bigoplus_k (\psi^{-1}(N_{k-1}) \cap M_k)/M_{k-1},$$
$$G(\mathrm{Im}'\psi) = \bigoplus_k (\psi(M) \cap N_k)/(\psi(M) \cap N_{k-1}),$$
$$\mathrm{Im}\,\overline{\psi} = \bigoplus_k \psi(M_k)/(\psi(M_k) \cap N_{k-1}).$$

Thus a) and b) are obvious.

c) Let $\bar{v} \in G(\text{Im}'\psi)$. We may assume $\bar{v}$ homogeneous. Let $v \in \psi(M) \cap N_k$ such that $\sigma(v) = \bar{v}$. Thus $v = \psi(u)$ for some $u \in M$ of order $k' \geqslant k$. We argue by induction on $k' - k$, and assume $k' - k > 0$. Then $\bar{\psi}(\sigma(u)) = 0$, and by the hypothesis there exists $w \in M$, with $\psi(w) = 0$, $\sigma(w) = \sigma(u)$. Then $\psi(u - w) = v$, $\text{ord}(u - w) < k'$.

d) Let $\bar{u} \in \text{Ker}\,\bar{\psi}$. We may assume $\bar{u}$ is homogeneous. Let $u \in M_k$, with $\sigma(u) = \bar{u}$. Then $\psi(u) \in N_{k-1}$. By assumption there exists $u_1 \in M_{k-1}$ and $v_2 \in N_{k-2}$ such that $\psi(u) = \psi(u_1) + v_2$. By induction we find $u_r \in M_{k-r}$ and $v_r \in N_{k-r}$ such that:

$$\psi(u - u_1 - \ldots - u_{r-1}) = v_r.$$

Thus:

$$\psi(u) \in \bigcap_l (\psi(M_{k-1}) + N_l).$$

Since $\psi(M_{k-1})$ is closed in FN, we find that $\psi(u) \in \psi(M_{k-1})$, and there exists $u' \in M_{k-1}$ such that $\psi(u - u') = 0$. This proves d) since $\sigma(u - u') = \sigma(u)$. $\square$

Now we shall give sufficient conditions for a filtered ring to be zariskian.

Proposition 1.1.5. *Let* FA *be a filtered ring. Assume:*
   i) GA *is a (left) noetherian graded ring.*
   ii) *Any* A_0-*submodule of* A_0^n ($n \in \mathbb{N}$), *is closed for the topology induced by* $(FA)^n$.
   Then FA *is zariskian.*

Proof. a) Let $a \in A_{-1}$. Since:

$$(1 - a)(1 + a + \ldots + a^k) = 1 - a^{k+1},$$

we get:

$$1 \in \bigcap_l ((1 - a)A_0 + A_l).$$

But $(1 - a)A_0$ is supposed to be closed in A_0, thus $(1 - a)$ is invertible.

b) It is enough to check that any submodule of a finite free FA-module is of finite type over FA for the induced filtration.

Let FN be a submodule of $FL = \bigoplus_{j=1}^{r} FA_{[k_j]}$, endowed with the induced filtration. Let $u_1, \ldots, u_d$ be elements of FN of order $l_1, \ldots, l_d$ whose symbols $\sigma(u_1), \ldots, \sigma(u_d)$ generate GN, (remark that GN is of finite type by the hypothesis).

Set $N_k = N \cap L_k$, and $N'_k = \sum_j A_{k-l_j} u_j$. Of course N'_k is contained in N_k and we have to prove the equality.

Let $u \in N$, u of order k. Then $\sigma(u) = \sum_j \bar{a}_j \sigma(u_j)$ for some $\bar{a}_j \in A_{k-l_j}/A_{k-l_j-1}$. Choose $a_j \in A_{k-l_j}$, with $\sigma(a_j) = \bar{a}_j$. We get:

$$u = \sum_j a_j u_j + v, \quad v \in N_{k-1}.$$

Applying the same procedure to v, we get:

(1.1.16) $$N'_k \subset N_k \subset \bigcap_l (N'_k + N_l)$$

For $k \ll 0$, N'_k is contained in A'_0 and thus $N'_k = N_k$ for k small enough, but this implies $N'_k = N_k \ \forall k$ by (1.1.16). $\square$

There is also a useful criterion to ensure that a filtered ring is noetherian.

Definition 1.1.6. Let FA be a filtered ring. The graded ring

$$A' = \bigoplus_{k \in \mathbb{Z}} A_k T^k,$$

(with the relations $T^k T^l = T^{k+l}$), will be called the formal graded ring associated to FA.

Remark that we may recover GA from A':

(1.1.17) $$GA \cong A'/A' \cdot T.$$

Proposition 1.1.7. *Let* FA *be a filtered ring. Then the associated formal graded ring* A' *is noetherian (as a graded ring) if and only if* FA *is noetherian. In particular if* FA *is noetherian then* GA *is (graded) noetherian.*

(For the notion of graded noetherian ring cf. the Appendix B.3).

Proof. To an FA-module FM we associate an A'-module M' by setting:

$$M' = \bigoplus_k M_k T^k,$$

and to an FA-linear map ψ from FM to FN we associate the A'-linear map ψ' from M' to N', in an obvious way.

First assume A' is noetherian, and consider a finite free 0-presentation $FL_0 \to FM \to 0$, and the associated finite free 0-presentation of M': $L'_0 \to M' \to 0$.

By hypothesis this presentation extends to a 1-presentation:

(1.1.18) $$K_1 \underset{\psi}{\longrightarrow} L'_0 \to M' \to 0.$$

The graded linear map ψ comes from a linear map of finite free FA-modules $FL_1 \underset{\phi}{\longrightarrow} FL_0$, that is, $K_1 = L'_1$, $\psi = \phi'$, and the exactness of (1.1.18) implies the exactness of the sequence:

$$FL_1 \to FL_0 \to FM \to 0.$$

To prove the converse we remark that any graded submodule of a finite free A'-module comes from an FA-submodule of a finite free FA-module by the above procedure.

Finally if A' is noetherian, then so is GA by (1.1.17).   □

Proposition 1.1.8. *Let* FA *be a filtered ring. Assume:*
   i) FA_0 *is a noetherian filtered ring.*
   ii) GA *is a noetherian graded ring.*
   Then FA *is noetherian.*

Proof. As in the proof of Proposition 1.1.5 let $u_1, \ldots, u_d$ be elements of order $k_1, \ldots, k_d$ of FN, where FN is a submodule of the finite free FA-module $FL = \bigoplus_{j=1}^{r} FA_{[k_j]}$, such that $\sigma(u_1), \ldots, \sigma(u_d)$ generate GN. We may assume that each u_j is contained in A_0^r, and the u_j's generate FN_0. Then the proof goes as for Proposition 1.1.5.   □

Remark 1.1.9. If FA is a filtered ring, if A_0 is noetherian (as a ring) and if GA is graded noetherian, we find by a similar proof that A is a noetherian ring.

Remark 1.1.10. Let $GB = \bigoplus_{k} B_k$ be a graded ring, and let us denote by B the underlying (non graded) ring. It is immediate that B noetherian implies GB noetherian. The converse is true if we assume $B_k = 0$ for $k < 0$. In fact we endow GB with the filtration:

$$F_k GB = \bigoplus_{j \leqslant k} B_j .$$

Since $GGB = GB$ and $F_0 GB$ is a noetherian ring, we find, by Remark 1.1.9, that B is noetherian.

1.2. Homological Properties

Proposition 1.2.1. *Assume* FA *noetherian.*
   a) *Any* FA-*module* FM *of finite type admits finite free* ∞-*presentations.*
   b) *Assume* FA *zariskian, and that any* GA-*module of finite type admits a finite free presentation of finite length. Then any finite free* FA-*module admits a finite free presentation of finite length.*

Proof. a) is obvious
   b) consider a finite free ∞-presentation of FM

$$(1.2.1) \qquad \ldots FL_{-1} \xrightarrow[\psi_{-1}]{} FL_0 \xrightarrow[\psi_0]{} FM \to 0,$$

and the associated graded complex, which will be a finite free ∞-presentation of GM:

(1.2.2) $$\ldots GL_{-1} \xrightarrow[\overline{\psi}_{-1}]{} GL_0 \xrightarrow[\overline{\psi}_0]{} GM \to 0.$$

Let FK_{-n} be the kernel of ψ_{-n} endowed with the induced filtration by FL_{-n}, and consider the exact sequence of FA-modules obtained by replacing FL_{-j} by 0 for $-j < -n-1$ and FL_{-n-1} by FK_{-n}. The associad graded sequence will be exact (Proposition 1.1.4), and GK_{-n} will be projective for n large enough, by the hypothesis. By Lemma B.2.2 the GA-module GK_{-n} is stably finite free (the lemma is proved on the category of A-modules, but the proof for graded modules over a graded ring would be the same).

Let GL be a finite free GA-module such that $GK_{-n} \oplus GL$ is free. The module GL is the graded module of a finite free FA-module FL. Consider the exact sequence:

$$0 \to FK_{-n} \oplus FL \to FL_{-n} \oplus FL \to FL_{-n+1} \to \ldots \to FL_0 \to FM \to 0.$$

It is a finite free presentation of FM, since the graded of $FK_{-n} \oplus FL$ being free, this FA-module is free by Proposition 1.1.3. $\square$

Proposition 1.2.2. *Assume* FA *is left and right zariskian. Let* FM *be a (left)* FA-*module of finite type, and assume that for some* $j \in \mathbb{N}$, $\mathrm{Ext}^j_{GA}(GM, GA)$ $= 0$. *Then* $\mathrm{Ext}^j_A(M, A) = 0$.

Proof. To a left FA-module FN we associate the right FA-module $FN^* = \mathrm{Hom}_A(N, A)$, with the filtration:

(1.2.3) $$N^*_k = \{u \in \mathrm{Hom}_A(N, A); \; u(N_l) \subset N_{k+l} \quad \forall l\}.$$

Then we have for a free FA-module N:

(1.2.4) $$GN^* = (GN)^*,$$

where $(GN)^* = \mathrm{Hom}_{GA}(GN, GA)$.

Now consider a finite free ∞-presentation of FM, as (1.2.1), and the associated presentation of GM, (1.2.2), and replace FM (resp. GM) by zero in (1.2.1) (resp. in (1.2.2)). We get:

(1.2.5) $$\ldots \to FL_{-1} \to FL_0 \to 0,$$

(1.2.6) $$\ldots \to GL_{-1} \to GL_0 \to 0.$$

Apply the functor $*$ to (1.2.5) and to (1.2.6). The group $\mathrm{Ext}^j_A(M, A)$ is the j-th cohomology group of the sequence (1.2.5)* obtained by applying the functor $*$ to (1.2.5) and forgetting the filtration. Thus the result follows from (1.2.4) and Proposition 1.1.3. $\square$

Remark 1.2.3. We have in fact proved more than simply the vanishing of $\mathrm{Ext}^j_A(M, A)$, since the exactness of the sequence (1.2.5)* is in the sense of FA-modules, not only of A-modules. This suggests that it would be useful to develop a theory of the derived functors of $\mathrm{Hom}_{FA}(\cdot, \cdot)$ but we shall not do this here (cf. Houzel [2]). Now we study the tensor product.

If FL is a finite free FA-module, $FL = \bigoplus_j FA_{[k_j]}$, and FB a right FA-module, we denote by FBL the filtered FB-module $\bigoplus_j FB_{[k_j]}$. We shall not use filtrations over $B \otimes_A L$ in more general cases.

Proposition 1.2.4. *Let* FA *and* FB *be two filtered rings, with* $FA \subset FB$, *the filtration on* FA *being induced by that of* FB. *We consider* FB *as a right* FA-*module, and we assume:*

i) FA *is zariskian.*

ii) FB *is zariskian and* GB *is flat (resp. faithfully flat) over* GA.

Then:

B *is flat (resp. faithfully flat) over* A, *and if* I *is a submodule of a finite free* FA-*module* FL, *we have:*

$$(1.2.7) \qquad\qquad G(BI) = GB \cdot GI.$$

Proof. Let $u_1, \ldots, u_r$ be elements of I of order $k_1, \ldots, k_r$ whose principal symbols generate GI. Set $FL_{-1} = \bigoplus_{j=1}^{r} FA_{[k_j]}$, and let ψ be the FA-linear map which associate $\sum_j a_j u_j$ to $(a_1, \ldots, a_r) \in FL_{-1}$.

We may extend ψ to an exact sequence of FA-modules:

$$(1.2.8) \qquad\qquad FL_{-2} \xrightarrow[\phi]{} FL_{-1} \xrightarrow[\psi]{} FL_0$$

We may consider the (exact) graded associated sequence, or we may tensor (1.2.8) by $B \otimes_A \cdot$ and take the graded sequence. We obtain:

$$(1.2.9) \qquad\qquad GL_{-2} \xrightarrow[\bar{\phi}]{} GL_{-1} \xrightarrow[\bar{\psi}]{} GL_0,$$

$$(1.2.10) \qquad\qquad FBL_{-2} \xrightarrow[\phi_B]{} FBL_{-1} \xrightarrow[\psi_B]{} FBL_0,$$

$$(1.2.11) \qquad\qquad GBL_{-2} \xrightarrow[\bar{\phi}_B]{} GBL_{-1} \xrightarrow[\bar{\psi}_B]{} GBL_0.$$

Since $GBL_i = GB \otimes_{GA} GL_i$, $(i = -2, -1, 0)$, and GB is flat over GA, the sequence (1.2.11) is exact, and:

$$(1.2.12) \qquad\qquad \mathrm{Ker}\,\bar{\psi}_B = GB \otimes_{GA} \mathrm{Ker}\,\bar{\psi}.$$

Now we use Proposition 1.1.4 and its notation. We have by construction $G(\mathrm{Im}'\psi) = \mathrm{Im}\,\bar{\psi}$. Thus $G(\mathrm{Ker}\,\psi) = \mathrm{Ker}\,\bar{\psi}$, and by (1.2.12), $G(\mathrm{Ker}(\psi_B) = \mathrm{Ker}\,\bar{\psi}_B$, hence $G(\mathrm{Im}'\psi_B) = \mathrm{Im}\,\bar{\psi}_B$. But $G(\mathrm{Im}'\psi_B) = G(BI)$, and $\mathrm{Im}\,\bar{\psi}_B = GB \cdot GI$, which proves (1.2.7).

Now we use the hypothesis that FB is zariskian. It implies that (1.2.10) is exact, and $\mathrm{Ker}\,\psi_B = B \otimes_A \mathrm{Ker}\,\psi$, which proves that B is flat over A. Finally assume I is an ideal of A and $BI = B$. Then $GB \cdot GI = GB$, and GB being faithfully flat over GA, $GI = GA$. But FA is zariskian, and we get $I = A$. $\square$

1.3. Characteristic Ideal

If B is a noetherian ring (or a noetherian graded ring) we shall denote by $\mathfrak{B}^f$ the abelian category of (left) B-modules of finite type.

Let FA be a filtered ring and let $\mathfrak{C}$ be an abelian subcategory of $\mathfrak{G}\mathfrak{A}$ such that the GA-modules of finite type are objects of $\mathfrak{C}$. Let Γ be a commutative group, and χ an additive function from $\mathfrak{C}$ to Γ (cf. Appendix D). We shall assume that $\mathfrak{C}$ is stable by the shift, and that we have $\chi(GM) = \chi(GM_{[+1]})$ for $GM \in Ob(\mathfrak{C})$.

Proposition 1.3.1. a) *Let M be an A-module of finite type*, FM *a good filtration on M. Then* $\chi(GM)$ *depends only on M. We set* $\chi(M) = \chi(GM)$.
   b) *Assume* FA *noetherian. Then the function* $\chi(\cdot)$ *on* $\mathfrak{A}^f$ *is additive.*

Proof. We follow the same argument as Kashiwara [5].
   a) Let $(M_k)_k$ and $(M'_k)_k$ be two good filtrations on M. Replacing M_k by $M_k + M'_k$, we may assume that for some $l \geqslant 0$ we have:

$$M'_k \subset M_k \subset M'_{k+l} \quad \forall k \in \mathbb{Z}.$$

i) Assume $l = 1$, and consider the exact sequences of groups:

$$0 \to M'_k/M_{k-1} \to M_k/M_{k-1} \to M_k/M'_k \to 0,$$
$$0 \to M_{k-1}/M'_{k-1} \to M'_k/M'_{k-1} \to M'_k/M_{k-1} \to 0.$$

We introduce:

$$\overline{L} = \bigoplus_k M'_k/M_{k-1}, \quad \overline{N} = \bigoplus_k M_k/M'_k, \quad \overline{N}_{[+1]} = \bigoplus_k M_{k-1}/M'_{k-1}.$$

We obtain exact sequences of GA-modules:

$$(1.3.1) \qquad\qquad 0 \to \overline{L} \to GM \to \overline{N} \to 0,$$

$$(1.3.2) \qquad\qquad 0 \to \overline{N}_{[+1]} \to GM' \to \overline{L} \to 0.$$

Thus $\overline{L}$ and $\overline{N}$ are GA-modules of finite type, and $\overline{N}$ being isomorphic to $\overline{N}_{[+1]}$ as a GA-module, $\chi(\overline{N}) = \chi(\overline{N}_{[+1]})$, hence $\chi(GM) = \chi(GM')$.
   ii) Set $M''_k = M_k + M'_{k+1}$. We have:

$$M_k \subset M''_k \subset M_{k+1},$$
$$M'_{k+1} \subset M''_k \subset M'_{k+l}.$$

Then $\chi(GM) = \chi(GM'')$, and we obtain $\chi(GM) = \chi(GM')$ by induction on l.
   b) Consider an exact sequence of A-modules of finite type:

$$(1.3.3) \qquad\qquad 0 \to L \to M \to N \to 0.$$

We may endow M with a good filtration FM and L with the induced filtration (which is a good filtration since FA is noetherian), N with the image filtration. The associated graded sequence will be exact, and the result follows. □

From now on and until the end of § 1.3. we shall assume GA commutative. If we denote by $[A_k, A_l]$ the subgroup of A generated by the brackets $[a, b]$, with $a \in A_k$, $b \in A_l$ the commutativity of GA is equivalent to:

(1.3.4) $[A_k, A_l] \subset A_{k+l-1}$ $\forall k, l$.

Then we endow GA with a new product, the "Poisson bracket", defined as follows.

Let $\bar{a} \in A_k/A_{k-1}$, $\bar{b} \in A_l/A_{l-1}$, and choose a and b in A such that $\sigma(a) = \bar{a}, \sigma(b) = \bar{b}$. One sets:

(1.3.5) $\{\bar{a}, \bar{b}\} = \sigma_{k+l-1}([a, b])$,

and one extends this formula to GA by bilinearity.

The following relations hold for $\bar{a}, \bar{b}, \bar{c} \in GA$:

(1.3.6) $\begin{cases} \{\bar{a}, \bar{b}\} = -\{\bar{b}, \bar{a}\}, \\ \{\bar{a}, \bar{b}\,\bar{c}\} = \bar{b}\{\bar{a}, \bar{c}\} + \bar{c}\{\bar{a}, \bar{b}\}, \\ \{\bar{a}, \{\bar{b}, \bar{c}\}\} + \{\bar{b}, \{\bar{c}, \bar{a}\}\} + \{\bar{c}, \{\bar{a}, \bar{b}\}\} = 0. \end{cases}$

Now we recall that if B is a commutative ring, and M a B-module, then the annihilator of M is the ideal I_M of B defined by:

$$x \in I_M \Leftrightarrow xu = 0 \quad \forall u \in M,$$

and the radical $\sqrt{I}$ of an ideal I of B is defined by:

$$x \in \sqrt{I} \Leftrightarrow \exists k \geqslant 1, \quad x^k \in I.$$

If we have an exact sequence of B-modules, $0 \to L \to M \to N \to 0$, then one checks immediately:

(1.3.7) $\sqrt{I_M} = \sqrt{I_L} \cap \sqrt{I_N} = \sqrt{I_L \cdot I_N}$.

We shall use these notions when $B = GA$. In that case we remark that the radical of a graded ideal is still a graded ideal.

Proposition 1.3.2. a) *Let* M *be an* A-*module of finite type,* FM *a good filtration on* M. *Then* $\sqrt{I_{GM}}$ *depends only on* M. *We set:*

$$\mathrm{Icar}(M) = \sqrt{I_{GM}}.$$

b) *Assume* FA *noetherian, and consider the exact sequence (1.3.3) of* A-*modules of finite type. Then:*

(1.3.8) $\mathrm{Icar}(M) = \mathrm{Icar}(L) \cap \mathrm{Icar}(N)$.

Proof. The proof of Proposition 1.3.1 applies in this situation, by considering formula (1.3.7) and the exact sequences (1.3.1) and (1.3.2).   □

Remark 1.3.3. We can give another proof of Proposition 1.3.2.

Let $\bar{a}$ be a homogeneous element of $\sqrt{I_{GM}}$. Then $\bar{a}^q \in I_{GM}$ for some $q \geqslant 1$. Choose $a \in A_p$ with $\sigma(a) = \bar{a}$. Then:

$$\bar{a}^q \in I_{GM} \Leftrightarrow a^q \, M_k \subset M_{k+pq-1} \quad \forall k,$$

hence:

$$a^{lq} \, M_k \subset M_{k+lpq-l} \quad \forall k,$$

and we get $a^{lq} \, M'_k \subset M'_{k+lpq-l}$ for $l \gg 0$, if $\{M_k'\}$ is another good filtration on M.

Now we recall the fundamental result of O. Gabber [1] und refer to his paper for the proof.

Theorem 1.3.4. *Let* FA *be a filtered ring, and assume* GA *is a commutative noetherian* $\mathbb{Q}$-*algebra. Then if* M *is an* A-*module of finite type,* Icar(M) *is involutive, that is:*

$$\{Icar(M), Icar(M)\} \subset Icar(M).$$

Remark 1.3.5. It is usefull to remark that Proposition 1.3.1 may be generalized by assuming that Γ is only a commutative monoid with unit as for example the set $\{\sqrt{I} ; I$ ideal of a commutative ring B$\}$, for the operation $\sqrt{I}, \sqrt{J} \mapsto \sqrt{I} \cap \sqrt{J}$, or else $\Gamma = \{$the subsets of a given set$\}$, for the operation of union. In fact we already applied this Remark for the proof of Proposition 1.3.2.

Remark 1.3.6. In Chapter III § 1 we shall need to consider characteristic ideals after a change of base rings. For that purpose let us recall some elementary and well known results (cf. Serre [1, Chap. I, c]).

Let B be a commutative noetherian ring, J an ideal of B, C the ring B/J. If M is a B-module of finite type we denote as before by $\sqrt{I_M}$ the radical of the annihilator of M in B, and we denote by $\sqrt{I_M^C}$ the radical of the annihilator in C of the C-module M/JM. First we remark:

$$(1.3.9) \qquad \sqrt{I_{M/JM}} = \sqrt{I_M + J}.$$

In fact the inclusion $\sqrt{I_M + J} \subset \sqrt{I_{M/JM}}$ is obvious. To prove the converse, let us choose a system of generators $(e_1, \ldots, e_r)$ of M. Let $x \in B$ such that $x^N M \subset JM$. Then there exists $(z_{i,j})_{i,j} (1 \leqslant i,j \leqslant r)$ in J with:

$$x^N e_i = \sum_j z_{i,j} e_j \quad \forall i.$$

Proceeding by elimination, we find some $N' \geqslant N$, and some $z \in J$ with

$$(x^{N'} + z) e_i = 0 \quad \forall i.$$

We deduce immediately from (1.3.9) and (1.3.7) that:

$$(1.3.10) \qquad \sqrt{I_M^C} = \sqrt{I_M + J}/J,$$

$$(1.3.11) \qquad \sqrt{I_M^C} = \sqrt{I_L^C} \cap \sqrt{I_N^C},$$

(where $0 \to L \to M \to N \to 0$ is an exact sequence of finitely generated B-modules).

Now let us make the assumption that for any finitely generated B-module M we have:

$$\text{Tor}_j^B(B/J, M) = 0 \quad \forall j \gg 0.$$

Let $\chi_C(\cdot)$ be an additive function on the category of finitely generated C-modules (with values in some commutative group Γ). Then we may associate to $\chi_C(\cdot)$ an additive function $\chi_B(\cdot)$ on the category of finitely generated B-modules, by setting for M a finitely generated B-module:

$$(1.3.12) \qquad \chi_B(M) = \sum_j (-1)^j \chi_C(\text{Tor}_j^B(B/J, M)).$$

(Recall that $\text{Tor}_j^B(B/J, M)$ are finitely generated C-modules). The additivity of $\chi_B(\cdot)$ follows immediately from the long exact sequence of $\text{Tor}_j(\cdot, \cdot)$ (Appendix B.2.2).

Remark 1.3.7. Let FA be a zariskian filtered ring such that GA is commutative and without zero divisors. Let $(a) = (a_{i,j})$ be an $(r \times r)$-matrix with entries in A, and assume that there exists $m_1, \ldots, m_r$ and $m'_1, \ldots, m'_r$ belonging to $\mathbb{Z}$ such that:

$$(1.3.13) \qquad \begin{cases} \text{ord}(a_{i,j}) \leqslant m_i - m'_j, \\ \det(\sigma_{m_i - m'_j}(a_{i,j})) \neq 0. \end{cases}$$

Then one says that the matrix (a) is "normal", in the sense of Leray-Volevic (cf. Andronikof [1]). We set:

$$FL_1 = \bigoplus_{j=1}^r A_{[m_j]}, \qquad FL_0 = \bigoplus_{j=1}^r A_{[m'_j]}$$

and consider (a) as a filtered morphism from FL_1 to FL_0 ((a) operates on the right).

Let FM be the cokernel of (a), endowed with the image filtration:

$$(1.3.14) \qquad FL_1 \xrightarrow{(a)} FL_0 \to FM \to 0.$$

Passing to the graded modules we obtain:

$$(1.3.15) \qquad GL_1 \xrightarrow{(\bar a)} GL_0 \to GM \to 0,$$

where $GL_1 = \bigoplus_{j=1}^r GA_{[m_j]}$, $GL_0 = \bigoplus_{j=1}^r GA_{[m'_j]}$ and $(\bar a)$ is the matrix $(\sigma_{m_i - m'_j}(a_{i,j}))$. By (1.3.13), $(\bar a)$ is injective, and so is (a) (Proposition 1.1.3). Then we have:

$$(1.3.16) \qquad \text{Icar}(M) = \sqrt{\det(\bar a) \cdot GA},$$

where $\det(\bar a)$ is the determinant of the matrix $(\bar a)$. In fact (1.3.16) is an easy exercise in commutative algebra, since GA is embedded in its field of fractions.

When the matrix (a) no longer satisfies condition (1.3.13), it is still possible to associate to (a) a "determinant" which gives $\mathrm{Icar}(A^r/A^r \cdot (a))$, but we do not expose this theory here and refer to Sato-Kashiwara [1] and Adjamagbo [1].

Remark 1.3.8. Let $F A$ be a noetherian filtered ring, and assume $G A$ commutative and without zero divisors. Then A is a Ore domain, that is for any non zero elements a and b in A, $Aa \cap Ab \neq \{0\}$. In fact we may even find an element $c \in Aa \cap Ab$ such that $\sigma(c) = (\sigma(a) \cdot \sigma(b))^n$ for some $n \in \mathbb{N}$. This follows immediately from the exact sequence:

$$0 \to A/Aa \cap Ab \to A/Aa \oplus A/Ab \to A/(Aa + Ab) \to 0$$

and Proposition 1.3.2.

1.4. Sheaves of Filtered Modules

We refer to Appendix C for basic notions on sheaves.

Let X be a topological space. We define a sheaf of filtered rings, or a "filtered Ring", $F\mathscr{A}$, as a sheaf of rings $\mathscr{A}$ endowed with a family of subsheaves of groups $\mathscr{A}_k$, $k \in \mathbb{Z}$, such that:

(1.4.1)
$$\begin{cases} \mathscr{A}_k \subset \mathscr{A}_{k+1}, \quad \mathscr{A}_l \cdot \mathscr{A}_k \subset \mathscr{A}_{l+k}, \\ 1 \text{ is a section of } \mathscr{A}_0, \\ \mathscr{A} = \bigcup_k \mathscr{A}_k, \text{ where the term on the right hand side} \\ \text{stands for the sheaf associated to the presheaf,} \\ U \mapsto \bigcup_k \mathscr{A}_k(U). \end{cases}$$

Remark that if u is a section of $\mathscr{A}$ on an open set $U \subset X$, the function on U, $x \mapsto \mathrm{ord}(u_x)$, where u_x is the image of u in $\mathscr{A}_x$, is upper semi-continuous.

We denote by $G\mathscr{A}$ the sheaf associated to the presheaf $U \to \bigoplus_k \mathscr{A}_k(U)/\mathscr{A}_{k-1}(U)$. Then $(G\mathscr{A})_x = G(\mathscr{A}_x)$. Moreover the natural morphism $\mathscr{A}_k \to \mathscr{A}_k/\mathscr{A}_{k-1} \to G\mathscr{A}$ define the "principal symbol" morphism, $\mathscr{A} \to G\mathscr{A}$ (this is not a morphism of sheaves of groups but only a morphism of sheaves of sets).

An $F\mathscr{A}$-module $F\mathscr{M}$ is an $\mathscr{A}$-module $\mathscr{M}$ endowed with a family of sheaves of subgroups $\mathscr{M}_k$ such that:

(1.4.2)
$$\begin{cases} \mathscr{M}_k \subset \mathscr{M}_{k+1}, \\ \mathscr{A}_l \cdot \mathscr{M}_k \subset \mathscr{M}_{l+k}, \\ \mathscr{M} = \bigcup_k \mathscr{M}_k, \end{cases}$$

where $\bigcup_k \mathscr{M}_k$ is the sheaf associated to the presheaf $U \to \bigcup_k \mathscr{M}_k(U)$. One define the graded associated Module $G\mathscr{M}$ as the sheaf associated to the presheaf $U \mapsto \bigoplus_k \mathscr{M}_k(U)/\mathscr{M}_{k-1}(U)$.

A morphism ψ of $F\mathscr{A}$-modules from $F\mathscr{M}$ to $F\mathscr{N}$ is a morphism of $\mathscr{A}$-modules from $\mathscr{M}$ to $\mathscr{N}$ such that $\psi(\mathscr{M}_k) \subset \mathscr{N}_k$, $\forall k$. Then the notion of complexes of $F\mathscr{A}$-modules is clear. A complex of $F\mathscr{A}$-modules:

$$F\mathscr{L} \underset{\phi}{\to} F\mathscr{M} \underset{\psi}{\to} F\mathscr{N}$$

is exact if for each k, $(\mathrm{Im}\phi)_k = (\mathrm{Ker}\psi)_k$, that is if $\phi(\mathscr{L}_k) = (\mathrm{Ker}\psi) \cap \mathscr{M}_k$. Thus the sequence is exact iff for any $x \in X$ the complex:

$$F\mathscr{L}_x \to F\mathscr{M}_x \to F\mathscr{N}_x$$

is an exact sequence of $F\mathscr{A}_x$-modules.

An $F\mathscr{A}$-module $F\mathscr{L}$ is finite free if it is isomorphic to a finite direct sum of $F\mathscr{A}_{[r]}$, for some shifts r. Then the notion of finite free s-presentation is clear, and an $F\mathscr{A}$-module is of finite type iff it admits a finite free 0-presentation.

It is also useful to define a "good filtration" on an $\mathscr{A}$-module $\mathscr{M}$ as a filtration $F\mathscr{M}$ which is locally of finite type.

It would be possible to develop a theory of coherent filtered rings and modules, but we shall not need it, thanks to the following:

Proposition 1.4.1. *Let $F\mathscr{A}$ be a filtered Ring. Assume that for any $x \in X$, $F\mathscr{A}_x$ is zariskian, and that the graded Ring $G\mathscr{A}$ is coherent. Then:*

a) *The Ring $\mathscr{A}$ is coherent*

b) *Let $F\mathscr{M}$ be an $F\mathscr{A}$-module locally of finite type. Then $\mathscr{M}$ is coherent iff $G\mathscr{M}$ is coherent (as a graded Module).*

c) *Let $F\mathscr{M}$ be an $F\mathscr{A}$-module locally of finite type, such that $\mathscr{M}$ is coherent. Let $\mathscr{N}$ be a coherent sub-Module of $\mathscr{M}$ and let us endow $\mathscr{N}$ with $F\mathscr{N}$, the induced filtration. Then the filtration $F\mathscr{N}$ is good, (i.e.: locally of finite type).*

d) *Let $F\mathscr{M}$ be as in c). Then for any $r \geqslant 0$, $F\mathscr{M}$ locally admits finite free r-presentations.*

e) *Assume moreover the graded Ring $G\mathscr{A}$ noetherian. Then the Ring $\mathscr{A}$ is noetherian.*

Proof. a) Let ψ be an $\mathscr{A}$-linear morphism from $\mathscr{L}_1$ to $\mathscr{L}_0$, where $\mathscr{L}_1$ and $\mathscr{L}_0$ are free $\mathscr{A}$-modules of finite type on an open set U. We may locally endow $\mathscr{L}_1$ and $\mathscr{L}_0$ with finite free filtrations such that ψ becomes an $F\mathscr{A}$-linear morphism:

(1.4.3) $$F\mathscr{L}_1 \underset{\psi}{\to} F\mathscr{L}_0.$$

Let us endow the Module $\mathrm{Ker}\,\psi$ with the induced filtration. We shall show that $\mathrm{Ker}\,\psi$ is locally of finite type over $F\mathscr{A}$. Let $\mathrm{Im}'\psi$ be the $\mathscr{A}$-module

Im ψ endowed with the induced filtration and let $\overline{\psi}$ be the graded morphism from $G\mathscr{L}_1$ to $G\mathscr{L}_0$ associated to ψ. Let $x \in U$. First we assume:

(1.4.4) $\qquad\qquad\qquad G(\text{Im}'\psi)_x = (\text{Im}\,\overline{\psi})_x.$

Applying Proposition 1.1.4 we get:

(1.4.5) $\qquad\qquad\qquad G(\text{Ker}\,\psi)_x = (\text{Ker}\,\overline{\psi})_x.$

Since $\text{Ker}\,\overline{\psi}$ is coherent, and $G(\text{Ker}\,\psi)$ is contained in $\text{Ker}\,\overline{\psi}$, we find that $G(\text{Ker}\,\psi) = \text{Ker}\,\overline{\psi}$ in a neighborhood of x. Thus $G(\text{Ker}\,\psi)$ is of finite type and by Proposition 1.1.3 we find that $\text{Ker}\,\psi$ is of finite type in a neighborhood of x. In the general case we may choose a finite free $F\mathscr{A}$-module $F\mathscr{L}_2$ and a morphism:

$$\phi \colon F\mathscr{L}_2 \to F\mathscr{L}_0,$$

such that $\text{Im}\,\phi \subset \text{Im}\,\psi$, but $(\text{Im}\,\overline{\phi})_x = G(\text{Im}'\psi)_x$. In fact choose $v_1, \ldots, v_s$ of order $l_1, \ldots, l_s$ in $\text{Im}'\psi$ whose principal symbols generate $G(\text{Im}'\psi)_x$. Set:

$$F\mathscr{L}_2 = \bigoplus_{i=1}^{s} F\mathscr{A}_{[l_i]},$$

$$\phi(a_1, \ldots, a_s) = \sum_i a_i v_i.$$

Denote by θ the map $\phi + \psi$ from $F\mathscr{L}_1 \oplus F\mathscr{L}_2$ to $F\mathscr{L}_0$. Then $\text{Im}\,\theta = \text{Im}\,\psi$, and we have just proved that $\text{Ker}\,\theta$, endowed with the induced filtration, is of finite type. Now it is easy to construct an isomorphism:

$$(\text{Ker}\,\psi) \oplus (F\mathscr{L}_1 \oplus F\mathscr{L}_2) \cong (\text{Ker}\,\theta) \oplus F\mathscr{L}_1,$$

which proves that $\text{Ker}\,\psi$ is of finite type over $F\mathscr{A}$ in a neighborhood of x. In particular $\text{Ker}\,\psi$ is of finite type over $\mathscr{A}$, which proves the coherency of $\mathscr{A}$.

b) Consider a finite free 0-presentation of $F\mathscr{M}$ on an open set $U \subset X$:

$$F\mathscr{L}_0 \to F\mathscr{M} \to 0$$

and let $F\mathscr{S}$ be the kernel of this map, endowed with the induced filtration. Assuming $G\mathscr{M}$ is coherent we find that $G\mathscr{S}$ is coherent, and $\Gamma\mathscr{S}$ will be locally of finite type by Proposition 1.1.3. Hence $\mathscr{M}$ is coherent.

Conversely assume $\mathscr{M}$ is coherent. Then $\mathscr{S}$ is locally of finite type and we may find a morphism ψ as (1.4.3) such that $F\mathscr{S} = \text{Im}'\psi$. Moreover since we are only interested in $\text{Im}\,\psi$, we may assume (1.4.4) at $x \in U$. Now the argument used for the proof of a) gives that $G(\text{Ker}\,\psi) = \text{Ker}\,\overline{\psi}$, thus $\text{Im}\,\overline{\psi} = G(\text{Im}'\psi)$ in a neighborhood of x, (Proposition 1.1.4), and $G\mathscr{S}$ is of finite type.

c) Set $\mathscr{H} = \mathscr{M}/\mathscr{N}$, and endow $\mathscr{H}$ with the image filtration. We get an exact sequence of $G\mathscr{A}$-modules:

$$0 \to G\mathscr{N} \to G\mathscr{M} \to G\mathscr{H} \to 0.$$

Since $F\mathscr{K}$ is locally of finite type we get by b) that $G\mathscr{K}$ is coherent. Thus $G\mathscr{N}$ is also coherent, and the result follows from Proposition 1.1.3.

d) and e) follow immediately from c).   □

Proposition 1.4.2. *Let* $F\mathscr{A}$ *be a filtered Ring. Assume:*

i) $\mathscr{A}_0$ *is a noetherian Ring, and* $\mathscr{A}$ *is coherent.*

ii) *The* $\mathscr{A}_0$*-modules* $\mathscr{A}_k$ *are coherent for* $k \geqslant 0$.

Then every coherent $\mathscr{A}$*-module* $\mathscr{M}$ *is* $\mathscr{A}_0$*-pseudo-coherent.*

Proof. Let $\mathscr{N}$ be a sub-$\mathscr{A}_0$-module of $\mathscr{M}$ of finite type. Since $\mathscr{A}$ is coherent, $\mathscr{A} \cdot \mathscr{N}$ is coherent, and replacing $\mathscr{M}$ by $\mathscr{A} \cdot \mathscr{N}$ we may assume $\mathscr{A} \cdot \mathscr{N} = \mathscr{M}$.

Locally there exists an exact sequence:

$$0 \to \mathscr{S} \to \mathscr{A}^n \to \mathscr{M} \to 0$$

and we have to prove that $\mathscr{A}_0^n \cap \mathscr{S}$ is $\mathscr{A}_0$-coherent.

Let $\mathscr{S}_0$ be an $\mathscr{A}_0$-module of finite type which generate $\mathscr{S}$

$$\mathscr{A}_0^n \cap \mathscr{S} = \bigcup_k (\mathscr{A}_0^n \cap (\mathscr{A}_k \cdot \mathscr{S}_0)).$$

Since $\mathscr{A}_k \cdot \mathscr{S}_0$ is contained in some $\mathscr{A}_l^n$ and is of finite type over $\mathscr{A}_0$, it is $\mathscr{A}_0$-coherent, and so is its intersection with $\mathscr{A}_0^n$. Hence $\mathscr{A}_0^n \cap \mathscr{S}$ being an increasing union of $\mathscr{A}_0$-coherent submodules of $\mathscr{A}_0^n$, is coherent since $\mathscr{A}_0$ is noetherian.   □

In Chapter III, § 1.4, we shall encounter non zariskian filtered rings (cf. also Example 1.5.5 below), and we shall need the following result.

Proposition 1.4.3. *Let* $F\mathscr{A}$ *be a filtered Ring. Assume:*

I) $G\mathscr{A}$ *is a noetherian graded Ring.*

ii) *For each* $k \in \mathbb{Z}$, $\mathscr{A}_k$ *is finitely generated over* $\mathscr{A}_0$.

iii) *For each* $l \geqslant 0$, *the Ring* $\mathscr{A}_0/\mathscr{A}_{-l}$ *is coherent.*

iv) *For each* $k \in \mathbb{Z}$, *each* $l \geqslant 0$, *each* $m \geqslant 0$ *with* $m \leqslant l$, *the* $\mathscr{A}_0/\mathscr{A}_{-l}$*-module* $\mathscr{A}_k/\mathscr{A}_{k-m}$ *is coherent.*

Let $F\mathscr{L}$ *be a finite free* $F\mathscr{A}$*-module,* $\mathscr{M}$ *an* $\mathscr{A}$*-module of finite type contained in* $F\mathscr{L}$, $F\mathscr{M}$ *the Module* $\mathscr{M}$ *endowed with the induced filtration by* $F\mathscr{L}$, $G\mathscr{M}$ *the associated graded Module. Then* $G\mathscr{M}$ *is* $G\mathscr{A}$*-coherent.*

Proof. The proof will be similar to that of Sato-Kashiwara-Kawaï [1, p. 386]. Let $x \in X$, and let $u_1, \ldots, u_N$ be sections of $\mathscr{M}$ which generate $\mathscr{M}$ on a neighborhood of x. Let k_j be the order of u_j at x, for the induced filtration. We set (on a fixed neighborhood of x):

$$\mathcal{M}_k = \mathcal{M} \cap \mathcal{L}_k,$$

$$\mathcal{M}_k' = \sum_j \mathcal{A}_{k-k_j} u_j,$$

$$\mathcal{M}_k^l = \mathcal{M}_{k+l}' \cap \mathcal{M}_k.$$

For each $l \geqslant 0$, we have the following exact sequence of $\mathcal{A}_0/\mathcal{A}_{-l-1}$-modules:

$$0 \to \mathcal{M}_k^l/\mathcal{M}_{k-1}^l \to \mathcal{M}_{k+1}^l/\mathcal{M}_{k-1}^l \to \mathcal{M}_{k+1}^l/\mathcal{M}_k^l \to 0.$$

By hypothesis ii), $\mathcal{M}_{k+l}'$ is finitely generated over $\mathcal{A}_0$. Thus $\mathcal{M}_{k+l}'/\mathcal{M}_{k-1}'$ is finitely generated over $\mathcal{A}_0/\mathcal{A}_{-l-1}$, and this sheaf being contained in $\mathcal{L}_{k+l}/\mathcal{L}_{k-1}$ which is coherent by the hypothesis iv), is itself coherent. Similarly $\mathcal{M}_{k+1}'/\mathcal{M}_k'$ is coherent over $\mathcal{A}_0/\mathcal{A}_{-l-1}$, and we find that $\mathcal{M}_k^l/\mathcal{M}_{k-1}^l$ is coherent over this Ring, then locally finitely generated over $\mathcal{A}_0/\mathcal{A}_{-1}$. For each $j \in \mathbb{N}$, the $\mathcal{A}_0/\mathcal{A}_{-1}$-module $\bigoplus_{|k| \leqslant j} \mathcal{M}_k^l/\mathcal{M}_{k-1}^l$ is finitely generated and contained in $\mathrm{G}\mathcal{L}$. Set:

$$\mathcal{K}_j^l = \mathrm{G}\mathcal{A}\left(\bigoplus_{|k| \leqslant j} \mathcal{M}_k^l/\mathcal{M}_{k-1}^l\right).$$

Then $\mathcal{K}_j^l$ is a coherent $\mathrm{G}\mathcal{A}$-module, and we have:

$$\mathrm{G}\mathcal{M} = \bigcup_{l \geqslant 0, j \geqslant 0} \mathcal{K}_j^l.$$

By hypothesis i), this sequence is locally stationnary, and $\mathrm{G}\mathcal{M}$ is coherent. □

Corollary 1.4.4. *Under the hypotheses of Proposition 1.4.3, and assuming moreover that $\mathcal{A}$ is coherent, let $\mathcal{N}$ be a coherent $\mathcal{A}$-module, $\mathrm{F}\mathcal{N}$ a good filtration on $\mathcal{N}$. Then $\mathrm{G}\mathcal{N}$ is $\mathrm{G}\mathcal{A}$-coherent.*

Proof. Consider a finite free 0-presentation of $\mathrm{F}\mathcal{N}$:

$$\mathrm{F}\mathcal{L} \xrightarrow{\psi} \mathrm{F}\mathcal{N} \to 0$$

and let $\mathcal{Y}$ be the kernel of the (non filtered) morphism $\psi: \mathcal{L} \to \mathcal{N}$. Then $\mathcal{Y}$ is coherent, and if $\mathrm{F}\mathcal{Y}$ is the induced filtration on $\mathcal{Y}$ by $\mathrm{F}\mathcal{L}$, $\mathrm{G}\mathcal{Y}$ is coherent by the preceding Proposition. Since we have an exact sequence $0 \to \mathrm{G}\mathcal{Y} \to \mathrm{G}\mathcal{L} \to \mathrm{G}\mathcal{N} \to 0$, we find that $\mathrm{G}\mathcal{N}$ is coherent. □

1.5. Examples

Example 1.5.1. Let E be a n-dimensional complex vector space. The ring of differential operators with polynomial coefficients on E is called "the Weyl algebra" on E. We shall denote it by W_n. We may endow this ring with the filtration given by the order of the operators. Then $W_n(k) = 0$ for $k < 0$, and

$\mathrm{gr}(W_n)$ is isomorphic to the (commutative) ring of polynomials on $E \times E^*$. Thus W_n is right and left zariskian.

We refer to the work of I.N. Bernstein [1], [2] and to the book of Björk [1] for a detailed study of this algebra.

Example 1.5.2. Let X be a complex manifold, $\mathscr{D}_X$ the Ring of differential operators on X, endowed with the filtration given by the order. We have:

$$\mathscr{D}_X(k) = 0 \quad \text{for} \quad k < 0,$$

$$\mathrm{gr}(\mathscr{D}_X) \cong \pi_* \left(\bigoplus_{j \geq 0} \mathscr{O}_{T^*X}(j) \right).$$

Since $\pi_*(\mathscr{O}_{T^*X}(j))$ is isomorphic to the restriction $\mathscr{O}_{T^*X}(j)|_{T^*_X X}$, we find that $\mathrm{gr}(\mathscr{D}_X)$ is a noetherian sheaf. Since $\mathscr{D}_{X,x}$ is right and left zariskian, we may apply Proposition 1.4.1 and find that the Ring $\mathscr{D}_X$ is right and left noetherian.

Example 1.5.3. The Ring $\mathscr{E}_X$ on T^*X will be studied in the next section.

Example 1.5.4. Let $f: X \to Y$ be a smooth holomorphic map ($\forall x \in X, df(x)$ is surjective, that is, f is a submersion at each $x \in X$). We define the "relative filtration", $F^f \mathscr{D}_X$ on $\mathscr{D}_X$ by induction, setting:

(1.5.1)
$$\begin{cases} F^f_k \mathscr{D}_X = 0 & k < 0, \\ [F^f_k \mathscr{D}_X, f^{-1} \mathscr{O}_Y] \subset F^f_{k-1} \mathscr{D}_X. \end{cases}$$

We shall denote $F^f_0 \mathscr{D}_X$ by $\mathscr{D}_{X/Y}$. Thus $\mathscr{D}_{X/Y}$ is the Ring of differential operators on X generated by the "vertical" vector fields, that is, the vector fields tangent to the fibers of f (i.e.: the vector fields which annihilate $f^{-1} \mathscr{O}_Y$).

We define the relative cotangent bundle $T^*(X/Y)$ by the exact sequence of bundles on X:

(1.5.2)
$$0 \to X \underset{Y}{\times} T^*Y \to T^*X \to T^*(X/Y) \to 0.$$

Let us denote by $\pi_* \mathscr{O}_{[T^*(X/Y)]}$ the subsheaf of $\pi_* \mathscr{O}_{T^*(X/Y)}$ of sections which are polynomial in the fibers of π. Then if we endow $\mathscr{D}_{X/Y}$ with the induced filtration by $\mathscr{D}_X$, we have a natural isomorphism:

(1.5.3)
$$\mathrm{gr}(\mathscr{D}_{X/Y}) \cong \pi_* \mathscr{O}_{[T^*(X/Y)]}$$

and we get:
  – $\forall x \in X$, the filtered ring $\mathscr{D}_{X/Y,x}$, is right and left zariskian,
  – the Ring $\mathscr{D}_{X/Y}$ is right and left noetherian.

Remark that locally on X, f is isomorphic to a projection $f: Z \times Y \to Y$, and in that case we have: $\mathscr{D}_{Z \times Y/Y} \cong \mathscr{O}_{Z \times Y} \otimes_{g^{-1} \mathscr{O}_Z} g^{-1} \mathscr{D}_Z$ where g is the projection on Z. We shall come back to this example in Chapter III, § 1.3.

Example 1.5.5. Let Y be a submanifold of X, and let $\mathscr{J}_Y$ be the Ideal of $\mathscr{O}_X$ of sections of $\mathscr{O}_X$ vanishing on Y, that is, the defining Ideal of Y in X.

Following Kashiwara [6] we define the filtration $\mathsf{F}^Y \mathscr{D}_X$ on $(\mathscr{D}_X)|_Y$ by setting for $k \in \mathbb{Z}$:

$$(1.5.4) \qquad \mathsf{F}^Y_k \mathscr{D}_X = \{ P \in (\mathscr{D}_X)|_Y ;\; P \mathscr{J}_Y^l \subset \mathscr{J}_Y^{l-k} \quad \forall l \}.$$

If one chooses coordinates $(y_1, \ldots, y_p, t_1, \ldots, t_q) = (y, t)$ on X such that $Y = \{(y, t);\ t = 0\}$, then y_i, D_{y_j} are of order 0, t_k of order -1 and D_{t_l} of order $+1$. Let $T_Y X$ be the normal bundle to Y in X: $T_Y X = (Y \underset{X}{\times} TX)/TY$. We may naturally identify $\mathsf{G}^Y \mathscr{D}_X$, the graded Ring of $\mathsf{F}^Y \mathscr{D}_X$, to the Ring $\mathscr{D}_{[T_Y X]}$ of differential operators on the bundle $T_Y X$ whith coefficients polynomials in the fibers of the bundle. In the system of coordinates (y, t), a section P of $\mathscr{D}_{[T_Y X]}$ is written:

$$(1.5.5) \qquad P = \sum_{\alpha, \beta, \gamma} a_{\alpha, \beta, \gamma}(y)\, t^\alpha D_t^\beta D_y^\gamma,$$

where $\alpha \in \mathbb{N}^q$, $\beta \in \mathbb{N}^q$, $\gamma \in \mathbb{N}^p$, $a_{\alpha, \beta, \gamma} \in \mathscr{O}_Y$. This Ring is graded by the subgroups of sections P as in (1.5.5) for which $|\beta| - |\alpha| = k$.

Remark that $\mathsf{G}^Y \mathscr{D}_X$ is not commutative, but noetherian. It is easy to prove, using the associated formal graded ring that $\mathsf{F}^Y \mathscr{D}_{X,x}$ is noetherian, but this ring is not zariskian in general. For example assume codim $Y = 1$. Then $1 + t^2 D_t$ has 1 as principal symbol, but is not invertible.

We shall come back to this example in Chapter III, § 1.4.

Exercices to II. 1

Ex. 1.1. Let A be a commutative noetherian ring, I an ideal of A. Set $A_{-k} = I^k$, $A_k = A$ for $k \geqslant 0$. Prove that $(A_k)_k$ defines a filtration $\mathsf{F}A$ on A, and that $\mathsf{F}A$ is noetherian. Prove that if I is contained in the radical of A, $\mathsf{F}A$ is zariskian (cf. Atiyah-Macdonald [1]).

Ex. 1.2. Let $\mathsf{F}A$ be a separated noetherian filtered ring. Set

$$\mathsf{F}\hat{A}_k - \varprojlim_{l \geqslant 0} A_k / A_{k-l}, \qquad \mathsf{F}\hat{A} - \bigcup_k \mathsf{F}\hat{A}_k$$

a) Prove that $\mathsf{F}A$ is a subring of $\mathsf{F}\hat{A}$.

b) Prove that $\mathsf{F}\hat{A}$ is zariskian.

c) Prove that $\mathsf{F}\hat{A}$ is flat (as an A-module) over A, and that if $\mathsf{F}A$ is zariskian, $\mathsf{F}\hat{A}$ is faithfully flat over A. (cf. Bourbaki [1]).

d) When $A = \mathscr{E}_{X,x}$, the ring of germs of microdifferential operators at $p \in T^* X$, compare $\mathsf{F}\hat{A}$ and $\hat{\mathscr{E}}_{X,p}$.

Ex. 1.3. Let t be a coordinate on $\mathbb{C}$. Find a non zero operator P in $\mathscr{D}_\mathbb{C} \cdot D_t^2 \cap \mathscr{D}_\mathbb{C} \cdot t^2$.

Ex. 1.4. Let $p \in T^*X$. Prove that $\mathscr{E}_{X,p}$ is a simple ring, that is any two-sided ideal $\mathscr{I}$ is $\{0\}$ or the whole of $\mathscr{E}_{X,p}$ (Hint: choose coordinates $(x_1, \ldots, x_n)$ and remark that $[x_i, P]$ and $[D_i, P]$ belong to I, $\forall i$ if $P \in \mathscr{I}$. If P is non zero, construct an invertible operator in $\mathscr{I}$ by this remark); (cf. Sato-Kashiwara-Kawaï [1, p. 456], or Björk [1] for another proof in the case of Weyl algebras).

Ex. 1.5. Let FA be a filtered noetherian ring, with GA commutative. Let a and b be two $(r \times r)$-matrices with entries in A. Set $L = A^r/A^r \cdot a$, $N = A^r/A^r \cdot b$, $M = A^r/A^r(b \cdot a)$, and assume that a, acting from the right on A^r, is injective. Prove that:

$$\mathrm{Icar}(M) = \mathrm{Icar}(L) \cap \mathrm{Icar}(N)$$

Ex. 1.6. Let $(x_1, \ldots, x_n)$ be coordinates on $\mathbb{C}^n$, $p = (0; dx_n)$, $n > 1$, and let P and Q be two sections of $\mathscr{E}_{\mathbb{C}^n,p}$ with $\sigma(P) = \xi_1^l$, $\sigma(Q) = x_1^k \xi_n^r$. Prove that the left ideal generated by P and Q is the whole of $\mathscr{E}_{\mathbb{C}^n,p}$.
    (Hint: use the involutivity theorem).

Ex. 1.7. Let $FA \subset FB$ be two filtered rings such that the filtration on B induces the filtration on A, and $GA = GB$. Assume FA noetherian and FB zariskian. Prove that B is flat over A.

§ 2. Structure of the Ring $\mathscr{E}_X$

2.1. The Ring $\mathscr{E}_X(0)$

Let X be a complex manifold. First we shall study the filtered ring $FE = \mathscr{E}_{X,p}(0)$ of germs at $p \in T^*X$ of microdifferential operators of order at most 0. The ring FE is filtered by the order, thus $E_k = E_0 = E$ for $k \geqslant 0$. We have:

$$GE = \bigoplus_{j \leqslant 0} \mathscr{O}_{T^*X,p}(j).$$

Let us show that this graded ring is noetherian. If $p = (x, 0)$ belongs to the zero section $T^*_X X$, $GE = \mathscr{O}_{X,x}$, the ring of germs at $x \in X$ of holomorphic functions, and this ring is noetherian. If p belongs to $\dot{T}^*X$, let p' be its image in $P^*X = \dot{T}^*X/\mathbb{C}^\times$. Then we have an isomorphism:

$$GE \cong \mathscr{O}_{P^*X,p'}[T^{-1}],$$

where the term on the right-hand side is the ring of polynomials in the variable T^{-1} with coefficients in $\mathscr{O}_{P^*X,p'}$, and this graded ring is noetherian.

Theorem 2.1.1. *Let $p \in T^*X$. The filtered ring $\mathscr{E}_{X,p}(0)$ is right and left zariskian.*

Proof. The right and left structures on $\mathscr{E}_{X,p}(0)$ being isomorphic, it is enough to prove the theorem for the left structure. Applying Proposition 1.1.5, it is enough to check that the submodules of $\mathsf{F}\,\mathsf{E}^r (r \in \mathbb{N})$ are closed for the topology defined by the filtration.

We shall follow the proof of Sato-Kashiwara-Kawaï [1, Th. 3.4.1, Chap. II], with slight modifications. In fact their proof uses the sheaf $\hat{\mathscr{E}}_X(0)$, but we prefer to introduce the ring $\hat{\mathsf{E}}$, the completion of $\mathsf{E} = \mathscr{E}_{X,p}(0)$ for the $\mathscr{E}_{X,p}(-1)$-adic topology:

$$\hat{\mathsf{E}} = \varprojlim_{k \geqslant 0} \mathsf{E}/\mathsf{E}_{-k}.$$

We endow $\hat{\mathsf{E}}$ with the filtration by the order, $\mathsf{F}\hat{\mathsf{E}}$.

If we choose a chart on X on a neighborhood of $\pi(p)$, an element of $\hat{\mathsf{E}}$ is a series $\sum_{j \leqslant 0} p_j$, where $p_j \in \mathscr{O}_{T^*X,p}(j)$. Recall that such a series only defines an element of $\hat{\mathscr{E}}_{X,p}(0)$ if all the p_j's are holomorphic in a common neighborhood of p.

The filtered ring $\mathsf{F}\hat{\mathsf{E}}$ is clearly zariskian, and it is enough to prove that $\hat{\mathsf{E}}$ is faithfully flat over E. We remark that it follows from the construction of $\hat{\mathsf{E}}$ that the division theorems of (I, § 2) extend to $\hat{\mathsf{E}}$ so that if ψ is a contact transformation and $\hat{\psi}$ a quantized contact transformation above ψ, $\hat{\psi}$ extends naturally as an isomorphism of $\mathsf{F}\hat{\mathsf{E}}$.

Let $f_1, \ldots, f_k$ be holomorphic functions on X, near $x_0 = \pi(p)$, with $df_1 \wedge \ldots \wedge df_k(x_0) \neq 0$. We denote by A (resp. $\hat{\mathsf{A}}$) the subring of E (resp. $\hat{\mathsf{E}}$) of elements P such that:

$$[P, f_i] \equiv 0 \qquad \forall i = 1, \ldots, k.$$

It is immediately verified that $\mathsf{F}\hat{\mathsf{A}}$ is zariskian (extend $(f_1, \ldots, f_k)$ as a system of local coordinates $(f_1, \ldots, f_n)$). We shall prove by induction that $\hat{\mathsf{A}}$ is faithfully flat over A.

Let I be a finite type ideal of A with $0 \neq \mathsf{I} \neq \mathsf{A}$. We have to prove:

$$\begin{cases} \operatorname{Tor}_1^{\mathsf{A}}(\hat{\mathsf{A}}, \mathsf{A}/\mathsf{I}) - 0, \\ \hat{\mathsf{A}}\,\mathsf{I} \neq \hat{\mathsf{A}}. \end{cases}$$

a) Assume that for a system of local coordinates $(x_1, \ldots, x_n)$ on X near $0 = \pi(p)$, with $p = (0; 0, \ldots, 1)$, $f_j = x_j$, $j = 1, \ldots, k < n-1$, there exists $P \in \mathsf{I}$ such that:

(2.1.1) $\sigma(P)(0; 0, \ldots, \xi_{n-1}, 1) \not\equiv 0$.

Let us denote by A' (resp. $\hat{\mathsf{A}}'$) the subring of A (resp. $\hat{\mathsf{A}}$) of elements which commute with x_{n-1}, and assume we have already proved that $\hat{\mathsf{A}}'$ is faithfully flat over A', and in particular that A' is noetherian. For every $Q \in \mathsf{A}$ there exists $S \in \mathsf{A}$ such that:

(2.1.2) $\begin{cases} S \circ Q \in A \cdot P, \\ \sigma(S)(0; 0, \dots, \xi_{n-1}, 1) \neq 0. \end{cases}$

In fact it follows from the division theorem that $A/A \cdot P$ is isomorphic to a direct sum $\displaystyle\bigoplus_{j=0}^{m-1} A' \cdot (D_{n-1} \circ D_n^{-1})^j$, where m is the order of the zero of $\sigma(P)(0; 0, \dots, \xi_{n-1}, 1)$ at p. Let u be the image of $1 \in A$ in $A/A \cdot P$. Since A' is noetherian, the sequence of A'-modules $\left\{ \left(\displaystyle\sum_{j=1}^{N} A' \cdot (D_{n-1} \circ D_n^{-1})^j \circ Qu \right) \right\}_N$ in $A/A \cdot P$ is stationary, and there exists an integer N with:

(2.1.3) $(D_{n-1} \circ D_n^{-1})^N \circ Qu = \displaystyle\sum_{j=0}^{N-1} S_j \circ ((D_{n-1} \circ D_n^{-1})^j \circ Qu),$

the S_j's not depending on D_{n-1}. Thus $S \circ Q = 0 \mod A \cdot P$, where:

$$S = (D_{n-1} \circ D_n^{-1})^N - \sum_j S_j \circ (D_{n-1} \circ D_n^{-1})^j,$$

which proves (2.1.2).

Now let $Q_1, \dots, Q_r$ be a system of generators of I (we have made the assumption that I is finitely generated), and for all $j = 1, \dots, r$, let P_j be such that $\sigma(P_j)(0; 0, \dots, \xi_{n-1}, 1) \neq 0$, $P_j Q_j \in A \cdot P$. Consider the morphism ψ_j from $A/A \cdot P_j$ to $A/A \cdot P$ given by:

$$\psi_j (R \mod A \cdot P_j) = R \circ Q_j \mod A \cdot P.$$

Set $G = \displaystyle\bigoplus_j A/A \cdot P_j$, $\psi = \displaystyle\sum_j \psi_j$, $N = \operatorname{Ker} \psi$. We have an exact sequence of A-modules:

$$0 \to N \to G \to A/AP \to A/I \to 0.$$

Let us apply both functors $\hat{A}' \otimes_A \cdot$ and $\hat{A} \otimes_A \cdot$ to this sequence. We get a commutative diagramm:

$$\begin{array}{ccccccccc} 0 \to & \hat{A}' \otimes_A N & \to & \hat{A}' \otimes_A G & \to & \hat{A}' \otimes_A A/A \cdot P & \to & \hat{A}' \otimes_A A/I & \to 0 \\ & \downarrow & & \downarrow & & \downarrow & & \downarrow & \\ & \hat{A} \otimes_A N & \to & \hat{A} \otimes_A G & \to & \hat{A} \otimes_A A/A \cdot P & \to & \hat{A} \otimes_A A/I & \to 0. \end{array}$$

Since $A/A \cdot P \cong A'^m$ and $\hat{A}/\hat{A} \cdot P \cong \hat{A}'^m$ by the division theorem (applied both for A' and for $\hat{A}'$), and since we have a similar result for the operators P_j $(j = 1, \dots, r)$, the two vertical arrows in the middle are isomorphisms. Since the first line is exact (we assumed $\hat{A}'$ flat over A') the second line is also exact, and we get the exact sequence:

$$0 \to \hat{A} \otimes_A G/N \to \hat{A} \otimes_A A/A \cdot P \to \hat{A} \otimes_A A/I \to 0.$$

Since $\operatorname{Tor}_1^A (\hat{A}, A/A \cdot P) = 0$ (that is, $P : \hat{A} \to \hat{A}$ is injective), we get:

$$\operatorname{Tor}_1^A (\hat{A}, A/I) = 0.$$

Now assume $\hat{A} I = \hat{A}$. In that case the arrow $\hat{A} \otimes_A G \to \hat{A} \otimes_A A/A \cdot P$ is surjective, and $\hat{A}' \otimes_{A'} A/I = 0$. But $\hat{A}'$ being faithfully flat over A' we get $A/I = 0$.

b) Assume now that for a system of local coordinates $(x_1, \ldots, x_n)$ on X near $0 = \pi(p)$, such that $p = (0; 0, \ldots, 1)$, $f_j = x_j$, $j = 1, \ldots, k$, there exists $P \in I$ with:

$$(2.1.4) \qquad \sigma(P)(x_1, 0, \ldots, 0; 0, \ldots, 1) \not\equiv 0.$$

Let A_1 (resp. $\hat{A}_1$) be the subring of A (resp. $\hat{A}$) of operators which commute with D_1, and assume we have already proved that $\hat{A}_1$ is faithfully flat over A_1. Then the same argument as for A' proves that $\hat{A}$ is faithfully flat over A.

c) Assume $k > 0$. Then there exist coordinates $(x_1, \ldots, x_n)$ which satisfy the conditions in b). In fact for $P \neq 0$, there exist coordinates (x, ξ) such that if $p = (x_0, \xi_0)$ we have $\sigma(P)(x, \xi_0) \not\equiv 0$. Moreover we may find a submanifold Y in X of dimension k passing through x_0 such that:

$$\begin{cases} \sigma(P)(x, \xi_0)|_Y \not\equiv 0, \\ df_1 \wedge \ldots \wedge df_k|_Y \neq 0. \end{cases}$$

One completes $(f_1, \ldots, f_k)$ as a system of local coordinates $(f_1, \ldots, f_n)$, with $Y = \{x; f_{k+1} = \ldots = f_n = 0\}$, and one replaces the f_j's $(j = 1, \ldots, k)$ by a linear combination of the f_j's.

d) By c) and b) we may assume $k = 0$. Now we proceed by induction on $\dim X$. Using a quantized contact transformation, by a) the problem is reduced to the case where $\dim X = 1$. Then by b) we only have to study the ring:

$$A = \{P \in \mathscr{E}_{\mathbb{C},(0;1)}(0); [D_1, P] = 0\},$$

but this filtered ring is commutative, and one proves that A is noetherian using the division theorem, as for the case of $\mathscr{O}_{\mathbb{C},0}$ (cf. Hervé [1]). Thus $\hat{A}$ is faithfully flat over A, in that case, which completes the proof.   □

Remark 2.1.2. A similar proof would work for the ring $\hat{\mathscr{E}}_{X,p}(0)$.

Corollary 2.1.3. *The Ring $\mathscr{E}_X(0)$ is right and left noetherian (hence coherent).*

Proof. Applying Proposition 1.4.1 and Theorem 2.1.1 we only have to prove that the graded Ring $\mathrm{gr}(\mathscr{E}_X(0))$ is noetherian. Locally on $\dot{T}^*X$ we have an isomorphism:

$$(2.1.5) \qquad \mathrm{gr}(\mathscr{E}_X(0)) \cong \gamma^{-1} \mathscr{O}_{P^*X}[T^{-1}]$$

Since $\mathscr{O}_{P^*X}$ is noetherian, the same property holds for $\mathrm{gr}(\mathscr{E}_X(0))$ on $\dot{T}^*X$.

Let us prove that $\mathrm{gr}(\mathscr{E}_X(0))$ is coherent in the neighborhood of T_X^*X. Let $(u_1, \ldots, u_r)$ be sections of this sheaf defined in a neighborhood of $(x, 0) \in T_X^*X$, and let ψ be the morphism from $\mathrm{gr}(\mathscr{E}_X(0))^r$ to $\mathrm{gr}(\mathscr{E}_X(0))$ given by:

$$\psi(a_1, \ldots, a_r) = \sum_{j=1}^r a_j u_j.$$

Let $K = \mathrm{Ker}\,\psi$. Since $u_1, \ldots, u_r$ are in fact holomorphic functions they define a morphism ϕ from $\mathscr{O}_X^r$ to $\mathscr{O}_X$ whose kernel $\mathscr{N}$ is an $\mathscr{O}_X$-module of finite type in a neighborhood of x. But $\mathrm{gr}(\mathscr{E}_X(0))$ being flat over $\pi^{-1}\mathscr{O}_X$, we have:

$$\mathscr{K} = \mathrm{gr}(\mathscr{E}_X(0)) \otimes_{\pi^{-1}\mathscr{O}_X} \pi^{-1}\mathscr{N}$$

and $\mathscr{K}$ is of finite type in a neighborhood of $(x, 0)$.

Finally let $\{\mathscr{I}_\alpha\}_\alpha$ be an increasing sequence of graded Ideals of $\mathrm{gr}(\mathscr{E}_X(0))$, in a neighborhood of $(x, 0) \in T_X^* X$. Then there exists a neighborhood U of x in X such that this family is stationary on U and also on $T^*U \setminus T_U^* U$, by (2.1.5). $\square$

Remark 2.1.4. There exists another proof of Theorem 2.1.1 by L. Boutet de Monvel [1], using an extension of the division theorem for ideals of $\mathscr{E}_{X,p}$.

2.2. Main Properties of $\mathscr{E}_X$

Let T be an indeterminate, $\mathbb{C}[T]$ the ring of polynomials on $\mathbb{C}$, $\mathbb{C}[T^{-1}, T]$ the ring of generalized polynomials with degree in $\mathbb{Z}$:

$$P \in \mathbb{C}[T^{-1}, T] \Leftrightarrow P = \sum_{i \in I} a_i T^i,$$

where I is a finite subset of $\mathbb{Z}$, with the obvious relations $T^i \cdot T^j = T^{i+j}$.

The module $\mathbb{C}[T^{-1}, T]$ over $\mathbb{C}[T]$ is flat since it is an inductive limit of free $\mathbb{C}[T]$-modules. (The same result holds when replacing T by T^{-1}). Now let p belong to $T^* X$. If $p = (x; 0) \in T_X^* X$, we have:

(2.2.1) $$\mathrm{gr}(\mathscr{E}_{X,p}) \cong \mathscr{O}_{X,x}[\xi_1, \ldots, \xi_n].$$

If $p \in \dot{T}^* X$, let $p' = \gamma(p) \in P^* X$. Then:

(2.2.2) $$\mathrm{gr}(\mathscr{E}_{X,p}) \cong \mathscr{O}_{P^* X, p'}[T^{-1}, T].$$

In all cases we find that $\mathrm{gr}(\mathscr{E}_{X,p})$ is noetherian. Thus applying Proposition 1.1.5 and Theorem 2.1.1 we get:

Proposition 2.2.1. *Let $p \in T^* X$. The filtered ring $\mathscr{E}_{X,p}$ is right and left zariskian.*

The following Lemma will be of constant use (cf. Malgrange [2]).

Let $E \xrightarrow{\pi} X$ be a complex vector bundle over X (e.g.: $E = T^* X$). We denote by $\mathscr{O}_E(j)$ the subsheaf of $\mathscr{O}_E$ of sections which are homogeneous of degree j, and set $\mathscr{O}_E^h = \bigoplus_{j \in \mathbb{Z}} \mathscr{O}_E(j)$.

Lemma 2.2.2. *Let* $p \in E$. *The functor* $\mathscr{O}_{E,p} \otimes_{\mathscr{O}_{E,p}^h} (\cdot)$ *from the category of graded* $\mathscr{O}_{E,p}^h$-*modules to the category of* $\mathscr{O}_{E,p}$-*modules, is exact and faithful.*

Proof. Set $A = \mathscr{O}_{E,p}^h$, $B = \mathscr{O}_{E,p}$.

First assume $p = (x; 0) \in X$ and choose coordinates $(x_1, \ldots, x_n; \xi_1, \ldots, \xi_l)$ in a neighborhood of x. Then:

$$A \cong \mathscr{O}_{\mathbb{C}^n, 0}[\xi_1, \ldots, \xi_l], \qquad B \cong \mathscr{O}_{\mathbb{C}^n \times \mathbb{C}^l, (x; 0)}.$$

Let $\mathfrak{m}_A$ (resp. $\mathfrak{m}_B$) be the ideal in A (resp. B) generated by $(\xi_1, \ldots, \xi_l)$. The rings A and B have the same completion for the $\mathfrak{m}_A$-adic (resp. $\mathfrak{m}_B$-adic) topology. Since A and B are noetherian, it is well known that this implies B flat over A (cf. Bourbaki [1]).

Now let I be a graded ideal of A, $J = BI$. We have to prove that $A \cap J = I$. To a section $f(x, \xi)$ of B, let us associate $\sigma_m(f)$, the homogeneous part of order m of the Taylor expansion of f along X. That is, if $f(x, \xi) = \sum_{\alpha \in \mathbb{N}^n} a_\alpha(x) \xi^\alpha$,

$$\sigma_m(f)(x, \xi) = \sum_{|\alpha| = m} a_\alpha(x) \xi^\alpha.$$

Let $u_1, \ldots, u_r$ be homogeneous generators of I, of order $k_1, \ldots, k_r$. Let $f \in J \cap A$:

$$f = \sum_{j=1}^r b_j u_j, \qquad b_j \in B.$$

For each $m \in \mathbb{N}$, we get:

$$\sigma_m(f) = \sum_{j=1}^r \sigma_{m - k_j}(b_j) u_j.$$

Since f belongs to A, f is a finite sum of $\sigma_k(f)$, $k \in \mathbb{N}$, and thus f belongs to I.

Now assume $p \in \dot{T}^* X$. Taking coordinates on X, we have isomorphisms:

$$A \cong \mathscr{O}_{\mathbb{C}^N, z_0}[t^{-1}, t], \qquad B \cong \mathscr{O}_{\mathbb{C}^N \times \mathbb{C}, (z_0, 1)},$$

where $N = n + l - 1$, t is a coordinate on $\mathbb{C}$, $z \in \mathbb{C}^N$.

The ring B is flat over $\mathscr{O}_{\mathbb{C}^N, z_0}[t]$, and A is the localization of this last ring by the ideal generated by t. Thus B is flat over A.

Finally let I be a graded ideal of A, $u_1, \ldots, u_r$ be homogeneous generators of I:

$$u_j = a_j(z) t^{k_j}.$$

Let $f = \sum_{j=1}^r b_j u_j$, $b_j \in B$, and assume $f \in A$. By taking the Taylor expansion of $b_j(z, t)$ at $t = 1$, we find that f belongs to I, as in the first part of the proof of. $\square$

Remark 2.2.3. Because of the preceding Lemma, we sometimes identify a graded gr($\mathscr{E}_X$)-Module $\mathscr{M}$, and the Module $\mathscr{O}_{T^*X} \otimes_{\mathrm{gr}(\mathscr{E}_X)} \mathscr{M}$.

Proposition 2.2.4. *The Ring $\mathscr{E}_X$ is right and left noetherian (hence coherent).*

Proof. Applying Proposition 1.4.1 and 2.2.1, it is enough to prove that gr($\mathscr{E}_X$) is a noetherian Ring. But this follows immediately from Lemma 2.2.2 since $\mathscr{O}_{T^*X}$ is noetherian over T^*X. $\square$

Proposition 2.2.5. *Let $p \in T^*X$. Then:*
   a) *$\mathscr{E}_{X,p}$ is flat over $\mathscr{E}_{X,p}(0)$ and over $(\pi^{-1}\mathscr{D}_X)_p = \mathscr{D}_{X,\pi(p)}$.*
   b) *$\hat{\mathscr{E}}_{X,p}$ is faithfully flat over $\mathscr{E}_{X,p}$.*

Proof. Applying Proposition 1.2.4 and Theorem 2.1.1 (or Remark 2.1.2) we have to prove similar results for the graded rings. Then b) follows since $\mathrm{gr}\,\mathscr{E}_X = \mathrm{gr}\,\hat{\mathscr{E}}_X$. To prove a) we remark that $\mathscr{O}_{X,x}[\xi_1, \ldots, \xi_n]$ is flat over $\mathscr{O}_{X,x}$ (the case where $p = (x; 0) \in T_X^* X$), and $\mathscr{O}_{\mathbb{C}^N, z}[t^{-1}, t]$ is flat over $\mathscr{O}_{\mathbb{C}^N, z}[t^{-1}]$ (the case where $p \in \dot{T}^*X$, with the notations of the proof of Lemma 2.2.2). $\square$

Remark 2.2.6. Thanks to Formula (C. 2.7) we may state Proposition 2.2.5 simply as "$\mathscr{E}_X$ is flat over $\mathscr{E}_X(0)$ and over $\pi^{-1}\mathscr{D}_X$, and $\hat{\mathscr{E}}_X$ is faithfully flat over $\mathscr{E}_X$".

Proposition 2.2.7. *Let U be an open set of $\dot{T}^*X$. A coherent $\mathscr{E}_X$-module on U is pseudo-coherent over $\mathscr{E}_X(0)$.*

Proof. Applying Proposition 1.4.2 we only have to check that $\mathscr{E}_X(k)$ is coherent over $\mathscr{E}_X(0)$. But this Module is locally free of rank one on $\dot{T}^*X$. In fact if T is an invertible operator of order one, it defines an isomorphism of left $\mathscr{E}_X(0)$-modules from $\mathscr{E}_X(0)$ to $\mathscr{E}_X(k)$:

$$P \to P \circ T^k.   \square$$

Remark 2.2.8. Proposition 2.2.7 does not extend all over T^*X. In fact $\mathscr{E}_X(1)$ is locally of finite type over $\mathscr{E}_X(0)$ on T^*X, but is not coherent, since it is free of rank $n+1$ at each point of $T_X^* X$ but not outside $T_X^* X$.

Remark 2.2.9. On an open set $U \subset \dot{T}^*X$, it is equivalent to endow a coherent $\mathscr{E}_X$-module $\mathscr{M}$ with a good filtration, or with a sub-$\mathscr{E}_X(0)$-module $\mathscr{M}_0$ locally of finite type, which generates $\mathscr{M}$. This follows immediately from the local existence of invertible operators of order one.

2.3. Characteristic Cycle

Let $\mathscr{M}$ be a coherent left $\mathscr{E}_X$-module on an open set $U \subset T^*X$. Locally on U we may find a matrix (P_0) of microdifferential operators such that $\mathscr{M}$ admits the presentation:

$$\mathscr{E}_X^{N_1} \xrightarrow[(P_0)]{} \mathscr{E}_X^{N_0} \to \mathscr{M} \to 0.$$

When (P_0) is a single operator, (a (1×1)-matrix), then the support of $\mathscr{M}$ in U is the set where the principal symbol $\sigma(P_0)$ vanishes (I, Proposition 1.3.5). Then it is natural to call the support of $\mathscr{M}$ the "characteristic variety" of $\mathscr{M}$. We shall write:

$$\text{support of } \mathscr{M} = \text{supp}(\mathscr{M}) = \text{char}(\mathscr{M}).$$

Moreover we also say that $\mathscr{M}$ is "a system of microdifferential equations".

Theorem 2.3.1. *Let $\mathscr{M}$ be a coherent $\mathscr{E}_X$-module on an open set $U \subset T^*X$. Then* char$(\mathscr{M})$ *is a closed analytic subset of U, conic for the action of $\mathbb{C}^\times$ on T^*X, involutive for the symplectic structure of T^*X.*

Proof. The theorem being local on U, we may endow $\mathscr{M}$ with a good filtration $(\mathscr{M}_k)_k$. Let gr$(\mathscr{M})$ be the associated graded Module. We set:

$$(2.3.1) \qquad \overline{\text{gr}}(\mathscr{M}) = \mathscr{O}_{T^*X} \otimes_{\text{gr}\,\mathscr{E}_X} \text{gr}(\mathscr{M})$$

We know by Proposition 1.4.1 that gr$(\mathscr{M})$ is coherent over gr$(\mathscr{E}_X)$. Applying Lemma 2.2.2 we find that $\overline{\text{gr}}(\mathscr{M})$ is $\mathscr{O}_{T^*X}$-coherent, and

$$(2.3.2) \qquad \text{supp}(\overline{\text{gr}}(\mathscr{M})) = \text{supp}(\text{gr}(\mathscr{M}))$$

Clearly the support of gr$(\mathscr{M})$ is $\mathbb{C}^\times$-conic and the support of $\overline{\text{gr}}(\mathscr{M})$ is analytic.

The involutivity of supp(gr$(\mathscr{M})$) follows from Theorem 1.3.4. since the two Poisson brackets on gr$(\mathscr{E}_X)$, one coming from the symplectic structure on T^*X, the other from the filtration on $\mathscr{E}_X$, are in fact the same. Finally it is enough to prove the equality of supp(gr$(\mathscr{M})$) and supp$(\mathscr{M})$, but this follows from Proposition 1.1.3. $\square$

In the preceding construction, we may not only consider the set supp(gr$(\mathscr{M})$), but also the analytic cycle [supp$(\overline{\text{gr}}(\mathscr{M}))$] (cf. Appendix D), which depends only on $\mathscr{M}$. In fact let W be an irreducible analytic set containing char$(\mathscr{M})$; applying Proposition 1.3.1 and Proposition D.1.2 of the Appendix, we find that mult$_W(\overline{\text{gr}}(\mathscr{M}))$ depends only on $\mathscr{M}$.

Definition 2.3.2. Let $\mathscr{M}$ be a coherent $\mathscr{E}_X$-module on an open set $U \subset T^*X$. The characteristic cycle of $\mathscr{M}$, denoted [char$(\mathscr{M})$], is the cycle [supp$(\overline{\text{gr}}(\mathscr{M}))$], where gr$(\mathscr{M})$ is the graded Module associated to any (locally defined) good filtration on $\mathscr{M}$, and $\overline{\mathrm{gr}}(\mathscr{M})$ is given by (2.3.1). If W is an irreducible analytic set containing char$(\mathscr{M})$, we also set:

$$\mathrm{mult}_W(\mathscr{M}) = \mathrm{mult}_W(\overline{\mathrm{gr}}(\mathscr{M})).$$

Proposition 2.3.3. *Consider an exact sequence of coherent $\mathscr{E}_X$-modules on U:*

(2.3.3) $$0 \to \mathscr{L} \to \mathscr{M} \to \mathscr{N} \to 0.$$

Then if codim$(\mathrm{char}(\mathscr{M})) \geqslant d$, *we have:*

(2.3.4) $$[\mathrm{char}(\mathscr{M})]_d = [\mathrm{char}(\mathscr{L})]_d + [\mathrm{char}(\mathscr{N})]_d.$$

In particular if W is an irreducible analytic set containing char$(\mathscr{M})$:

(2.3.5) $$\mathrm{mult}_W(\mathscr{M}) = \mathrm{mult}_W(\mathscr{L}) + \mathrm{mult}_W(\mathscr{N}).$$

Remark that the cycle $[\mathrm{char}(\mathscr{M})]$ is positive, and its support is nothing but supp$(\mathscr{M})$. Remark also that $\mathrm{mult}_W(\mathscr{M}) = 0$ iff dim$(\mathrm{char}(\mathscr{M})) < \dim W$.

Example 2.3.4. Let t be a coordinate on $\mathbb{C}$, $\mathscr{M} = \mathscr{D}_\mathbb{C}/\mathscr{D}_\mathbb{C}(D_t^2, tD_t - 1)$. Then $\mathscr{M}$ is isomorphic to $\mathscr{O}_\mathbb{C}$, since the annihilator of $t \in \mathscr{O}_\mathbb{C}$ is exactly the left ideal $\mathscr{J}$ generated by D_t^2 and $(tD_t - 1)$, and t generates $\mathscr{O}_\mathbb{C}$ since $D_t \cdot t = 1$ in $\mathscr{O}_\mathbb{C} = \mathscr{D}_\mathbb{C}/\mathscr{D}_\mathbb{C} \cdot D_t$. Thus char$(\mathscr{M}) = T_\mathbb{C}^* \mathbb{C}$, and the multiplicity of $\mathscr{M}$ along $T_\mathbb{C}^* \mathbb{C}$ is 1. However the symbol Ideal of $\mathscr{J}$ is generated by τ^2 and $t\tau$ and is not reduced, (cf. I, Remark 6.2.4).

Now we shall study the right $\mathscr{E}_X$-modules $\mathscr{E}xt^j_{\mathscr{E}_X}(\mathscr{M}, \mathscr{E}_X)$.

Proposition 2.3.5. a) *For each $j \in \mathbb{N}$, $\mathscr{E}xt^j_{\mathscr{E}_X}(\mathscr{M}, \mathscr{E}_X)$ is a right coherent $\mathscr{E}_X$-module.*

b) $\mathscr{E}xt^j_{\mathscr{E}_X}(\mathscr{M}, \mathscr{E}_X) = 0$ *for $j < $* codim$(\mathrm{char}\, \mathscr{M})$.

c) codim$(\mathrm{char}(\mathscr{E}xt^j_{\mathscr{E}_X}(\mathscr{M}, \mathscr{E}_X))) \geqslant j$.

Proof. a) For any $r \geqslant 0$, $\mathscr{M}$ admits locally a finite free r-presentation. Replacing $\mathscr{M}$ by 0 in such a complex and applying the functor $\mathscr{H}om_{\mathscr{E}_X}(\cdot, \mathscr{E}_X)$, we find a finite free complex whose j-th cohomology group is $\mathscr{E}xt^j_{\mathscr{E}_X}(\mathscr{M}, \mathscr{E}_X)$ for $j < r$. Then it is clear by induction on j that these Modules are coherent.

b, c) Let us endow $\mathscr{M}$ with a good filtration, and let $\overline{\mathrm{gr}}(\mathscr{M})$ be the Module defined in (2.3.1). We know that char$(\mathscr{M}) = $ supp$(\overline{\mathrm{gr}}(\mathscr{M}))$, and we know by Proposition 1.2.3 and Lemma 2.2.2 that

$$\mathrm{supp}(\mathscr{E}xt^j_{\mathscr{E}_X}(\mathscr{M}, \mathscr{E}_X)) \subset \mathrm{supp}(\mathscr{E}xt^j_{\mathscr{O}_{T^*X}}(\overline{\mathrm{gr}}(\mathscr{M}), \mathscr{O}_{T^*X})).$$

Thus the results follow from the corresponding ones for $\mathscr{O}_{T^*X}$-modules (cf. Appendix D). $\square$

Proposition 2.3.6. *Let $\mathscr{M}$ be a coherent $\mathscr{E}_X$-module on $U \subset T^*X$, and let $\dim X = n$.*

a) *For any $\mathscr{E}_X$-module $\mathscr{N}$ we have*

$$\mathscr{E}\kern-1pt xt^j_{\mathscr{E}_X}(\mathscr{M}, \mathscr{N}) = 0 \quad \forall j > n.$$

b) *Assume $\mathscr{E}\kern-1pt xt^j_{\mathscr{E}_X}(\mathscr{M}, \mathscr{E}_X) = 0 \ \forall j > p$. Then $\mathscr{M}$ locally admits a finite free presentation of length $\sup(1, p)$.*

c) *Assume $\mathscr{M}$ is projective. Then $\mathscr{M}$ is locally stably finite-free, i.e. there exists locally a finite free $\mathscr{E}_X$-module $\mathscr{L}$ such that $\mathscr{M} \oplus \mathscr{L}$ is free.*

Proof. a) Since $\mathscr{M}$ is coherent, we have for $p \in U$:

$$\mathscr{E}\kern-1pt xt^j_{\mathscr{E}_X}(\mathscr{M}, \mathscr{N})_p = \operatorname{Ext}^j_{\mathscr{E}_{X,p}}(\mathscr{M}_p, \mathscr{N}_p).$$

Moreover we know that $\mathscr{E}\kern-1pt xt^j_{\mathscr{E}_X}(\mathscr{M}, \mathscr{E}_X) = 0$ for $j > n$, since the support of this Module has to be both involutive and of codimension $\geqslant j$. In order to apply Lemma B.2.1 to the $\mathscr{E}_{X,p}$-module $\mathscr{M}_p$, we use Proposition 1.2.1, by remarking that a $\operatorname{gr}(\mathscr{E}_{X,p})$-module of finite type admits a finite free presentation of finite length (this is well known for $\mathscr{O}_{T^*X,p}$-modules, and we may use Lemma 2.2.2. Remark that projective graded $\operatorname{gr}(\mathscr{E}_{X,p})$-modules of finite type are free).

b), c) We may apply Lemma B.2.2 to $\mathscr{M}_p$, and $\mathscr{M}$ being coherent, the results follow. $\square$

Remark 2.3.7. Let $\mathscr{M}$ be a coherent $\mathscr{E}_X$-module- Set:

$\mathscr{M}_r =$ the sub-Module of $\mathscr{M}$ generated by the sections u such that

$$\operatorname{codim}(\operatorname{char}(\mathscr{E}_X u)) \geqslant r.$$

M. Kashiwara has proved that $\mathscr{M}_r$ is a coherent $\mathscr{E}_X$-module, and other important results of this kind. We refer to Kashiwara [3].

2.4. Holonomic Modules

Definition 2.4.1. A holonomic Module $\mathscr{M}$ on an open set $U \subset T^*X$ is a coherent $\mathscr{E}_X$-module such that $\operatorname{char}(\mathscr{M})$ is Lagrangean in U.

Remark that if $\mathscr{M}$ is holonomic, $\operatorname{char}(\mathscr{M})$ is purely dimensional.

The category of (left) holonomic $\mathscr{E}_X$-modules on U is an abelian subcategory of the category of coherent $\mathscr{E}_X$-modules. If we consider an exact sequence of coherent $\mathscr{E}_X$-modules as (2.3.3), then $\mathscr{M}$ is holonomic iff $\mathscr{L}$ and $\mathscr{N}$ are holonomic. Let $n = \dim X$. Applying Proposition 2.3.5 and the involutivity theorem, we find for a holonomic Module $\mathscr{M}$:

(2.4.1) $$\mathscr{E}\kern-1pt xt^j_{\mathscr{E}_X}(\mathscr{M}, \mathscr{E}_X) = 0 \quad j \neq n,$$

and $\mathscr{E}\kern-1pt xt^n_{\mathscr{E}_X}(\mathscr{M}, \mathscr{E}_X)$ is a right holonomic $\mathscr{E}_X$-module.

We define the functor * on the category of holonomic Modules by setting:

(2.4.2)
$$\mathscr{M}^* = \mathscr{E}xt^n_{\mathscr{E}_X}(\mathscr{M}, \mathscr{E}_X).$$

Applying the functor $\mathscr{H}om_{\mathscr{E}_X}(\cdot, \mathscr{E}_X)$ to the short exact sequence (2.3.3) we find an exact sequence of right holonomic Modules:

$$0 \to \mathscr{N}^* \to \mathscr{M}^* \to \mathscr{L}^* \to 0,$$

since all the groups $\mathscr{E}xt^j_{\mathscr{E}_X}(\mathscr{K}, \mathscr{E}_X)$ are zero for $j \neq n$, $\mathscr{K} = \mathscr{L}, \mathscr{M}, \mathscr{N}$. To better understand the functor *, consider a bounded complex of finite free $\mathscr{E}_X$-modules quasi-isomorphic to $\mathscr{M}$ (Proposition 2.3.6):

(2.4.3)
$$0 \to \mathscr{E}_X^{N_{-n}} \xrightarrow{P_{1-n}} \ldots \to \mathscr{E}_X^{N_{-1}} \xrightarrow{P_0} \mathscr{E}_X^{N_0} \to 0.$$

That is, the complex (2.4.3) is exact, except at step 0, and we have:

(2.4.4)
$$\mathscr{M} \cong \mathscr{E}_X^{N_0} / \mathscr{E}_X^{N_{-1}} \cdot P_0.$$

Remark that the P_j's are $(N_j \times N_{j-1})$-matrices of microdifferential operators satisfying:

(2.4.5)
$$P_j \circ P_{j-1} = 0.$$

Applying the functor $\mathscr{H}om_{\mathscr{E}_X}(\cdot, \mathscr{E}_X)$ to the complex (2.4.3), we find a complex of right $\mathscr{E}_X$-modules:

(2.4.6)
$$0 \to \mathscr{E}_X^{N_0'} \xrightarrow{P_0} \mathscr{E}_X^{N_1'} \to \ldots \xrightarrow{P_{n-1}} \mathscr{E}_X^{N_n'} \to 0,$$

where we have set:

$$\mathscr{E}_X^{N_i'} = \mathscr{H}om_{\mathscr{E}_X}(\mathscr{E}_X^{N_{-i}}, \mathscr{E}_X),$$

(thus $N_i' = N_{-i}$), and $P_i = P_{-i}$, but now P_i operates on the left. Thus the relation (2.4.5) still holds.

By (2.4.1) the complex (2.4.6) is exact, except at step n, and we have:

(2.4.7)
$$\mathscr{M}^* \cong \mathscr{E}_X^{N_n'} / P_{n-1} \cdot \mathscr{E}_X^{N_{n-1}'}.$$

Since char$(\mathscr{M}^*)$ is contained in char$(\mathscr{M})$, $\mathscr{M}^*$ is holonomic. We may define a similar functor, still denoted *, on the category of right holonomic Modules, and the preceding discussion shows that $** = I$. We summarize:

Proposition 2.4.2. *The functor * from the category of left (resp. right) holonomic Modules on an open set $U \subset T^*X$, to the category of right (resp. left) holonomic Modules on U, is a contravariant exact functor which satisfies $** = I$.*

Of course char$(\mathscr{M})$ = char$(\mathscr{M}^*)$, and it could be easily proved that this equality still holds for the characteristic cycles.

Now recall that an object in an abelian category is simple if it has no sub-object other than $\{0\}$ or itself. In particular a holonomic Module $\mathscr{M}$ is simple if the only sub-Modules of $\mathscr{M}$ are $\{0\}$ and $\mathscr{M}$.

An object M in an abelian category is of finite length if it admits a composition sequence:

$$\{0\} = M_0 \subset M_1 \subset \ldots \subset M_{l-1} \subset M_l = M,$$

such that each quotient M_j/M_{j-1} is simple $(0 < j \leqslant l)$.

Proposition 2.4.3. *A holonomic Module $\mathscr{M}$ is locally of finite length.*

Proof. Locally char$(\mathscr{M})$ has a finite number of irreducible components, and the multiplicity on each component is finite. Thus the result follows from Proposition 2.3.3. □

Remark that we can give another proof. In fact $\mathscr{E}_X$ is noetherian, and any increasing sequence of coherent sub-Modules of a coherent Module is locally stationary. Applying Proposition 2.4.2 we find that a decreasing sequence of holonomic Modules is locally stationary.

Example 2.4.4. We have:

$$(2.4.8) \qquad\qquad \mathscr{O}_X^* = \Omega_X.$$

If we choose coordinates $(x_1, \ldots, x_n)$ on X,

$$\mathscr{O}_X \cong \mathscr{D}_X / \mathscr{D}_X(D_1, \ldots, D_n),$$

$$\Omega_X \cong \mathscr{D}_X / (D_1, \ldots, D_n) \mathscr{D}_X,$$

where $\mathscr{D}_X(D_1, \ldots, D_n)$ (resp. $(D_1, \ldots, D_n)\mathscr{D}_X$) is the left (resp. right) Ideal generated by $D_1, \ldots, D_n$.

The holonomic Module $\mathscr{O}_X$ is sometimes called the "De Rham system" on X.

Proposition 2.4.5. *Let $\mathscr{M}$ be a coherent $\mathscr{E}_X$-module on the open set $U \subset T^*X$. Assume that $\mathscr{E}\kern-1pt xt^j_{\mathscr{E}_X}(\mathscr{M}, \mathscr{E}_X) = 0$ for $j < n = \dim X$. Then $\mathscr{M}$ is holonomic.*

Proof. Set $\mathscr{M}^* = \mathscr{E}\kern-1pt xt^n_{\mathscr{E}_X}(\mathscr{M}, \mathscr{E}_X)$. Then $\mathscr{M}^*$ is holonomic by Proposition 2.3.5, and $\mathscr{M} = \mathscr{M}^{**}$ is itself holonomic. □

Remark 2.4.6. As an application of Proposition 2.4.3 one can prove, by purely algebraic arguments, that holonomic Modules are cyclic, that is locally generated by only one section (a theorem of Stafford, cf. Björk [1], ch. 1, § 7]).

Remark 2.4.7. We do not pretend to have treated the theory of holonomic Modules in these few lines. This theory is a fundamental branch of Analysis, including the theory of "regular holonomic Modules" (Kashiwara-Kawaï [1]), which has itself deep connections with Topology, through the so-called "Riemann-Hilbert correspondence" (Kashiwara [7], Mebkhout [1]). We shall study the holomorphic solutions of holonomic Modules in Chapter III.

2.5. Adjunction of a Dummy Variable

As we have noticed, it may be difficult to work with $\mathscr{E}_X$-modules in a neighborhood of the zero section $T^*_X X$ of $T^* X$. It is also difficult to work with general involutive manifolds but easy to manipulate regular involutive manifolds (Appendix A). In order to overcome this kind of difficulties we expose the trick of the dummy variable.

Let t be a coordinate on $\mathbb{C}$, (t, τ) the associated coordinates on $T^* \mathbb{C}$. Consider the map:

$$\psi: T^* X \times \dot{T}^* \mathbb{C} \to T^* X, \quad ((x, \xi), (t, \tau)) \to (x, \xi/\tau).$$

If U is an open set of $T^* X$, $\psi^{-1}(U)$ is a conic open set, and if V is an involutive manifold of $T^* X$, $\psi^{-1}(V)$ is regular involutive.

To a left coherent $\mathscr{E}_X$-module $\mathscr{M}$ on $U \subset T^* X$, we associate:

$$\mathscr{M} \hat{\otimes} \mathscr{E}_{\mathbb{C}} \underset{\mathrm{def}}{=} \psi^{-1} \mathscr{M} \otimes_{\psi^{-1}\mathscr{E}_X} \mathscr{E}_{X \times \mathbb{C}}$$

Remark that $\mathscr{M} \hat{\otimes} \mathscr{E}_{\mathbb{C}}$ is a left $\mathscr{E}_{X \times \mathbb{C}}$-module, and also a right $\mathscr{E}_{\mathbb{C}}$-module.

Lemma 2.5.1. *The Ring $\mathscr{E}_{X \times \mathbb{C}}$ is faithfully flat over $\psi^{-1}\mathscr{E}_X$.*

Proof. Applying Propositions 1.2.4 and 2.2.1 it is enough to prove that $\mathrm{gr}(\mathscr{E}_{X \times \mathbb{C}})$ is faithfully flat over $\psi^{-1}(\mathrm{gr}(\mathscr{E}_X))$, but this follows from Lemma 2.2.2 and the fact that $\mathscr{O}_{T^*(X \times \mathbb{C})}$ is faithfully flat over $\psi^{-1}\mathscr{O}_{T^*X}$. $\square$

Now consider the two morphisms on $\mathscr{M} \hat{\otimes} \mathscr{E}_{\mathbb{C}}$:

$$v \mapsto \mathrm{ad}_t(v) = [t, v] = tv - vt$$

$$v \mapsto \mathrm{ad}_{D_t}(v) = [D_t, v] = D_t v - v D_t$$

Those two morphisms are well defined since $\mathscr{M} \hat{\otimes} \mathscr{E}_{\mathbb{C}}$ is also a right $\mathscr{E}_{\mathbb{C}}$-module, and they commute with each together. In fact we have:

$$[t, [D_t, v]] = t D_t v - t v D_t - D_t v t + v D_t t$$

$$[D_t, [t, v]] = D_t t v - D_t v t - t v D_t + v t D_t$$

then:

$$(\mathrm{ad}_t \circ \mathrm{ad}_{D_t} - \mathrm{ad}_{D_t} \circ \mathrm{ad}_t) v = t D_t v + v D_t t - D_t t v - v t D_t$$

$$= [t, D_t] v - v [t, D_t] = 0.$$

Thus we can construct the Koszul complex associated to these morphisms:

$$(2.5.1) \qquad 0 \to \psi^{-1} \mathscr{M} \underset{v}{\to} \mathscr{M} \hat{\otimes} \mathscr{E}_{\mathbb{C}} \underset{\lambda}{\to} (\mathscr{M} \hat{\otimes} \mathscr{E}_{\mathbb{C}})^2 \underset{\mu}{\to} \mathscr{M} \hat{\otimes} \mathscr{E}_{\mathbb{C}} \to 0,$$

where $v(u) = u \otimes 1$, $\lambda(u) = ([t, u], [D_t, u])$, $\mu(v, w) = [D_t, v] - [t, w]$.

Proposition 2.5.2. a) *The complex of left $\psi^{-1}(\mathscr{E}_X)$-modules (2.5.1) is exact.*

b) *Assume $\mathscr{M}$ is coherent on $U \subset T^*X$. Then $\mathscr{M} \hat{\otimes} \mathscr{E}_{\mathbb{C}}$ is coherent on $\psi^{-1}(U)$ and $\mathrm{char}(\mathscr{M} \hat{\otimes} \mathscr{E}_{\mathbb{C}}) = \psi^{-1}(\mathrm{char}(\mathscr{M}))$. Moreover if $\mathscr{M}$ has simple characteristics along an involutive manifold V and u is a simple generator of $\mathscr{M}$ on V, then $\mathscr{M} \hat{\otimes} \mathscr{E}_{\mathbb{C}}$ has simple characteristic along $\psi^{-1}(V)$ with simple generator $u \otimes 1$.*

c) *Assume now that $\mathscr{M} \hat{\otimes} \mathscr{E}_{\mathbb{C}}$ is coherent on $\psi^{-1}(U)$. Then $\mathscr{M}$ is coherent on U.*

Proof. a) by Lemma 2.5.1 it is enough to prove the exactness of (2.5.1) when $\mathscr{M} = \mathscr{E}_X$. We remark that:

- $\psi^{-1}\mathscr{E}_X = \{P \in \mathscr{E}_{X \times \mathbb{C}}; \ [t, P] = [D_t, P] = 0\}$
- If $\sum_j p_j$ is the total symbol of P for a choice of local coordinates on X,

then the total symbol of $[t, P]$ is $-\dfrac{\partial}{\partial \tau}\left(\sum_j p_j\right)$ and the total symbol of $[D_t, P]$

is $\dfrac{\partial}{\partial t}\left(\sum_j p_j\right)$. Thus $\mathrm{ad}_t(\cdot)$ is surjective on $\mathscr{E}_{X \times \mathbb{C}}$ and $\mathrm{ad}_{D_t}(\cdot)$ is surjective as an endomorphism of $\mathrm{Ker}(\mathrm{ad}_t(\cdot))$. Thus the sequence $(\mathrm{ad}_t(\cdot), \mathrm{ad}_{D_t}(\cdot))$ is a regular sequence of $\psi^{-1}(\mathscr{E}_X)$-linear morphisms of $\mathscr{E}_{X \times \mathbb{C}}$, and the exactness of (2.5.1) follows from Proposition B.4.2.

b) follows from Lemma 2.5.1 and Proposition 1.2.4.

c) let $v_1, \ldots, v_N$ be generators of $\mathscr{M} \hat{\otimes} \mathscr{E}_{\mathbb{C}}$ on an open set $V \subset \psi^{-1}(U)$. We may assume $v_j = u_j \otimes 1$. Consider on $\psi(V)$ the morphism α from $\mathscr{E}_X^N$ to $\mathscr{M}$, $\alpha((P_j)_{j=1}^N) = \sum_j P_j u_j$. Since the corresponding morphism from $\mathscr{E}_{X \times \mathbb{C}}^N$ to $\mathscr{M} \hat{\otimes} \mathscr{E}_{\mathbb{C}}$ is surjective, and the functor $\mathscr{M}_p \to \mathscr{M}_p \hat{\otimes} \mathscr{E}_{\mathbb{C},(0,dt)}$ is faithful, α is surjective. Let $\mathscr{N} = \mathrm{Ker}\,\alpha$. The sequence:

$$0 \to \mathscr{N} \hat{\otimes} \mathscr{E}_{\mathbb{C}} \to \mathscr{E}_{X \times \mathbb{C}}^N \to \mathscr{M} \hat{\otimes} \mathscr{E}_{\mathbb{C}} \to 0$$

is exact, and $\mathscr{M} \hat{\otimes} \mathscr{E}_{\mathbb{C}}$ being coherent, $\mathscr{N} \hat{\otimes} \mathscr{E}_{\mathbb{C}}$ is coherent, and $\mathscr{N}$ is locally of finite type by the preceding argument.   □

Remark 2.5.3. There is another method to add a dummy variable, (cf. Kashiwara-Kawaï [1]).

Let p denote the projection from $T^*X \times \dot{T}^*\mathbb{C}$ to T^*X. To a left $\mathscr{E}_X$-module $\mathscr{M}$ we associate:

$$\mathscr{M} \hat{\otimes} \delta_t \underset{\mathrm{def}}{=} p^{-1}\mathscr{M} \otimes_{p^{-1}\mathscr{E}_X} (\mathscr{E}_{X \times \mathbb{C}}/\mathscr{E}_{X \times \mathbb{C}} \cdot t).$$

Since $\mathscr{E}_{X \times \mathbb{C}}/\mathscr{E}_{X \times \mathbb{C}} \cdot t$ is faithfully flat over $p^{-1}\mathscr{E}_X$, the analogous to Proposition 2.5.2 a) holds. We have an exact sequence of $p^{-1}\mathscr{E}_X$-modules:

$$0 \to p^{-1}\mathscr{M} \to \mathscr{M} \hat{\otimes} \delta_t \underset{t}{\to} \mathscr{M} \hat{\otimes} \delta_t \to 0$$

and $\mathscr{M}$ is coherent iff $\mathscr{M} \hat{\otimes} \delta_t$ is coherent.

Remark that $\mathcal{M} \hat{\otimes} \delta_t$ is supported by $\{(x, t; 0, \tau); t=0\}$ and the functor:

$$\mathcal{M} \to \mathcal{M} \hat{\otimes} \delta_t|_{t=0,\, \tau=1}$$

is exact and faithful.

2.6. $\mathcal{D}_X$-modules

It is possible to study $\mathcal{D}_X$-modules directly, without introducing the sheaf $\mathcal{E}_X$, by using the filtration on $\mathcal{D}_X$.

Recall that the filtration on $\mathcal{D}_{X,x}$ is (right and left) zariskian (§ 1.5), and the sheaf $\mathcal{D}_X$ is (right and left) noetherian.

Let $\mathcal{M}$ be a coherent $\mathcal{D}_X$-module endowed with a good filtration $(\mathcal{M}_k)_k$. Then locally we have for some $k_0 \geqslant 0$:

$$(2.6.1) \quad \begin{cases} \mathcal{M}_k = 0 \quad \text{for } k \ll 0, \\ \mathcal{D}_X(l)\, \mathcal{M}_k = \mathcal{M}_{k+l} \quad \forall l \geqslant 0, \; \forall k \geqslant k_0, \\ \bigcup_k \mathcal{M}_k = \mathcal{M}, \\ \text{the } \mathcal{O}_X\text{-modules } \mathcal{M}_k \text{ are locally of finite type.} \end{cases}$$

Conversely assume the coherent Module $\mathcal{M}$ endowed with a filtration $(\mathcal{M}_k)_k$ satisfying (2.6.1). Then this filtration is good. In fact choose locally k_0 and k_1 such that $\mathcal{M}_k = 0$ for $k \leqslant k_0$, and $\mathcal{D}_X(l)\, \mathcal{M}_k = \mathcal{M}_{k+l}$ for $k \geqslant k_1 \; \forall l$, and choose generators $(u_j^l)_{1 < j < N_l}$ of $\mathcal{M}_l$. We have for any $k \in \mathbb{Z}$:

$$\mathcal{M}_k = \sum_{k_0 \leqslant l \leqslant k_1} \sum_{j=1}^{N_l} \mathcal{D}_X(k-l)\, u_j^l.$$

Now we shall study the relation between a coherent $\mathcal{D}_X$-module $\mathcal{M}$, and the $\mathcal{E}_X$-module $\mathcal{E}_X \otimes_{\pi^{-1}\mathcal{D}_X} \pi^{-1}\mathcal{M}$.

To a coherent $\mathcal{D}_X$-module $\mathcal{M}$ we may associate a characteristic cycle in $T^* X$, which, for the time being we denote by $[\mathrm{Char}(\mathcal{M})]$, namely the cycle $[\mathrm{supp}(\mathcal{O}_{T^*X} \otimes_{\pi^{-1}\mathrm{gr}(\mathcal{D}_X)} \pi^{-1}\mathrm{gr}(\mathcal{M}))]$, for a good filtration on $\mathcal{M}$. In particular we may define the characteristic variety of $\mathcal{M}$, $\mathrm{Char}(\mathcal{M})$, and the multiplicity of $\mathcal{M}$ along an irreducible analytic set W containing (locally) $\mathrm{Char}(\mathcal{M})$, $\mathrm{Mult}_W(\mathcal{M})$, exactly as we did for $\mathcal{E}_X$-modules.

Proposition 2.6.1. a) *For a coherent $\mathcal{D}_X$-module $\mathcal{M}$, let $\mathcal{N} = \mathcal{E}_X \otimes_{\pi^{-1}\mathcal{D}_X} \pi^{-1}\mathcal{M}$. Then $\mathcal{N}$ is a coherent $\mathcal{E}_X$-module and we have:*

$$[\mathrm{char}(\mathcal{N})] = [\mathrm{Char}(\mathcal{M})].$$

In particular $\mathrm{char}(\mathcal{N}) = \mathrm{Char}(\mathcal{M})$, and if W is an irreducible analytic set containing $\mathrm{Char}(\mathcal{M})$, then

$$\mathrm{mult}_W(\mathcal{N}) = \mathrm{Mult}_W(\mathcal{M}).$$

b) *Let $\mathscr{N}$ be a coherent $\mathscr{E}_X$-module defined in all of T^*X. Set $\mathscr{M} = \mathscr{N}|_{T_X^*X}$. Then $\mathscr{M}$ is a coherent $\mathscr{D}_X$-module and we have a natural isomorphism:*

$$\mathscr{N} \cong \mathscr{E}_X \otimes_{\pi^{-1}\mathscr{D}_X} \pi^{-1}\mathscr{M}.$$

Proof. Locally we may represent $\mathscr{M}$ by an exact sequence:

(2.6.2) $$0 \to \mathscr{S} \to \mathscr{D}_X^N \to \mathscr{M} \to 0,$$

where $\mathscr{S}$ is a coherent $\mathscr{D}_X$-module. We endow $\mathscr{M}$ with the image filtration, and $\mathscr{S}$ with the induced filtration, and take the associated graded exact sequence:

(2.6.3) $$0 \to \mathrm{gr}(\mathscr{S}) \to \mathrm{gr}(\mathscr{D}_X^N) \to \mathrm{gr}(\mathscr{M}) \to 0.$$

We take the inverse image by π of (2.6.2) (resp. (2.6.3)) and tensorize by $\mathscr{E}_X$ (resp. $\mathrm{gr}(\mathscr{E}_X)$). Since $\mathscr{E}_X$ (resp. $\mathrm{gr}(\mathscr{E}_X)$) is flat over $\pi^{-1}\mathscr{D}_X$ (resp. $\pi^{-1}\mathrm{gr}(\mathscr{D}_X)$) we get the exact sequences:

(2.6.4) $$0 \to \mathscr{E}_X \otimes_{\pi^{-1}\mathscr{D}_X} \pi^{-1}\mathscr{S} \to \mathscr{E}_X^N \to \mathscr{E}_X \otimes_{\pi^{-1}\mathscr{D}_X} \pi^{-1}\mathscr{M} \to 0,$$

(2.6.5) $$0 \to \mathrm{gr}(\mathscr{E}_X) \otimes_{\pi^{-1}\mathrm{gr}(\mathscr{D}_X)} \pi^{-1}\mathrm{gr}(\mathscr{S})$$
$$\to \mathrm{gr}(\mathscr{E}_X^N) \to \mathrm{gr}(\mathscr{E}_X) \otimes_{\pi^{-1}\mathrm{gr}(\mathscr{D}_X)} \pi^{-1}\mathrm{gr}(\mathscr{M}) \to 0$$

Applying Proposition 1.2.4 we get:

(2.6.6) $$\mathrm{gr}(\mathscr{E}_X) \otimes_{\pi^{-1}\mathrm{gr}(\mathscr{D}_X)} \pi^{-1}\mathrm{gr}(\mathscr{M}) = \mathrm{gr}(\mathscr{N}).$$

Thus:

$$\mathscr{O}_{T^*X} \otimes_{\pi^{-1}\mathrm{gr}(\mathscr{D}_X)} \pi^{-1}\mathrm{gr}(\mathscr{M}) = \mathscr{O}_{T^*X} \otimes_{\mathrm{gr}(\mathscr{E}_X)} \mathrm{gr}(\mathscr{N}),$$

which proves a).

b) We represent $\mathscr{N}$ locally by an exact sequence defined in a neighborhood of T_X^*X:

$$0 \to \mathscr{J} \to \mathscr{E}_X^N \to \mathscr{N} \to 0.$$

Then $\mathscr{J}$ is generated by differential operators, and if we set $\mathscr{S} = \mathscr{J}|_{T_X^*X}$, then $\mathscr{J} \cong \mathscr{E}_X \otimes_{\pi^{-1}\mathscr{D}_X} \pi^{-1}\mathscr{S}$. The corresponding result for $\mathscr{N}$ follows. □

Remark 2.6.2. We shall often identify a $\mathscr{D}_X$-module $\mathscr{M}$ with $\mathscr{E}_X \otimes_{\pi^{-1}\mathscr{D}_X} \pi^{-1}\mathscr{M}$, and from now on we shall write $\mathrm{char}(\mathscr{M})$ and $\mathrm{mult}_V(\mathscr{M})$ instead of $\mathrm{Char}(\mathscr{M})$ and $\mathrm{Mult}_V(\mathscr{M})$.

Remark 2.6.3. Let $E \xrightarrow{\pi} X$ be a complex vector bundle over X. We have already introduced the subsheaf $\mathscr{O}_E^h = \bigoplus_{j \in \mathbb{Z}} \mathscr{O}_E(j)$ of $\mathscr{O}_E$, and now we introduce $\mathscr{O}_{[E]}$ the subsheaf of $\mathscr{O}_E$ of sections which are polynomials in the fiber of π, that is $\mathscr{O}_{[E]} = \pi^{-1}\pi_* \left(\bigoplus_{j \geq 0} \mathscr{O}_E(j) \right)$.

To a coherent graded $\pi_* \mathcal{O}_{[E]}$-module $\mathcal{M}$, let us associate the coherent $\mathcal{O}_E$-module $\tilde{\mathcal{M}} \underset{\mathrm{def}}{=} \mathcal{O}_E \otimes_{\mathcal{O}_{[E]}} \pi^{-1} \mathcal{M}$. Then:

a) $[\mathrm{supp}\, \tilde{\mathcal{M}}]$ is a closed conic analytic cycle of E. In fact $\mathcal{O}_E$ being faithfully flat over $\mathcal{O}_E^h$ considered as a graded Ring (Lemma 2.2.2), we have $\mathrm{supp}\, \tilde{\mathcal{M}} = \mathrm{supp}\,(\mathcal{O}_E^h \otimes_{\mathcal{O}_{[E]}} \mathcal{M})$.

b) If $0 \to \mathscr{L} \to \mathcal{M} \to \mathcal{N} \to 0$ is an exact sequence of graded $\pi_* \mathcal{O}_{[E]}$-modules, then $\mathrm{supp}\, \tilde{\mathcal{M}} = \mathrm{supp}\, \tilde{\mathscr{L}} \cup \mathrm{supp}\, \tilde{\mathcal{N}}$ and if V is an irreducible analytic set in E containing locally $\mathrm{supp}\, \tilde{\mathcal{M}}$, then $\mathrm{mult}_V(\tilde{\mathcal{M}}) = \mathrm{mult}_V(\tilde{\mathscr{L}}) + \mathrm{mult}_V(\tilde{\mathcal{N}})$. This follows from the fact that $\mathcal{O}_E$ is flat over $\mathcal{O}_{[E]}$ (Lemma 2.2.2).

c) We have:
$$\mathrm{supp}\,(\tilde{\mathcal{M}}) = \{\theta \in E;\ f(\theta) = 0 \quad \forall f \in \sqrt{\mathscr{I}_M}\}$$

where $\sqrt{\mathscr{I}_M}$ is the radical of the annihilator Ideal of $\mathcal{M}$ in $\pi_* \mathcal{O}_E$. In particular if $\mathcal{M} = \pi_* \mathcal{O}_{[E]}/\mathscr{I}$, for a coherent graded Ideal $\mathscr{I}$ of $\pi_* \mathcal{O}_{[E]}$, then in that case:

$$\mathrm{supp}\,(\tilde{\mathcal{M}}) = \{\theta \in E;\ f(\theta) = 0 \quad \forall f \in \mathscr{I}\}.$$

This follows immediately from the fact that for each $p \in E$, $\mathcal{O}_{E, p}$ is faithfully flat over the localization of $\pi_* \mathcal{O}_{[E]}$ at p.

We may summarize our results on the characteristic varieties of coherent $\mathscr{D}_X$-modules.

Proposition 2.6.4. a) *Let $\mathcal{M}$ be a coherent $\mathscr{D}_X$-module. Then $[\mathrm{char}(\mathcal{M})]$ is a closed conic analytic involutive positive cycle of $T^* X$.*

b) *If $0 \to \mathscr{L} \to \mathcal{M} \to \mathcal{N} \to 0$ is an exact sequence of coherent $\mathscr{D}_X$-modules, then $\mathrm{char}(\mathcal{M}) = \mathrm{char}(\mathscr{L}) \cup \mathrm{char}(\mathcal{N})$, and if V is an irreducible analytic set of $T^* X$ containing $\mathrm{char}(\mathcal{M})$ (locally on $T^* X$), then $\mathrm{mult}_V(\mathcal{M}) = \mathrm{mult}_V(\mathscr{L}) + \mathrm{mult}_V(\mathcal{N})$.*

c) *Let $\mathcal{M} = \mathscr{D}_X/\mathscr{I}$, for a coherent Ideal $\mathscr{I}$ of $\mathscr{D}_X$. Then:*

$$\mathrm{char}(\mathcal{M}) = \{\theta \in T^* X;\ \sigma(P)(\theta) = 0 \quad \forall P \in \mathscr{I}\}$$

Exercises to II.2

Ex. 2.1. Let P and Q be two sections of $\mathscr{E}_X$ in a neighborhood of $p \in T^* X$, with $P \neq 0$, $Q \neq 0$. Prove

$$\mathscr{E}_X P \cap \mathscr{E}_X Q \neq \{0\}.$$

Ex. 2.2. Let t be a coordinate on $\mathbb{C}$, and consider the $\mathscr{E}_{\mathbb{C}}(0)$-submodule of $\mathscr{E}_{\mathbb{C}}$, $\mathscr{L}(0) = \mathscr{E}_{\mathbb{C}}(0) + t \mathscr{E}_{\mathbb{C}}(1)$. Prove that the $\mathscr{L}(k) = \mathscr{E}_{\mathbb{C}}(k) \mathscr{L}(0)$ define a good filtration on $\mathscr{E}_{\mathbb{C}}$, but that the graded Module associated to this filtration is not free over $\mathrm{gr}(\mathscr{E}_{\mathbb{C}})$.

Ex. 2.3. Prove that the Ideal $\mathscr{I} = \mathscr{D}_{\mathbb{C}} \cdot D_t^2 + \mathscr{D}_{\mathbb{C}} \cdot (t D_t - 1)$ of $\mathscr{D}_{\mathbb{C}}$ is projective, but not free. (Hint: $\mathrm{mult}_{T^*\mathbb{C}}\, \mathscr{I} \leqslant 1$, but if $\mathscr{I}$ would be free, it would be of rank 2, since $\mathscr{I}$ does not admit a single generator).

Ex. 2.4. Calculate the characteristic cycle of $\mathscr{D}_{\mathbb{C}}/\mathscr{D}_{\mathbb{C}} \cdot (t^2 D_t + 1)$ on $T^*\mathbb{C}$.

Ex. 2.5. Prove that coherent projective Modules over $\mathscr{E}_X(0)$ are locally free.
   (Hint: $\mathscr{E}_{X,p}(0)$ is a local ring).

Ex. 2.6. Is the $\mathscr{E}_X(0)$-module $\mathscr{E}_X(-1)$ locally of finite type on T^*X?

Ex. 2.7. Calculate the characteristic cycle of $\mathscr{D}_{\mathbb{C}^2}/\mathscr{D}_{\mathbb{C}^2} \cdot (D_1^2, x_2^3)$.

Ex. 2.8. Prove that the $\mathscr{D}_X$-module $\mathscr{B}_{Z|X}$ is simple (assuming Z connected).

Ex. 2.9. Prove that a holonomic Module with simple characteristics on a connected Lagrangean manifold $\Lambda \subset \dot{T}^*X$ is simple (§ 2.4).

Ex. 2.10. Find locally a generator of the $\mathscr{D}_X$-module $\mathscr{O}_X^r$. Same problem for $\mathscr{B}_{Z|X}^r$.

Ex. 2.11. Let $\mathscr{M} = \mathscr{E}_X u$, where u satisfies $Pu = Qu = 0$, with $\sigma(P) = x_1^l$, $\sigma(Q) = \xi_1^k$. Prove that $u = 0$. (Hint: do not proceed by induction).

Ex. 2.12. Let Z be a single point, $\mathscr{M}$ a $\mathscr{D}_X$-module with $\mathrm{char}(\mathscr{M}) \subset T_Z^* X$, $\mathrm{mult}_{T_Z^* X}(\mathscr{M}) = r$. Prove that $\mathscr{M}$ is isomorphic to $\mathscr{B}_{Z|X}^r$.

Ex. 2.13. On $U = \{(x_1, x_2; \xi_1, \xi_2) \in T^*\mathbb{C}^2, \xi_2 \neq 0\}$ consider the $\mathscr{E}_X(0)$-module:

$$\mathscr{L} = \mathscr{E}_X(0)/\mathscr{E}_X(0)(x_1, D_1 D_2^{-1}).$$

Prove that $\mathrm{supp}(\mathscr{L})$ is not involutive.

Ex. 2.14. Calculate the characteristic variety of the $\mathscr{D}_{\mathbb{C}^2}$-module $\mathscr{M} = \mathscr{D}_{\mathbb{C}^2}/\mathscr{D}_{\mathbb{C}^2} \cdot P$ where

$$P = \begin{pmatrix} D_1^2 & x_1 \\ D_2 & x_2 D_2 \end{pmatrix},$$

(cf. Remark 1.3.7).

Ex. 2.15. On $X = \mathbb{C}^n$, let $\mathscr{M}$ be the $\mathscr{D}_X$-module $\mathscr{D}_X u$ with one generator u and relations: $\left(\sum_{j=1}^{n} D_j^2 + \lambda \right) u = 0$, $(x_i D_j - x_j D_i) u = 0$, $1 \leqslant i, j \leqslant n$. Prove that $\mathscr{M}$ is holonomic.

Ex. 2.16. On $X = \mathbb{C}^n$, let $P = \sum\limits_{i=1}^{r} D_i^2 - \sum\limits_{j=r+1}^{n} D_j^2$. Find a non zero holonomic $\mathscr{D}_X$-module $\mathscr{M}$, and a surjective morphism $\mathscr{D}_X / \mathscr{D}_X \cdot P \to \mathscr{M} \to 0$ (that is find operators $(Q_j)_{1 \leqslant j \leqslant r}$ such that the system $\mathscr{D}_X u$, with $Pu = Q_j u = 0, j = 1, \ldots, r$, is holonomic).

§ 3. Operations on $\mathscr{E}_X$-modules

3.1. Definitions

Let Y and X be two complex manifolds, V and U two open sets of $T^* Y$ and $T^* X$ respectively. We denote by p_1 and p_2 the projections from $Y \times X$ to Y and X respectively, and we keep the same notations for the projections from $T^*(Y \times X) = T^* Y \times T^* X$ to $T^* Y$ and $T^* X$.

Let $\mathscr{N}$ (resp. $\mathscr{M}$) be a left $\mathscr{E}_Y$-module on V (resp. $\mathscr{E}_X$-module on U).

Definition 3.1.1. One puts:

$$\mathscr{N} \hat{\otimes} \mathscr{M} = \mathscr{E}_{Y \times X} \otimes_{(p_1^{-1} \mathscr{E}_Y \otimes p_2^{-1} \mathscr{E}_X)} (p_1^{-1} \mathscr{N} \otimes p_2^{-1} \mathscr{M}).$$

One says that $\mathscr{N} \hat{\otimes} \mathscr{M}$ is the external product of $\mathscr{N}$ and $\mathscr{M}$.

Remark. When unspecified, the tensor product $\cdot \otimes \cdot$ means the tensor product over $\mathbb{C}$.

Now let f be a holomorphic map from Y to X. We identify Y with the graph of f in $Y \times X$, and $Y \times T^* X$ with $T^*_Y(Y \times X)$ (cf. I. § 4.3). Then the projections p_1 and p_2 induce the natural maps

$$T^* Y \xleftarrow{\ \rho\ } Y \underset{X}{\times} T^* X \xrightarrow{\ \varpi\ } T^* X.$$

Recall that we have put (I, § 4.3):

$$\mathscr{E}_{Y \to X} = \mathscr{C}_{Y|Y \times X} \otimes_{\varpi^{-1} \pi^{-1} \mathscr{O}_X} \varpi^{-1} \pi^{-1} \Omega_X$$

and this sheaf on $Y \underset{X}{\times} T^* X$ is naturally endowed with a structure of $(\rho^{-1} \mathscr{E}_Y, \varpi^{-1} \mathscr{E}_X)$-bi-Module. It is also useful to define the $(\varpi^{-1} \mathscr{E}_X, \rho^{-1} \mathscr{E}_Y)$-bi-Module:

$$\mathscr{E}_{X \leftarrow Y} = \mathscr{C}_{Y|Y \times X} \otimes_{\rho^{-1} \pi^{-1} \mathscr{O}_Y} \rho^{-1} \pi^{-1} \Omega_Y,$$

that is:

$$\mathscr{E}_{X \leftarrow Y} = \mathscr{E}_{Y \to X} \otimes_{\varpi^{-1} \pi^{-1} \mathscr{O}_X} \varpi^{-1} \pi^{-1} \Omega_X^{\otimes -1} \otimes_{\rho^{-1} \pi^{-1} \mathscr{O}_Y} \rho^{-1} \pi^{-1} \Omega_Y.$$

For the sake of simplicity we shall not always write the symbols π^{-1} or ρ^{-1}. Thus we write for example:

$$\mathscr{E}_{X \leftarrow Y} = \mathscr{E}_{Y \rightarrow X} \otimes_{f^{-1}\mathscr{O}_X} f^{-1} \Omega_X^{\otimes -1} \otimes_{\mathscr{O}_Y} \Omega_Y.$$

We also set:

$$\mathscr{D}_{X \leftarrow Y} = \mathscr{E}_{X \leftarrow Y} |_{Y \underset{X}{\times} T_X^* X}.$$

Definition 3.1.2. a) Let V be an open set of $T^* Y$, $\mathscr{M}$ a left coherent $\mathscr{E}_X$-module on an open set U of $T^* X$. One says that f is non characteristic for $\mathscr{M}$ on V if the map ρ is finite on $\rho^{-1}(V) \cap \overline{\omega}^{-1}(\operatorname{char}(\mathscr{M}))$.

In that case one defines the inverse image of $\mathscr{M}$, denoted $\mathscr{M}_Y$, by:

(3.1.1) $$\mathscr{M}_Y = \rho_*(\mathscr{E}_{Y \rightarrow X} \otimes_{\overline{\omega}^{-1}\mathscr{E}_X} \overline{\omega}^{-1}\mathscr{M}).$$

b) Let U be an open set of $T^* X$, $\mathscr{N}$ a right coherent $\mathscr{E}_Y$-module on an open set V of $T^* Y$. One says that f is non characteristic for $\mathscr{N}$ on U if the map $\overline{\omega}$ is finite on $\overline{\omega}^{-1}(U) \cap \rho^{-1}(\operatorname{char}(\mathscr{N}))$. In that case one defines the direct image of $\mathscr{N}$, denoted $\int_f \mathscr{N}$, by:

(3.1.2) $$\int_f \mathscr{N} = \overline{\omega}_*(\rho^{-1}\mathscr{N} \otimes_{\rho^{-1}\mathscr{E}_Y} \mathscr{E}_{Y \rightarrow X})$$

Remark 3.1.3. In Formula (3.1.1) ρ is in fact the restriction of ρ to $\rho^{-1}(V)$. Similarly in (3.1.2) $\overline{\omega}$ is the restriction of $\overline{\omega}$ to $\overline{\omega}^{-1}(U)$. Thus the notations $\mathscr{M}_Y$ or $\int_f \mathscr{N}$ are not very good, and if necessary we shall specifiy U or V.

Remark 3.1.4. A continuous map g from a locally compact space to another is proper if the inverse image of a compact set is compact, and g is finite if, moreover, the inverse image of a finite set is finite. In Definition 3.1.2, if ρ is proper over $\overline{\omega}^{-1}(\operatorname{char}(\mathscr{M}))$, ρ is finite over this set since $\overline{\omega}^{-1}(\operatorname{char}(\mathscr{M}))$ is an analytic set of $Y \underset{X}{\times} T^* X$, and the fibers of ρ are vector spaces. Of course a similar remark does not hold for $\overline{\omega}$.

Remark 3.1.5. If $\mathscr{M}$ is a $\mathscr{D}_X$-module, (and $V = T^* Y$), then f is non characteristic for $\mathscr{M}$ iff:

$$T_Y^* X \cap \overline{\omega}^{-1}(\operatorname{char}(\mathscr{M})) \subset Y \underset{X}{\times} T_X^* X,$$

where $T_Y^* X$ is the kernel of ρ.

In that case we find:

$$\mathscr{M}_Y = \mathscr{D}_{Y \rightarrow X} \otimes_{f^{-1}\mathscr{D}_X} f^{-1}\mathscr{M}.$$

If $\mathscr{N}$ is a $\mathscr{D}_Y$-module (and $U = T^* X$), then the hypothesis that f is non characteristic is very strong. It is equivalent to saying that f is finite on $\operatorname{supp}(\mathscr{N})$. In that case:

$$\int_f \mathscr{N} = f_*(\mathscr{N} \otimes_{\mathscr{D}_Y} \mathscr{D}_{Y \rightarrow X})$$

Remark 3.1.6. It is also possible to define the inverse image of a right $\mathscr{E}_X$-module, or the direct image of a left $\mathscr{E}_Y$-module, replacing $\mathscr{E}_{Y\to X}$ by $\mathscr{E}_{X\gets Y}$ in formulas (3.1.1) and (3.1.2), but such operations are not so natural, since $\mathscr{E}_{Y\to X}$ has a canonical section, $1_{Y\to X}$, but not $\mathscr{E}_{X\gets Y}$. For example one takes the restriction (inverse image) of a holomorphic function, but the integral (direct image) of a holomorphic form.

Remark 3.1.7. Of course one can write formula (3.1.1) or (3.1.2) without making the assumption that f is non characteristic, or even without assuming $\mathscr{M}$ or $\mathscr{N}$ coherent. But in that case it is necessary to consider the derived functors, and the language of derived categories (cf. Hartshorne [1]) is better adapted. Then one would define:

$$(3.1.3) \qquad \mathscr{M}_Y = \mathbb{R}\rho_* (\mathscr{E}_{Y\to X} \overset{\mathbb{L}}{\otimes}_{\bar{\omega}^{-1}\mathscr{E}_X} \bar{\omega}^{-1}\mathscr{M}),$$

$$(3.1.4) \qquad \int_f \mathscr{N} = \mathbb{R}\bar{\omega}_* (\rho^{-1}\mathscr{N} \overset{\mathbb{L}}{\otimes}_{\rho^{-1}\mathscr{E}_Y} \mathscr{E}_{Y\to X}).$$

We shall see later that under the assumptions made in Definition 3.1.2, formulas (3.1.1) and (3.1.3) (resp. (3.1.2) and (3.1.4)) are equivalent.

Remark 3.1.8. It is possible to generalize both inverse and direct images in a single construction. Let Λ be a conic Lagrangean manifold in $T^*(Y \times X)$, $\mathscr{L}$ a right holonomic system with simple characteristics on Λ, $\mathscr{M}$ a left $\mathscr{E}_X$-module on U such that the projection p_1 from $V \times U \subset T^*(Y \times X)$ to T^*Y is finite on $(V \times \mathrm{char}(\mathscr{M})) \cap \Lambda$. Then we may consider the left $\mathscr{E}_Y$-module on V, $p_{1*}(\mathscr{L} \otimes_{p_2^{-1}\mathscr{E}_X} p_2^{-1}\mathscr{M}) \otimes_{\mathscr{O}_Y} \Omega_Y$.

Now we shall prepare the next sections. Let Θ_Y be the sheaf on Y of holomorphic vector fields, that is, the sheaf of sections of the tangent bundle TY over Y, and let $\mathscr{M}$ be a left $\mathscr{D}_X$-module. One defines the action of Θ_Y on $\mathscr{O}_Y \otimes_{f^{-1}\mathscr{O}_X} f^{-1}\mathscr{M}$ as follows.

Let v be a section of Θ_Y. Then $f_*(v)$ is a section of the bundle $Y \underset{X}{\times} TX$, and locally it may be written:

$$f_*(v) = \sum_j a_j \otimes w_j,$$

where a_j is a section of $\mathscr{O}_Y$ and w_j a section of Θ_X.

Let $a \otimes u$ be a section of $\mathscr{O}_Y \otimes_{f^{-1}\mathscr{O}_X} f^{-1}\mathscr{M}$. One sets:

$$v(a \otimes u) = v(a) \otimes u + \sum_j a_j a \otimes w_j(u).$$

If we choose local coordinates $(x_1, \ldots, x_n)$ on X, $(y_1, \ldots, y_p)$ on Y so that $f = (f_1, \ldots, f_n)$, then we get:

$$D_{y_k}(a \otimes u) = D_{y_k}(a) \otimes u + \sum_{j=1}^n a \frac{\partial f_j}{\partial x_k} \otimes D_{x_j}(u).$$

The action of Θ_Y on $\mathcal{O}_Y \otimes_{f^{-1}\mathcal{O}_X} f^{-1}\mathcal{M}$ may be extended as an action of $\mathscr{D}_Y$, thanks to the following Lemma.

Lemma 3.1.9 *(Bernstein [2], Kashiwara [4]). Let $\mathscr{F}$ be an $\mathcal{O}_Y$-module, ψ an homomorphism from $\Theta_Y \otimes \mathscr{F}$ to $\mathscr{F}$ which satisfies:*

   i) $\psi(av \otimes s) = a\psi(v \otimes s)$ *(resp.* $\psi(av \otimes s) = \psi(v \otimes as)$).

   ii) $\psi(v \otimes as) = a\psi(v \otimes s) + v(a)\psi(v \otimes s)$, *(resp.*
$\psi(av \otimes s) = a\psi(v \otimes s) - v(a)\psi(v \otimes s)$), *where* $a \in \mathcal{O}_Y$, $v \in \Theta_Y$, $s \in \mathscr{F}$.

   iii) $\psi([v_1, v_2] \otimes s) = \psi(v_1 \otimes \psi(v_2 \otimes s)) - \psi(v_2 \otimes \psi(v_1 \otimes s))$, *where*
$v_1, v_2 \in \Theta_Y, s \in \mathscr{F}$.

   Then there exists a unique structure of left (resp. right) $\mathscr{D}_Y$-module on $\mathscr{F}$ such that $\psi(v \otimes s) = vs$ (resp. $\psi(v \otimes s) = sv$), and such that the induced structure of $\mathcal{O}_Y$-module is that of $\mathscr{F}$.

The proof is left to the reader.

As a particular case of the preceding considerations, we find that $\mathcal{O}_Y \otimes_{f^{-1}\mathcal{O}_X} f^{-1}\mathscr{D}_X$ is naturally endowed with the structure of a left $\mathscr{D}_Y$-module.

Proposition 3.1.10. *The morphism from $\mathcal{O}_Y \otimes_{f^{-1}\mathcal{O}_X} f^{-1}\mathscr{D}_X$ to $\mathscr{D}_{Y \to X}$:*

$$f \otimes P \to f \cdot 1_{Y \to X} \cdot P$$

is an isomorphism of $(\mathscr{D}_Y, f^{-1}\mathscr{D}_X)$-bimodules.

The proof is immediate by choosing coordinates on Y and X.

Let Z be another manifold, g a holomorphic map from Z to Y. One checks the isomorphism:

$$(\mathcal{O}_Z \otimes_{g^{-1}\mathcal{O}_Y} g^{-1}\mathscr{D}_Y) \otimes_{g^{-1}\mathscr{D}_Y} g^{-1}(\mathcal{O}_Y \otimes_{f^{-1}\mathcal{O}_X} f^{-1}\mathscr{D}_X)$$
$$\cong \mathcal{O}_Z \otimes_{(f \circ g)^{-1}\mathcal{O}_X} (f \circ g)^{-1}\mathscr{D}_X$$

or equivalently:

(3.1.5) $\qquad \mathscr{D}_{Z \to Y} \otimes_{g^{-1}\mathscr{D}_Y} g^{-1}(\mathscr{D}_{Y \to X}) = \mathscr{D}_{Z \to X}.$

Remark 3.1.11. As we shall see in Remark 3.3.4, Proposition 3.1.10 and Formula (3.1.5) do not extend to microdifferential operators without extra assumptions.

Remark 3.1.12. It follows from Proposition 3.1.10 that for a left $\mathscr{D}_X$-module $\mathscr{M}$ we have a canonical isomorphism for all j:

(3.1.6) $\qquad \mathscr{T}or_j^{f^{-1}\mathcal{O}_X}(\mathcal{O}_Y, f^{-1}\mathscr{M}) \cong \mathscr{T}or_j^{f^{-1}\mathscr{D}_X}(\mathscr{D}_{Y \to X}, f^{-1}\mathscr{M})$

and those sheaves are naturally endowed with a structure of left $\mathscr{D}_Y$-modules.

In fact the isomorphism follows from Proposition 3.1.10 since $\mathscr{D}_X$ is flat over $\mathscr{O}_X$: if one tensorizes a free projective resolution of $\mathscr{O}_Y$ over $\mathscr{O}_X$ by $\mathscr{D}_X$ we get a free projective resolution of $\mathscr{D}_{Y \to X}$ over $\mathscr{D}_X$.

The structure of a left $\mathscr{D}_Y$-module of the term on the right hand side of (3.1.6) is immediately obtained by replacing locally $\mathscr{M}$ by a complex of free $\mathscr{D}_X$-modules.

3.2. Operations on $\mathscr{B}_{S|X}$

We first give a new construction of the Modules $\mathscr{B}_{S|X}$ introduced in (I, § 4). Let S be a (complex) submanifold of codimension p. Let $\mathscr{J}_S$ be the defining Ideal of S in $\mathscr{O}_X$ generated by the sections vanishing on S.

Lemma 3.2.1. *Let $\mathscr{M}$ be a left $\mathscr{D}_X$-module. Then for all j, the groups*

$$(3.2.1) \qquad \mathscr{H}_{[S]}^j(\mathscr{M}) \underset{\mathrm{def}}{=} \varinjlim_{k>0} \mathscr{E}xt_{\mathscr{O}_X}^j(\mathscr{O}_X / \mathscr{J}_S^k, \mathscr{M})$$

are naturally endowed with a structure of left $\mathscr{D}_X$-module.

Proof. First we remark that if $\mathscr{F}$ is a coherent $\mathscr{O}_X$-module, then for all j there exists a canonical isomorphism:

$$(3.2.2) \qquad \mathscr{E}xt_{\mathscr{O}_X}^j(\mathscr{F}, \mathscr{M}) \cong \mathscr{E}xt_{\mathscr{D}_X}^j(\mathscr{D}_X \otimes_{\mathscr{O}_X} \mathscr{F}, \mathscr{M}).$$

In fact $\mathscr{D}_X$ being flat over $\mathscr{O}_X$, both terms in (3.2.2) may be calculated by replacing $\mathscr{F}$ by a bounded complex of finite free $\mathscr{O}_X$-modules.

Now replacing $\mathscr{M}$ in (3.2.1) by a complex of injective $\mathscr{D}_X$-modules it is enough to prove the lemma for $j = 0$.

If P is a differential operator of order l, the multiplication on the right by P, $Q \mapsto Q \circ P$, defines a left $\mathscr{D}_X$-linear morphism:

$$\mathscr{D}_X \mathscr{J}_S^k \to \mathscr{D}_X \mathscr{J}_S^{k-l}.$$

Thus we have an homomorphism:

$$\mathscr{H}om_{\mathscr{D}_X}(\mathscr{D}_X / \mathscr{D}_X \mathscr{J}_S^{k-l}, \mathscr{M}) \to \mathscr{H}om_{\mathscr{D}_X}(\mathscr{D}_X / \mathscr{D}_X \mathscr{J}_S^k, \mathscr{M}).$$

Taking the inductive limit with respect to k we obtain the left action of $\mathscr{D}_X$. $\square$

Proposition 3.2.2. *We have $\mathscr{H}_{[S]}^j(\mathscr{O}_X) = 0$ for $j \neq p$, and there exists a canonical isomorphism of $\mathscr{D}_X$-modules: $\mathscr{H}_{[S]}^p(\mathscr{O}_X) \cong \mathscr{B}_{S|X}$.*

Proof. For the proof we shall use the results of Appendix B.4.

For each $k \geq 1$, we have an exact sequence:

(3.2.3) $0 \to \mathscr{J}_S^{k-1}/\mathscr{J}_S^k \to \mathscr{O}_X/\mathscr{J}_S^k \to \mathscr{O}_X/\mathscr{J}_S^{k-1} \to 0.$

Moreover $\mathscr{J}_S^{k-1}/\mathscr{J}_S^k$ is locally isomorphic to a direct sum of $\mathscr{O}_X/\mathscr{J}_S$. Hence it is enough to prove the vanishing of $\mathscr{E}xt^j_{\mathscr{O}_X}(\mathscr{O}_X/\mathscr{J}_S^k, \mathscr{O}_X)$ for $k=1$ to get the result for any $k \geqslant 1$, by induction.

Let $(s) = (s_1, \ldots, s_p)$ be a system of equations defining S (i.e.: $ds_1 \wedge \ldots \wedge ds_p \neq 0$ on S). The sequence $(s_1, \ldots, s_p)$ is a regular sequence since s_i is injective on $\mathscr{O}_X/\mathscr{O}_X(s_1, \ldots, s_{i-1})$. Then the only non vanishing cohomology group of the Koszul complex below is the p-th group:

(3.2.4) $0 \to \mathscr{O}_X^{(0)} \xrightarrow[(s)]{} \mathscr{O}_X^{(1)} \to \cdots \xrightarrow[(s)]{} \mathscr{O}_X^{(p)} \to 0$

To the sequence $(s) = (s_1, \ldots, s_p)$ we have associated in (I, § 4) a generator $\delta_{(s)}$ of $\mathscr{B}_{S|X}$, and the left Ideal $\mathscr{I}_{(s)}$, the annihilator of $\delta_{(s)}$.

Now let $\tilde{\delta}_{(s)}$ be the class of $1 \mod(\mathscr{O}_X^{(p-1)} \cdot (s))$ in $\mathscr{E}xt^p_{\mathscr{O}_X}(\mathscr{O}_X/\mathscr{J}_S, \mathscr{O}_X)$. If one extends (s) to a system of local coordinates $(s_1, \ldots, s_n)$ on X it is immediately verified that $\tilde{\delta}_{(s)}$ is a generator of $\mathscr{H}^p_{[S]}(\mathscr{O}_X)$ over $\mathscr{D}_X$, and the annihilator of $\tilde{\delta}_{(s)}$ is exactly $\mathscr{I}_{(s)}$.

Thus the correspondence $\tilde{\delta}_{(s)} \to \delta_{(s)}$ gives the isomorphism of $\mathscr{D}_X$-modules we were looking for. $\square$

Remark 3.2.3. It follows from the results of (I, § 4) that the class $\tilde{\delta}_{(s)} ds_1 \wedge \ldots \wedge ds_p$ in $\mathscr{E}xt^p_{\mathscr{O}_X}(\mathscr{O}_X/\mathscr{J}_S, \Omega_X^{(p)})$ depends only on S. Of course this fact is wellknown, (cf. Grothendieck [1]).

Now we come back to the situation of Section 3.1, and consider a holomorphic map f from Y to X. One says that f is transversal to the submanifold S in X if the graph of f (still denoted by Y) is transversal to $Y \times S$ in $Y \times X$:

$$T_Y^*(Y \times X) \cap T_{Y \times S}^*(Y \times X) \subset T_{Y \times X}^*(Y \times X)$$

or equivalently:

(3.2.5) $T_Y^* X \cap \bar{\omega}^{-1} T_S^* X \subset Y \underset{X}{\times} T_X^* X,$

where $T_Y^* X$ is the kernel of ρ.

In other words f is transversal to S iff f is non characteristic for $\mathscr{B}_{S|X}$.

The transversality hypothesis implies that $f^{-1}(S)$ is a submanifold of codimension p in Y.

Proposition 3.2.4. *Assume f transversal to S. Then there exists a natural isomorphism of left $\mathscr{D}_Y$-modules:*

$$\mathscr{D}_{Y \to X} \otimes_{f^{-1}\mathscr{D}_X} f^{-1}\mathscr{B}_{S|X} \simeq \mathscr{B}_{f^{-1}(S)|Y}.$$

Proof. Set $T = f^{-1}(S)$. Let $(s) = (s_1, \ldots, s_p)$ be a system of equations defining S in X. Then $(s \circ f) = (s_1 \circ f, \ldots, s_p \circ f)$ is a system of equations of T in Y. Let $\delta_{(s)}$

be the generator of $\mathscr{B}_{S|X}$ associated to the system (s) and $\mathscr{I}_{(s)}$ the annihilator of $\delta_{(s)}$, and similarly for $\delta_{(s\circ f)}$ and $\mathscr{I}_{(s\circ f)}$ in $\mathscr{D}_Y$.

First we prove:

$$(3.2.6) \qquad \mathscr{I}_{(s\circ f)}\cdot 1_{Y\to X}\subset \mathscr{D}_{Y\to X}\cdot f^{-1}\mathscr{I}_{(s)},$$

(the term on the left-hand side is the image of $\mathscr{I}_{(s\circ f)}$ by the map $Q\to Q\cdot 1_{Y\to X}$, and the term on the right-hand side is the image of $\mathscr{D}_{Y\to X}\otimes_{f^{-1}\mathscr{D}_X}f^{-1}\mathscr{I}_{(s)}$ in $\mathscr{D}_{Y\to X}$).

To prove (3.2.6) we may choose coordinates on Y and X. We choose $(x_1,\ldots,x_n)$ on X, $(y_1,\ldots,y_m)$ on Y such that $(s)=(x_1,\ldots,x_p)$ and $y_i=x_i\circ f$ for $1\leqslant i\leqslant p$. Then:

$$y_i\cdot 1_{Y\to X}=1_{Y\to X}\otimes x_i \quad 1\leqslant i\leqslant p,$$

$$D_{y_i}\cdot 1_{Y\to X} = \sum_{j=1}^{n}\frac{\partial f_j}{\partial x_i}\cdot 1_{Y\to X}\otimes D_{x_j},$$

$$= \sum_{j=p+1}^{n}\frac{\partial f_j}{\partial x_i}\cdot 1_{Y\to X}\otimes D_{x_j} \quad \text{for} \quad p+1\leqslant i\leqslant m.$$

Now we may define a $\mathscr{D}_Y$-linear morphism from $\mathscr{B}_{T|Y}$ to $\mathscr{D}_{Y\to X}\otimes_{f^{-1}\mathscr{D}_X}f^{-1}\mathscr{B}_{S|X}$ by setting:

$$(3.2.7) \qquad \psi(P\delta_{(s\circ f)})=P\cdot 1_{Y\to X}\otimes\delta_{(s)}.$$

This morphism does not depend on the choice of (s). In fact if (s') is another system of equations defining S, we shall have $\delta_{(s')}=h\delta_{(s)}$ for a non vanishing function h, and $\delta_{(s'\circ f)}=(h\circ f)\delta_{(s\circ f)}$ (cf. I, Lemma 4.1.1). Thus:

$$\psi(\delta_{(s'\circ f)}) = \psi((h\circ f)\delta_{(s\circ f)})=(h\circ f)1_{Y\to X}\otimes\delta_{(s)}=1_{Y\to X}\otimes h\,\delta_{(s)}$$

$$=1_{Y\to X}\otimes\delta_{(s')}.$$

Finally to prove that ψ is an isomorphism it is possible to decompose f as an embedding and a projection and to consider separately each case. Recall that an embedding $f: Y\to X$ is a proper injective immersion in an open subset of X.

For example assume f is an embedding. Then we may even assume Y is a submanifold of X. We choose coordinates $(x)=(y,s,t)$ on X such that $S=\{(y,s,t); s=0\}$, $Y=\{(y,s,t); t=0\}$. Denoting by $\mathscr{D}_X(P_1,\ldots,P_r)$ the left Ideal of $\mathscr{D}_X$ generated by $P_1,\ldots,P_r$, and similarly for right Ideals, we have:

$$\mathscr{D}_{Y\to X}\otimes_{\mathscr{D}_X}\mathscr{B}_{S|X} \cong \mathscr{D}_X/(t)\,\mathscr{D}_X\otimes_{\mathscr{D}_X}\mathscr{D}_X/\mathscr{D}_X(s,D_y,D_t)$$

$$\cong \mathscr{D}_X/((t)\,\mathscr{D}_X+\mathscr{D}_X(s,D_y,D_t))$$

$$\cong \mathscr{D}_Y/\mathscr{D}_Y(s,D_y)$$

$$\cong \mathscr{B}_{S\cap Y|Y}. \qquad \square$$

Notation 3.2.5. We denote by $u|_Y$ the image in $\mathscr{B}_{T|Y}$ of $1_{Y\to X}\otimes u$, where u is a section of $\mathscr{B}_{S|X}$, and one says that $u|_Y$ is the inverse image, or the restriction, of u.

Remark 3.2.6. We can give another proof of the preceding result, by using Proposition 3.2.2. In fact we have the canonical isomorphism:

$$\mathscr{O}_Y\otimes_{f^{-1}\mathscr{O}_X}f^{-1}\mathscr{E}xt^j_{\mathscr{O}_X}(\mathscr{O}_X/\mathscr{J}_S^k,\mathscr{O}_X)\cong\mathscr{E}xt^j_{\mathscr{O}_Y}(\mathscr{O}_Y/\mathscr{J}_T^k,\mathscr{O}_Y).$$

Taking the inductive limit with respect to k we get:

$$\mathscr{O}_Y\otimes_{f^{-1}\mathscr{O}_X}f^{-1}\mathscr{B}_{S|X}\cong\mathscr{B}_{f^{-1}(S)|Y},$$

but one has to check the $\mathscr{D}_Y$-linearity of this morphism, which is left as an exercise.

Lemma 3.2.7. *Let S be a submanifold of X. We have a canonical isomorphism of right $\mathscr{D}_X$-modules:*

$$\mathscr{B}_{S|X}\otimes_{\mathscr{O}_X}\Omega_X\cong\Omega_S\otimes_{\mathscr{D}_S}\mathscr{D}_{S\to X}.$$

Proof. Let δ_S be the fundamental class of S, a section of $\mathscr{B}_{S|X}\otimes_{\mathscr{O}_X}\Omega_X\otimes_{\mathscr{O}_S}\Omega_S^{\otimes-1}$. Let us choose a volume element ds on S, and let us denote by $\mathscr{J}_{ds}$ the right Ideal of $\mathscr{D}_X$, the annihilator of $\delta_S\,ds$ in $\mathscr{B}_{S|X}\otimes_{\mathscr{O}_X}\Omega_X$, and by $\mathscr{J}_{ds}$ the right Ideal of $\mathscr{D}_S$, the annihilator of ds in Ω_S.

First we prove:

$$(3.2.8) \qquad\qquad 1_{S\to X}\cdot\mathscr{J}_{ds}\subset\mathscr{J}_{ds}\cdot\mathscr{D}_{S\to X}.$$

But this formula is obvious by choosing coordinates $(x)=(x',x'')$ on X such that $S=\{(x);\ x''=0\}$, $ds=dx'$.

By (3.2.8) we may define a $\mathscr{D}_X$-linear morphism ψ from $\mathscr{B}_{S|X}\otimes_{\mathscr{O}_X}\Omega_X$ to $\Omega_S\otimes_{\mathscr{D}_S}\mathscr{D}_{S\to X}$ by setting:

$$\psi(\delta_S\,ds\cdot P)=ds\otimes 1_{S\to X}\cdot P$$

and one proves as in Proposition 3.2.4 that ψ does not depend on the choice of ds.

Finally it is clear by considering the coordinates (x',x'') that ψ is an isomorphism. $\square$

Proposition 3.2.8. *Let T be a submanifold of Y, and assume f induces a closed embedding from T to X. Then there exists a natural isomorphism of right $\mathscr{D}_X$-modules:*

$$f_*(\mathscr{B}_{T|Y}\otimes_{\mathscr{O}_Y}\Omega_Y\otimes_{\mathscr{D}_Y}\mathscr{D}_{Y\to X})\cong\mathscr{B}_{f(T)|X}\otimes_{\mathscr{O}_X}\Omega_X.$$

Proof. Applying Lemma 3.2.7 we have:

$$\Omega_T \otimes_{\mathscr{D}_T} \mathscr{D}_{T \to Y} \cong \mathscr{B}_{T|Y} \otimes_{\mathscr{O}_Y} \Omega_Y$$

We tensorize both terms by $\mathscr{D}_{Y \to X}$ over $\mathscr{D}_Y$ and identify T with its image in X.

Since $\mathscr{D}_{T \to Y} \otimes_{\mathscr{D}_Y} \mathscr{D}_{Y \to X} = \mathscr{D}_{T \to X}$ (Formula (3.1.5)) we get the result, again by Lemma 3.2.7. $\square$

Notation 3.2.9. We shall denote by $\int v\, dy$ the image in $\mathscr{B}_{S|X} \otimes_{\mathscr{O}_X} \Omega_X$ of a section $v\, dy \otimes 1_{Y \to X}$, and say that $\int v\, dy$ is the direct image, or the integral, of $v\, dy$.

Proposition 3.2.10. *Let S and Z be two submanifolds of X, S and Z being transversal. There exists a natural $\mathscr{O}_X$-linear morphism on $S \cap Z$:*

$$\mathscr{B}_{S|X} \otimes_{\mathscr{O}_X} \mathscr{B}_{Z|X} \to \mathscr{B}_{(S \cap Z)|X}.$$

Proof. Let X' be another copy of X, Z' the image of Z in X'. We have natural morphisms:

$$\mathscr{B}_{S|X} \times \mathscr{B}_{Z'|X'} \to \mathscr{B}_{S \times Z'|X \times X'} \to \mathscr{O}_{X \underset{X}{\times} X'} \otimes_{\mathscr{O}_{X \times X'}} \mathscr{B}_{S \times Z'|X \times X'} \to \mathscr{B}_{(S \cap Z)|X}$$

The last morphism is defined by Proposition 3.2.4 and is in fact an isomorphism.

We have to check the $\mathscr{O}_X$-bilinearity of the composite morphism.

Let $f \in \mathscr{O}_X$, let f' be the image of f in X', and let $u \in \mathscr{B}_{S|X}$, $v \in \mathscr{B}_{Z'|X'}$. We have, in $\mathscr{O}_{X \underset{X}{\times} X'} \otimes_{\mathscr{O}_{X \times X'}} \mathscr{B}_{S \times Z'|X \times X'}$:

$$(f u, v) = (u, f' v),$$

since $\mathscr{D}_X$ and $\mathscr{D}_{X'}$ commute in $\mathscr{D}_{X \times X'}$. $\square$

Remark 3.2.11. In the preceding situation, let u_1 be a section of $\mathscr{B}_{S|X}$, u_2 a section of $\mathscr{B}_{Z|X}$. Then:

(3.2.9) $$u_1 \cdot u_2 = (-1)^{\operatorname{codim} S \cdot \operatorname{codim} Z} u_2 \cdot u_1.$$

To check this formula, choose local coordinates $(x) = (x', x'', t)$ on X such that $S = \{(x); x' = 0\}$, $Z = \{(x); x'' = 0\}$. We have:

$$\frac{dx' \wedge dx''}{x' \cdot x''} = \frac{dx'' \wedge dx'}{x'' \cdot x'} \quad \text{in} \quad \mathscr{B}_{(S \cap Z)|X} \otimes_{\mathscr{O}_X} \Omega_X^{\operatorname{codim}(S \cap Z)},$$

which proves formula (3.2.9) when $u_1 = 1/x'$, $u_2 = 1/x''$, thus when $u_1 = D_{x'}^\alpha(1/x')$, $u_2 = D_{x''}^\beta(1/x'')$. The general case follows by $\mathscr{O}_X$-linearity. (cf. Sato-Kashiwara-Kawaï [1, p. 327 and p. 331]).

3.3. Operations on $\mathscr{C}_{S|X}$

We keep the notations of section 3.1. For the sake of simplicity we often write $\mathscr{E}\!\mathit{xt}^j_{\mathscr{E}_X}(\cdot,\cdot)$ or $\mathscr{T}\!\mathit{or}^{\mathscr{E}_X}_j(\cdot,\cdot)$ instead of $\mathscr{E}\!\mathit{xt}^j_{\overline{\omega}^{-1}\mathscr{E}_X}(\cdot,\cdot)$ or $\mathscr{T}\!\mathit{or}^{\overline{\omega}^{-1}\mathscr{E}_X}_j(\cdot,\cdot)$. Similarly if $\mathscr{M}$ is an $\mathscr{E}_X$-module, we write $\mathscr{M}\otimes_{\mathscr{O}_X}\Omega_X$ instead of $\mathscr{M}\otimes_{\pi^{-1}\mathscr{O}_X}\pi^{-1}\Omega_X$, as we did previously.

Lemma 3.3.1. *Assume Y is a submanifold of X, of codimension p. Then:*
  a) $\mathscr{E}_{Y\to X}$ *is a coherent* $\overline{\omega}^{-1}\mathscr{E}_X$*-module.*
  b) $\mathscr{E}_{Y\to X}$ *is flat over* $\rho^{-1}\mathscr{E}_Y$.
  c) $\mathscr{E}\!\mathit{xt}^j_{\mathscr{E}_X}(\mathscr{E}_{Y\to X},\mathscr{E}_{Y\to X})=0$ *for* $j\neq0$, *and for* $j=0$ *this Ring is canonically isomorphic to* $\rho^{-1}\mathscr{E}_Y$.
  d) $\mathscr{T}\!\mathit{or}^{\mathscr{E}_X}_j(\mathscr{E}_{Y\to X},\mathscr{E}_{X\leftarrow Y})=0$ *for* $j\neq p$, *and for* $j=p$ *this Ring is canonically isomorphic to* $\rho^{-1}\mathscr{E}_Y$.

Proof. a) There exists a natural right $\overline{\omega}^{-1}\mathscr{E}_X$-linear morphism from $\mathscr{O}_Y\otimes_{\mathscr{O}_X}\overline{\omega}^{-1}\mathscr{E}_X$ to $\mathscr{E}_{Y\to X}$, defined by:

(3.3.1) $$f\otimes P\mapsto f\cdot 1_{Y\to X}\cdot P.$$

To prove it is an isomorphism we choose coordinates (y,t) on X such that $Y=\{(y,t); t=0\}$. Let (y') be another copy of the coordinates (y). We have:

$$\mathscr{C}_{Y|Y\times X}\cong\mathscr{E}_{Y\times X}/\mathscr{E}_{Y\times X}(y-y',D_y+D_{y'},t)$$
$$\cong\mathscr{E}_X/\mathscr{E}_X(t).$$

We find that this Module is coherent over $\mathscr{E}_X$, and isomorphic to $\mathscr{E}_X\otimes_{\mathscr{O}_X}\mathscr{O}_Y$ as a left $\mathscr{E}_X$-module. If we tensorize it by Ω_X over $\mathscr{O}_X$ we find $\mathscr{E}_{Y\to X}$.
  b) We may send $\rho^{-1}\mathscr{E}_Y$ into $\mathscr{E}_{Y\to X}$ by:

(3.3.2) $$P\mapsto P\cdot 1_{Y\to X}.$$

To prove that (3.3.2) is injective we choose the same coordinates (y,t) as previously. Then one may identify $\mathscr{E}_{Y\to X}$ with the subsheaf $\widetilde{\mathscr{E}}_X$ of sections P of $\mathscr{E}_X$ which do not depend on t, (i.e. $[P,D_{t_j}]=0\ \forall j=1,\ldots,p$). Since the structure of $\rho^{-1}\mathscr{E}_Y$-module of $\mathscr{E}_{Y\to X}$ is given by the structure of $\widetilde{\mathscr{E}}_X$ it is enough to prove that the Ring $\widetilde{\mathscr{E}}_X$ is flat over $\rho^{-1}\mathscr{E}_Y$, but this follows from Proposition 1.2.4.
  c), d) Let us choose a system of equations $(t)=(t_1,\ldots,t_p)$ of Y in X and consider the Koszul complex associated to the sequence $(t_1,\ldots,t_p)$, acting on the right $\mathscr{E}_X$-module $\mathscr{E}_{Y\to X}\cong\mathscr{E}_X/(t)\mathscr{E}_X$:

(3.3.3) $$0\to(\mathscr{E}_X/(t)\mathscr{E}_X)^{(0)}\xrightarrow{(t)}(\mathscr{E}_X/(t)\mathscr{E}_X)^{(1)}\to\ldots\to(\mathscr{E}_X/(t)\mathscr{E}_X)^{(p)}\to0,$$

For each i, $1\leqslant i\leqslant p$, the operator t_i is surjective as an operator over $(\operatorname{Ker}t_1)\cap\ldots\cap(\operatorname{Ker}t_{i-1})$, that is, as an endomorphism of the space $\{P\in\mathscr{E}_X/(t)\mathscr{E}_X;\ P\circ t_1=\ldots=P\circ t_{i-1}=0\}$.
  Applying the results of the Appendix B.4 we find that the sequence (3.3.3) is exact, except at step 0.

The j-th cohomology group of the complex (3.3.3) gives both the Groups $\mathcal{E}xt^j_{\mathcal{E}_X}(\mathcal{E}_{Y\to X}, \mathcal{E}_{Y\to X})$ and $\mathcal{T}or^{\mathcal{E}_X}_{-j}(\mathcal{E}_{Y\to X}, \mathcal{E}_{X\leftarrow Y})$. For $j=0$ we find the sub-Group of $\mathcal{E}_{Y\to X}$ defined by:

$$\{P\in\mathcal{E}_X/(t)\mathcal{E}_X;\ P\circ t_i=0\quad\forall i=1,\ldots,p\},$$

which is isomorphic to $\rho^{-1}\mathcal{E}_Y$.

It is immediately verified that the isomorphisms $\mathcal{H}om_{\mathcal{E}_X}(\mathcal{E}_{Y\to X}, \mathcal{E}_{Y\to X})$ $\cong\mathcal{T}or^{\mathcal{E}_X}_p(\mathcal{E}_{Y\to X}, \mathcal{E}_{X\leftarrow Y})\cong\rho^{-1}\mathcal{E}_Y$ do not depend on the choice of $(t)=(t_1,\ldots,t_p)$, since those three groups are naturally embedded into $\mathcal{E}_{Y\to X}$. $\square$

Lemma 3.3.2. *Let Y and S be two submanifolds of X, Y and S being transversal.*

a) *There exists a natural isomorphism of $\mathcal{E}_Y$-modules on T^*Y:*

$$\rho_*(\mathcal{E}_{Y\to X}\otimes_{\mathcal{E}_X}\overline{\omega}^{-1}\mathcal{C}_{S|X})\cong\mathcal{C}_{(S\cap Y)|Y}.$$

b) $\mathcal{T}or^{\overline{\omega}^{-1}\mathcal{E}_X}_i(\mathcal{E}_{Y\to X}, \overline{\omega}^{-1}\mathcal{C}_{S|X})=0\quad\forall i\neq 0.$

Proof. Remark that Y and S being transversal, ρ induces a bijection from $(Y\times_X T^*X)\cap T^*_S X$ to $T^*_{(S\cap Y)}Y$.

For the sake of simplicity we shall not write ρ^{-1} and $\overline{\omega}^{-1}$ in the next formulas.

First we notice that $\rho_*(\mathcal{E}_{Y\to X}\otimes_{\mathcal{E}_X}\mathcal{C}_{S|X})$ is $\mathcal{E}_Y$-coherent. To check it, we choose coordinates (y,s,t) on X (as in the proof of Proposition 3.2.4) such that $S=\{(y,s,t);\ s=0\}$, $Y=\{(y,s,t);\ t=0\}$. Then:

$$\mathcal{E}_{Y\to X}\otimes_{\mathcal{E}_X}\mathcal{C}_{S|X}\cong\mathcal{E}_X/((t)\mathcal{E}_X+\mathcal{E}_X(s,D_t,D_y))$$

$$\cong\mathcal{E}_Y/\mathcal{E}_Y(s,D_y).$$

The restriction to $T^*_Y Y$ of this coherent $\mathcal{E}_Y$-module coincides with $\mathcal{D}_{Y\to X}\otimes_{\mathcal{D}_X}\mathcal{B}_{S|X}$, and the restriction to $T^*_Y Y$ of $\mathcal{C}_{(S\cap Y)|Y}$ is $\mathcal{B}_{(S\cap Y)|Y}$. Thus a) follows from Propositions 3.2.4 and 2.6.1. To prove b) we keep the coordinates (y,s,t) and remark that for any i, $1\leqslant i\leqslant p=\mathrm{codim}\,Y$, t_i is injective on $\mathcal{E}_X/((t_1,\ldots,t_{i-1})\mathcal{E}_X+\mathcal{E}_X(s,D_t,D_y))$. $\square$

Let Z be another manifold, g a holomorphic map from Z to Y, $h=f\circ g$. Let us denote by $\rho_f,\overline{\omega}_f,\rho_g$, etc. ... the associated maps on the cotangent bundles. We have a diagram:

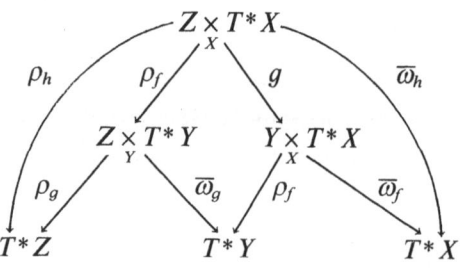

For the sake of simplicity we shall not always write the symbols $\rho_f^{-1}, \overline{\omega}_f^{-1}$, etc. ... in the formulas.

Proposition 3.3.3. *Assume g is an embedding. Then there exists a canonical isomorphism of $(\rho_h^{-1}\mathscr{E}_Z, \overline{\omega}_h^{-1}\mathscr{E}_X)$-bimodule on $Z \underset{X}{\times} T^*X$:*

$$\mathscr{E}_{Z \to Y} \otimes_{\mathscr{E}_Y} \mathscr{E}_{Y \to X} \cong \mathscr{E}_{Z \to Y},$$

and we have:

$$\mathscr{T}\!\mathit{or}_i^{\mathscr{E}_Y}(\mathscr{E}_{Z \to Y}, \mathscr{E}_{Y \to X}) = 0 \quad \forall i \neq 0.$$

Proof. Since $\mathscr{E}_{Z \to Y}$ is $\mathscr{E}_Y$-coherent, and $\mathscr{E}_{Y \to X}$ is flat over $\mathscr{E}_Y$, the canonical morphism:

$$\mathscr{E}_{Z \to Y} \otimes_{\mathscr{E}_Y} \mathscr{C}_{Y|Y \times X} \to \mathscr{E}_{Z \times X \to Y \times X} \otimes_{\mathscr{E}_{Y \times X}} \mathscr{C}_{Y|Y \times X}$$

is an isomorphism. But this Module is isomorphic to $\mathscr{C}_{Z|Z \times X}$ by the preceding lemma.

The proof for the vanishing of $\mathscr{T}\!\mathit{or}_i^{\mathscr{E}_Y}(\cdot, \cdot)$ is similar.   □

Remark 3.3.4. Proposition 3.3.3 is not true without the hypothesis that g is an embedding. For example take $\dim Z > 0$, $\dim X > 0$, $Y = \{y\}$, $y \in X$. Then:

$$\mathscr{E}_{Z \to Y} \otimes_{\mathscr{E}_Y} \mathscr{C}_{Y|Y \times X} \cong \mathscr{O}_Z \otimes \mathscr{C}_{\{y\}|X},$$

$$\mathscr{C}_{Z|Z \times X} = \mathscr{C}_{Z \times \{y\}|Z \times X} \cong \mathscr{O}_Z \hat{\otimes} \mathscr{C}_{\{y\}|X}$$

and these Modules are different outside of the zero section of $T^*_Z(Z \times X)$.

Proposition 3.3.5. a) *Let f and S be as in Proposition 3.2.4. Then there exists a natural isomorphism of left $\mathscr{E}_Y$-modules on T^*Y:*

$$\rho_*(\mathscr{E}_{Y \to X} \otimes_{\overline{\omega}^{-1}\mathscr{E}_X} \overline{\omega}^{-1}\mathscr{C}_{S|X}) \cong \mathscr{C}_{f^{-1}(S)|Y}.$$

b) *Let T and f be as in Proposition 3.2.8. Then there exists a natural isomorphism of right $\mathscr{E}_X$-modules on T^*X:*

$$\overline{\omega}_*(\rho^{-1}\mathscr{C}_{T|Y} \otimes_{\mathscr{O}_Y} \Omega_Y \otimes_{\rho^{-1}\mathscr{E}_Y} \mathscr{E}_{Y \to X}) \cong \mathscr{C}_{f(T)|X} \otimes_{\mathscr{O}_X} \Omega_X$$

Proof. a) By applying Propositions 2.6.1 and 3.2.4 it is enough to prove that $\rho_*(\mathscr{E}_{Y \to X} \otimes_{\overline{\omega}^{-1}\mathscr{E}_X} \overline{\omega}^{-1}\mathscr{C}_{S|X})$ is $\mathscr{E}_Y$-coherent. Applying Proposition 3.3.3 we may decompose f by the graph map, and prove the coherency separately when f is an embedding or a projection.

The case where f is an embedding is treated in Lemma 3.3.2. If f is the projection $Y \times X \to X$, then:

$$\mathscr{E}_{Y \times X \to X} \otimes_{\mathscr{E}_X} \mathscr{C}_{Z|X} = \mathscr{E}_Y \hat{\otimes} \mathscr{C}_{Z|X} = \mathscr{C}_{Y \times Z|Y \times X}.$$

b) The proof is similar.   □

Proposition 3.3.6. *Let S (resp. T) be a submanifold of X (resp. Y). There exists a natural isomorphism of left $\mathscr{E}_{Y \times X}$-modules:*

$$\mathscr{C}_{T|Y} \hat{\otimes} \mathscr{C}_{S|X} \cong \mathscr{C}_{T \times S|Y \times X}.$$

The proof is trivial.

Proposition 3.3.7. *In the situation of Proposition 3.2.10 let p be the natural map from $T_S^* X \underset{X}{\times} T_Z^* X$ to $T_{S \cap Z}^* X$. Then there exists a natural $\mathscr{O}_X$-linear morphism on $T_S^* X \underset{X}{\times} T_Z^* X$:*

$$\mathscr{C}_{S|X} \otimes_{\mathscr{O}_X} \mathscr{C}_{Z|X} \to p^{-1}(\mathscr{C}_{(S \cap Z)|X}).$$

The proof is formally the same as for Proposition 3.2.10, with the use of Proposition 3.3.5 a) instead of Proposition 3.2.4.

Remark 3.3.8. By Definition 3.1.2, we may translate Proposition 3.3.5 as follows: if f is non characteristic for $\mathscr{C}_{S|X}$, then the inverse image of this Module is $\mathscr{C}_{f^{-1}(S)|Y}$, and if f is non characteristic for $\mathscr{C}_{T|Y} \otimes_{\mathscr{O}_Y} \Omega_Y$, and f is an embedding on T, the direct image of this Module is $\mathscr{C}_{f(T)|X} \otimes_{\mathscr{O}_X} \Omega_X$.

3.4. Operations on $\mathscr{E}_X$-modules

Let Y, X, V, U, p_1, p_2 be as in Definition 3.1.1

Proposition 3.4.1. *Assume $\mathscr{M}$ (resp. $\mathscr{N}$) is $\mathscr{E}_X$-coherent on U (resp. $\mathscr{E}_Y$-coherent on V). Then:*
a) *The $\mathscr{E}_{Y \times X}$-module $\mathscr{N} \hat{\otimes} \mathscr{M}$ is coherent,*
b) $\mathrm{char}(\mathscr{N} \hat{\otimes} \mathscr{M}) = (\mathrm{char}(\mathscr{N})) \times (\mathrm{char}(\mathscr{M}))$ *and if S (resp. S') is an irreducible analytic set containing $\mathrm{char}(\mathscr{M})$ (resp. $\mathrm{char}(\mathscr{N})$) then:*

$$\mathrm{mult}_{S' \times S}(\mathscr{N} \hat{\otimes} \mathscr{M}) = \mathrm{mult}_{S'}(\mathscr{N}) \times \mathrm{mult}_S(\mathscr{M}).$$

The proof follows immediately from the fact that $\mathscr{E}_{Y \times X}$ (resp. $\mathrm{gr}(\mathscr{E}_{Y \times X})$) is faithfully flat over $p_1^{-1} \mathscr{E}_Y \otimes p_2^{-1} \mathscr{E}_X$ (resp. $p_1^{-1} \mathrm{gr}(\mathscr{E}_Y) \otimes p_2^{-1} \mathrm{gr}(\mathscr{E}_X)$) and Proposition 1.2.4.

Now we consider the situation of Definition 3.1.2 a): f is a holomorphic map from Y to X, $\mathscr{M}$ a left coherent $\mathscr{E}_X$-module on $U \subset T^*X$.

Theorem 3.4.2. *Assume f non characteristic for $\mathscr{M}$ on the open set $V \subset T^*Y$. Then:*
a) $\mathscr{T}or_i^{\varpi^{-1} \mathscr{E}_X}(\mathscr{E}_{Y \to X}, \varpi^{-1} \mathscr{M}) = 0 \quad \forall i \neq 0$
b) $\mathscr{M}_Y = \rho_*(\mathscr{E}_{Y \to X} \otimes_{\varpi^{-1} \mathscr{E}_X} \varpi^{-1} \mathscr{M})$ *is $\mathscr{E}_Y$-coherent on V.*
c) *Assume* $\mathrm{codim}(\mathrm{char}(\mathscr{M})) \geqslant d$. *Then:* $[\mathrm{char}(\mathscr{M}_Y)]_d = \rho_* \varpi^*[\mathrm{char}(\mathscr{M})]_d$.
d) $\mathrm{char}(\mathscr{M}_Y) = \rho \varpi^{-1}(\mathrm{char}(\mathscr{M}))$.

(The definition of the direct image or inverse image of an analytic cycle is given in Appendix D).

Proof. By decomposing f with the graph map, and applying Proposition 3.3.3, we may consider separately the case where f is the projection $Y \times X \to X$ and the case where f is an embedding.

The first case follows from Proposition 3.4.1, since:

$$\mathscr{E}_{Y \times X \to X} \otimes_{\mathscr{E}_X} \mathscr{M} = \mathscr{E}_Y \,\hat{\otimes}\, \mathscr{M}$$

and $\mathscr{E}_{Y \times X \to X}$ is flat over $\overline{\omega}^{-1}\mathscr{E}_X$. Now assume Y is a submanifold of X. Again by Proposition 3.3.3 we may argue by induction on the codimension of Y, and thus assume that Y is a hypersurface.

Let $(x_1, \ldots, x_n) = (x_1, x')$ be a system of local coordinates on X, such that $Y = \{x; \ x_1 = 0\}$. Since ρ is finite over $\overline{\omega}^{-1}(\mathrm{char}(\mathscr{M}))$, it is enough to argue locally in the neighborhood of each point $p \in (Y \underset{X}{\times} T^* X) \cap \mathrm{char}(\mathscr{M})$.

By the hypothesis that ρ is finite there exists a homogeneous function $f(x, \xi)$ defined in a neighborhood of p, of Weierstrass type with respect to ξ_1, vanishing on $\mathrm{char}(\mathscr{M})$.

Let u be a section of $\mathscr{M}$, $\mathscr{I}$ the annihilator Ideal of u in $\mathscr{E}_X$. Since $\mathrm{char}(\mathscr{E}_X u)$ is contained in $\mathrm{char}(\mathscr{M})$, f^l belongs to $\mathrm{gr}(\mathscr{I})$ for $l \gg 0$. Thus there exists an operator P, let us say of order m, with $\sigma(P) = f^l$ and $Pu = 0$. By the Weierstrass theorem for microdifferential operators, we may assume P of Weierstrass type with respect to D_1:

$$P = D_1^m + \sum_{j=0}^{m-1} A_j(x, D') D_1^j.$$

Now we prove a). Since $\mathscr{E}_{Y \to X} \cong \mathscr{E}_X / x_1 \mathscr{E}_X$, it is enough to prove that x_1 is injective on $\mathscr{M}$. Assume $x_1 u = 0$. We get:

$$\mathrm{ad}_{x_1}^m (P)(u) = m! \, u = 0$$

and $u = 0$.

Let us prove b). Let $(u_1, \ldots, u_r)$ be a system of generators of $\mathscr{M}$ and for each j, $1 \leqslant j \leqslant r$, choose an operator P_j of Weierstrass type with respect to D_1, of order m_j, such that $P_j u_j = 0$.

Let $\tilde{\mathscr{E}}_X$ be the sub-Ring of $\mathscr{E}_X$ of operators independ of D_1, that is $R \in \tilde{\mathscr{E}}_X$ iff $[x_1, R] = 0$, and set:

$$\tilde{\mathscr{E}}_X(k) = \tilde{\mathscr{E}}_X \cap \mathscr{E}_X(k).$$

By the division theorem, any operator $Q \in \mathscr{E}_X$ may be written:

$$Q = S P_j + \sum_{i=0}^{m_j - 1} R_i^j(x, D') D_1^i$$

with $R_i^j \in \tilde{\mathscr{E}}_X$, and hence:

$$\mathscr{E}_X(k) u_j \subset \sum_{i=0}^{m_j-1} \tilde{\mathscr{E}}_X(k-i) D_1^i u_j.$$

We set:

$$\mathscr{M}_k = \sum_{j=1}^{r} \sum_{i=0}^{m_j-1} \tilde{\mathscr{E}}_X(k-i) D_1^i u_j.$$

The $\mathscr{M}_k$'s define a good filtration on $\mathscr{M}$. Now we set:

$$\mathscr{N}_k = \mathscr{M}_k / (x_1 \mathscr{M} \cap \mathscr{M}_k)$$

$$= \sum_{j=1}^{r} \sum_{i=0}^{m_j-1} \mathscr{E}_Y(k-i) (1_{Y \to X} \otimes D_1^i u_j).$$

The $\mathscr{N}_k$'s define a good filtration on the $\rho^{-1}(\mathscr{E}_Y)$-module $\mathscr{N} \underset{\mathrm{def}}{=} \mathscr{M}/x_1 \mathscr{M}$, and the $\rho_*(\mathscr{N}_k)$'s a good filtration on the $\mathscr{E}_Y$-module $\mathscr{M}_Y$.

Let us denote by W the space $\mathbb{C}^n \times \mathbb{C}^{n-1}$ with coordinates $(x_1, x'; \xi')$ and by ϕ the projection $(x_1, x'; \xi_1, \xi') \mapsto (x_1, x'; \xi')$ from T^*X to W. We consider $\tilde{\mathscr{E}}_X$ as a sheaf over W. Remark that for $p \in W$, $\tilde{\mathscr{E}}_{X,p}$ is zariskian and $\tilde{\mathscr{E}}_X$ is notherian. We identify T^*Y with the subspace of W defined by the equation $\{x_1 = 0\}$. Then $\tilde{\mathscr{E}}_Y \simeq \tilde{\mathscr{E}}_X / \tilde{\mathscr{E}}_X \cdot x_1$.

The Module $\tilde{\mathscr{M}} \underset{\mathrm{def}}{=} \phi_* \mathscr{M}$ is finitely generated over $\tilde{\mathscr{E}}_X$. Moreover $\mathrm{gr}(\mathscr{M})$ is coherent over $\mathrm{gr}(\mathscr{E}_X)$, and $\phi_* \mathrm{gr}(\mathscr{M})$ is coherent over $\mathrm{gr}(\tilde{\mathscr{E}}_X)$, since ϕ is finite over $\mathrm{supp}(\mathrm{gr}(\mathscr{M}))$ (Appendix D.1). Since $\phi_* \mathrm{gr}(\mathscr{M}) = \mathrm{gr}(\phi_* \mathscr{M})$, we can apply Proposition 1.4.1 and obtain that $\tilde{\mathscr{M}}$ is $\tilde{\mathscr{E}}_X$-coherent.

Now consider the exact sequence of $\tilde{\mathscr{E}}_X$-modules:

$$(3.4.1) \qquad 0 \longrightarrow \tilde{\mathscr{M}} \xrightarrow{x_1} \tilde{\mathscr{M}} \longrightarrow \mathscr{M}_Y \longrightarrow 0.$$

Since $\mathscr{E}_Y \simeq \tilde{\mathscr{E}}/x_1 \tilde{\mathscr{E}}$, we find that $\mathscr{M}_Y$ is $\mathscr{E}_Y$-coherent.

Now we prove the multiplicity formula. First we remark that:

$$[\mathrm{supp}\, \phi_* \mathrm{gr}\, \mathscr{M}] = \phi_* [\mathrm{supp}\, \mathrm{gr}\, \mathscr{M}],$$

(Appendix, Proposition D.1.3), and since $\phi_* \mathrm{gr}\, \mathscr{M} = \mathrm{gr}\, \phi_* \mathscr{M}$, it remains to prove the formula for the $\mathrm{gr}(\tilde{\mathscr{E}}_X)$-module $\mathrm{gr}(\tilde{\mathscr{M}})$. Let $\tilde{\mathscr{M}}'$ be this Module endowed with the filtration:

$$(3.4.2) \qquad (\tilde{\mathscr{M}}')_k = \{u \in \tilde{\mathscr{M}}; \; x_1 u \in \phi_* \mathscr{M}_k\}.$$

This filtration is good over $\tilde{\mathscr{E}}_X$ since $\tilde{\mathscr{M}}$ is $\tilde{\mathscr{E}}_X$-coherent and x_1 is $\tilde{\mathscr{E}}_X$-linear. Set $\mathscr{F} = \mathrm{gr}(\tilde{\mathscr{M}})$, $\mathscr{F}' = \mathrm{gr}(\tilde{\mathscr{M}}')$, and denote by $\bar{x}_1$ (resp. $\bar{x}_1'$) the graded morphism associated to x_1 from $\tilde{\mathscr{M}}$ to $\tilde{\mathscr{M}}$ (resp. $\tilde{\mathscr{M}}'$ to $\tilde{\mathscr{M}}$). We define the graded Modules $\mathscr{N}$ and $\mathscr{H}$ by the exact sequences:

$$\begin{array}{ccccccc}
 & & & & 0 & & \\
 & & & & \uparrow & & \\
0 \longrightarrow \mathscr{F}' & \underset{\bar{x}_1'}{\longrightarrow} & \mathscr{F} & \longrightarrow & \mathscr{F}/\bar{x}_1' \mathscr{F}' & \longrightarrow 0 \\
 & & & & \uparrow & & \\
0 \longrightarrow \mathscr{N} \longrightarrow & \mathscr{F} & \underset{\bar{x}_1}{\longrightarrow} & \mathscr{F} & \longrightarrow \mathscr{F}/\bar{x}_1 \mathscr{F} & \longrightarrow 0 \\
 & & & & \uparrow & & \\
 & & & & \mathscr{K} & & \\
 & & & & \uparrow & & \\
 & & & & 0 & &
\end{array}$$

For simplicity we shall confuse a graded $\mathrm{gr}(\tilde{\mathscr{E}}_X)$-module and the Module obtained by tensorizing it over $\mathcal{O}_W$.

First we prove:

(3.4.3) $\qquad\qquad [\operatorname{supp} \mathscr{N}]_d = [\operatorname{supp} \mathscr{K}]_d$.

Since the filtration $\tilde{\mathscr{M}}'$ over $\tilde{\mathscr{M}}$ is good, and since any good filtrations are locally equivalent, there exists $r \geqslant 0$ such that:

$$\tilde{\mathscr{M}}_k \subset \tilde{\mathscr{M}}_k' \subset \tilde{\mathscr{M}}_{k+r} \qquad \forall k.$$

Now we have:

$$\mathscr{N} = \bigoplus_k \frac{\{u \in \tilde{\mathscr{M}}_k ;\ x_1 u \in \tilde{\mathscr{M}}_{k-1}\}}{\tilde{\mathscr{M}}_{k-1}}$$

$$\cong \bigoplus_k \frac{(x_1 \tilde{\mathscr{M}}_{k+1} \cap \tilde{\mathscr{M}}_k)}{x_1 \tilde{\mathscr{M}}_k},$$

$$\mathscr{K} = \bigoplus_k \frac{(x_1 \tilde{\mathscr{M}}_{k+r} \cap \tilde{\mathscr{M}}_k) + \tilde{\mathscr{M}}_{k-1}}{x_1 \tilde{\mathscr{M}}_k + \tilde{\mathscr{M}}_{k-1}}$$

$$= \bigoplus_k \frac{x_1 \tilde{\mathscr{M}}_{k+r} \cap \tilde{\mathscr{M}}_k}{(x_1 \tilde{\mathscr{M}}_{k+r-1} \cap \tilde{\mathscr{M}}_{k-1}) + x_1 \tilde{\mathscr{M}}_k},$$

since $(x_1 \tilde{\mathscr{M}} \cap \tilde{\mathscr{M}}_k) \cap (x_1 \tilde{\mathscr{M}}_k + \tilde{\mathscr{M}}_{k-1}) = x_1 \tilde{\mathscr{M}} \cap \tilde{\mathscr{M}}_{k-1} + x_1 \tilde{\mathscr{M}}_k$.

We argue by induction on $r \geqslant 1$, and set:

$$\mathsf{A}_k = x_1 \tilde{\mathscr{M}}_{k+r} \cap \tilde{\mathscr{M}}_k, \qquad \mathsf{B}_k = x_1 \tilde{\mathscr{M}}_{k+r-1} \cap \tilde{\mathscr{M}}_k,$$

$$\mathsf{C}_k = x_1 \tilde{\mathscr{M}}_k + (x_1 \tilde{\mathscr{M}}_{k+r-1} \cap \tilde{\mathscr{M}}_{k-1}),$$

$$\mathsf{D}_k = x_1 \tilde{\mathscr{M}}_k + (x_1 \tilde{\mathscr{M}}_{k+r-2} \cap \tilde{\mathscr{M}}_{k-1}),$$

$$\mathsf{A}/\mathsf{B} = \bigoplus_k \mathsf{A}_k/\mathsf{B}_k, \qquad \mathsf{C}/\mathsf{D} = \bigoplus_k \mathsf{C}_k/\mathsf{D}_k,$$

$$\mathsf{A}/\mathsf{C} = \bigoplus_k \mathsf{A}_k/\mathsf{C}_k, \qquad \mathsf{B}/\mathsf{C} = \bigoplus_k \mathsf{B}_k/\mathsf{D}_k.$$

The functor $[\operatorname{supp}(\cdot)]_d$ being additive:

$$[\operatorname{supp}(\mathsf{A}/\mathsf{B})]_d - [\operatorname{supp}(\mathsf{C}/\mathsf{D})]_d = [\operatorname{supp}(\mathsf{A}/\mathsf{C})]_d - [\operatorname{supp}(\mathsf{B}/\mathsf{D})]_d.$$

The term on the left hand side is zero since:

$$\frac{x_1 \tilde{\mathscr{M}}_k + (x_1 \tilde{\mathscr{M}}_{k+r-1} \cap \tilde{\mathscr{M}}_{k-1})}{x_1 \tilde{\mathscr{M}}_k + (x_1 \tilde{\mathscr{M}}_{k+r-2} \cap \tilde{\mathscr{M}}_{k-1})} = \frac{x_1 \tilde{\mathscr{M}}_{k+r-1} \cap \tilde{\mathscr{M}}_{k-1}}{x_1 \tilde{\mathscr{M}}_{k+r-2} \cap \tilde{\mathscr{M}}_{k-1}}$$

This proves (3.4.3).

To complete the proof of c) it remains to apply Proposition D.1.3 to the sheaf $\mathscr{F}$, by remarking that no irreducible component of $\operatorname{supp}(\mathscr{F})$ is contained in $Y \underset{X}{\times} W$, by the involutivity of $\operatorname{char}(\mathscr{M})$.

Finally we may reduce the proof of d) to the case where Y is a hypersurface of X and $\operatorname{char}(\mathscr{M})$ is purely codimensional. Then d) follows from c) since in that case, setting $S = \phi(\operatorname{char}(\mathscr{M}))$ we have: $\mathscr{T}\!or_j^{\mathscr{C}_W}(\mathscr{O}_{Y \underset{X}{\times} W}, \mathscr{O}_S) = 0$ for $j \neq 0$, and the inverse image of $[S]$ on $Y \underset{X}{\times} W$ is a positive cycle. $\square$

Examples 3.4.3. a) Take $\mathscr{M} = \mathscr{O}_X$. If Y is a submanifold of X, Y is always non characteristic for $\mathscr{M}$ since $\operatorname{char}(\mathscr{M}) = T_X^* X$, and we find $(\mathscr{O}_X)_Y = \mathscr{O}_Y$.

b) Let Z be another manifold and consider the "partial De Rham system", $\mathscr{O}_X \hat{\otimes} \mathscr{D}_Z$ on $X \times Z$. Let Y be an hypersurface on $X \times Z$, $\{f = 0\}$ an equation of Y, and assume Y non characteristic for $\mathscr{O}_X \hat{\otimes} \mathscr{D}_Z$, that is, $d_x f \neq 0$ on Y.

Let $\mathscr{J}$ be the left Ideal of $\mathscr{D}_{X \times Z}$ generated by the vector fields of $Z \times TX$ which annihilate f. It we choose coordinates $(x_1, \dots, x_n)$ on X, then $\mathscr{J}$ is generated by $\frac{\partial f}{\partial x_i} D_{x_j} - \frac{\partial f}{\partial x_j} D_{x_i}$. Let $\mathscr{J}_Y = \mathscr{O}_Y \otimes_{\mathscr{D}_{X \times Z}} \mathscr{J}$ be the left Ideal of $\mathscr{D}_Y$ generated by the restriction to Y of the preceding vector fields. Then:

$$\mathscr{O}_Y \otimes_{\mathscr{O}_{X \times Z}} (\mathscr{O}_X \hat{\otimes} \mathscr{D}_Z) \cong \mathscr{D}_Y / \mathscr{J}_Y$$

c) Let $\overline{X}$ be the anti-holomorphic manifold associated to X ($\overline{X}$ is a complex manifold, and $\overline{X} = X$ as a topological space, but the holomorphic functions on $\overline{X}$ are the anti-holomorphic functions on X). We identify X with the diagonal of $X \times \overline{X}$, and denote by $X^{\mathbb{R}}$ the underlying real analytic manifold of X. Then $X \times \overline{X}$ is a complexification of $X^{\mathbb{R}}$, and the Cauchy-Riemann system on $X^{\mathbb{R}}$ is the $\mathscr{D}_{X \times \overline{X}}$-module, $\mathscr{D}_X \hat{\otimes} \mathscr{O}_{\overline{X}}$. Let S be a real analytic hypersurface of $X^{\mathbb{R}}$, Y its complexification in $X \times \overline{X}$. The induced Cauchy-Riemann system on S is the $\mathscr{D}_Y$-module $\mathscr{O}_Y \otimes_{\mathscr{O}_{X \times \overline{X}}} (\mathscr{D}_X \hat{\otimes} \mathscr{O}_{\overline{X}})$.

If we choose a system of local holomorphic coordinates $(z_1, \dots, z_n)$ on X, and if $\{f(z, \overline{z}) = 0\}$ is an equation of Y in $X \times \overline{X}$, then this Module is just $\mathscr{D}_Y / \mathscr{J}_Y$, where $\mathscr{J}_Y$ is the left Ideal generated by the restriction to Y of the vector fields $\frac{\partial f}{\partial \overline{z}_i} D_{\overline{z}_j} - \frac{\partial f}{\partial \overline{z}_j} D_{\overline{z}_i}$ $(1 \leqslant i, j \leqslant n)$.

We can state a theorem similar to Theorem 3.4.2 for direct images. Consider the situation of Definition 3.1.2 b).

Theorem 3.4.4. *Assume f non characteristic for the right coherent $\mathscr{E}_Y$-module $\mathscr{N}$ on $U \subset T^* X$. Then:*

a) $\mathscr{T}\!or_i^{\rho^{-1}\mathscr{E}_Y}(\rho^{-1}\mathscr{N}, \mathscr{E}_{Y \to X}) = 0 \quad \forall i \neq 0$.

b) $\int_f \mathscr{N} = \overline{\omega}_* (\rho^{-1}\mathscr{N} \otimes_{\rho^{-1}\mathscr{E}_Y} \mathscr{E}_{Y \to X})$ *is $\mathscr{E}_X$-coherent on U.*

c) *Assume* codim$(\text{char}(\mathscr{N})) \geqslant d)$. *Then:*

$[\text{char}(\int_f \mathscr{N})]_d = \overline{\omega}_* \rho^* [(\text{char}\,\mathscr{N})]_d$.

d) char$(\int_f \mathscr{N}) = \overline{\omega}\rho^{-1}(\text{char}(\mathscr{N}))$.

Proof. Of course it is possible to prove Theorem 3.4.4 by the same method as for Theorem 3.4.2, but we prefer to give another method. First we may decompose f as an immersion and a submersion and consider separately each case, and since the first case is trivial we shall assume f is smooth (i. e.: df is everywhere surjective).

Let $V = Y \underset{X}{\times} T^* X \subset T^* Y$. Outside of $T^*_Y Y$ the manifold V is regular involutive, and a partial Legendre contact transformation interchanges V and $Z \underset{Y}{\times} \dot{T}^* Y$, for a submanifold Z of Y.

We may quantize this canonical transformation and interchange the systems with simple characteristics $\mathscr{E}_{Y \to X}$ on V and $\mathscr{E}_{Y \leftarrow Z}$ on $Z \underset{Y}{\times} T^* Y$.

Then Theorem 3.4.4 is a particular case of Theorem 3.4.2 since $\overline{\omega}^{-1}\mathscr{E}_X = \mathscr{E}nd_{\mathscr{E}_Y}(\mathscr{E}_{Y \to X})$ is interchanged with $\rho^{-1}\mathscr{E}_Z = \mathscr{E}nd_{\mathscr{E}_Y}(\mathscr{E}_{Y \leftarrow Z})$. Now to avoid the zero section $T^* Y$, we use the trick of the dummy variable (Proposition 2.5.2), and remark that:

$$(3.4.4) \qquad \int_f (\mathscr{N} \hat{\otimes} \mathscr{E}_{\mathbb{C}}) = (\int_f \mathscr{N}) \hat{\otimes} \mathscr{E}_{\mathbb{C}}. \qquad \square$$

Example 3.4.5. Let Y be an open set of $\mathbb{C}^n \times \mathbb{C}$, with coordinates $(x, t) = (x_1, \dots, x_n, t)$, and let f be the projection from Y to $X = \mathbb{C}^n$. Let $p(x, t)$ be a function on Y of Weierstrass type with respect to t. Then if $\mathscr{N}$ is a right coherent $\mathscr{D}_Y$-module supported by $\{(x, t); p(x, t) = 0\}$, the right $\mathscr{D}_X$-module:

$$\int_f \mathscr{N} = f_*(\mathscr{N}/\mathscr{N} D_t)$$

is coherent. Moreover the (right-)action of D_t on $\mathscr{N}$ is injective.

As an application of Proposition 3.4.1 and 3.4.2 we may study the tensor product over $\mathscr{O}_X$ of two left $\mathscr{D}_X$-modules $\mathscr{M}$ and $\mathscr{N}$, by noticing:

$$\mathscr{M} \otimes_{\mathscr{O}_X} \mathscr{N} = \mathscr{O}_{X \underset{X}{\times} X} \otimes_{\mathscr{O}_{X \times X}} (\mathscr{M} \hat{\otimes} \mathscr{N}).$$

In particular the term on the left hand side inherits a structure of a left $\mathscr{D}_X$-module.

Corollary 3.4.6. *Let $\mathscr{M}$ and $\mathscr{N}$ be two coherent left $\mathscr{D}_X$-modules and assume* char$(\mathscr{M}) \cap$ char$(\mathscr{N})$ *is contained in $T^*_X X$. Then the left $\mathscr{D}_X$-module $\mathscr{M} \otimes_{\mathscr{O}_X} \mathscr{N}$ is $\mathscr{D}_X$-coherent.*

In fact by hypothesis, the diagonal $X \underset{X}{\times} X$ is non characteristic for $\mathscr{M} \hat{\otimes} \mathscr{N}$, and we may apply Theorem 3.4.2.

Remark that Proposition 3.2.10 is an illustration of this result.

3.5. Complement on Inverse Images

When the Module $\mathscr{M}$ corresponds to "one equation with one unknown", that is, if $\mathscr{M} \cong \mathscr{E}_X / \mathscr{E}_X \cdot P$ we may refine Theorem 3.4.2.

Let $(x_1, \ldots, x_n) = (x_1, x')$ be a system of local coordinates on X, Y the hypersurface $\{x \in X; x_1 = 0\}$, U an open set of T^*X and P a section of $\mathscr{E}_X$ on U.

Proposition 3.5.1. *Assume that for any* $(x', \xi') \in \rho \overline{\omega}^{-1}(U)$, *the equation* $\sigma(P)(0, x', \xi_1, \xi') = 0$ *has* m *roots (counted with their multiplicities). Let* $\mathscr{M} = \mathscr{E}_X / \mathscr{E}_X P = \mathscr{E}_X u$. *Then* $\mathscr{M}_Y$ *is free of rank* m, *generated by* $1_{Y \to X} \otimes u$, ..., $1_{Y \to X} \otimes D_1^{m-1} u$.

Proof. We follow the proof of Sato-Kashiwara-Kawaï [1, p. 408].

First we remark that it is enough to prove that for each $(x', \xi') \in \rho \overline{\omega}^{-1}(U)$, $\mathscr{M}_{Y, (x', \xi')}$ is free of rank m over $\mathscr{E}_{Y, (x', \xi')}$, since $\mathscr{M}_Y$ is coherent. We fix (x', ξ') and set:

$$Z = \{(0, x'; \xi_1, \xi'); \ \sigma(P)(0, x'; \xi_1, \xi') = 0\}.$$

If Z reduces to one point, we get from the division theorem that any section Q of $\mathscr{E}_X$ may be written in a unique way on a neighborhood of Z:

$$Q = SP + \sum_{j=0}^{m-1} R_j(x', D_{x'}) D_1^j + x_1 R$$

and thus:

$$\mathscr{E}_X / (x_1 \mathscr{E}_X + \mathscr{E}_X P) \cong \mathscr{E}_Y^m.$$

Now we argue by induction on the number of points in Z, and set $Z = Z_1 \sqcup Z_2$, so that the equation $\sigma(P)(0, x', \xi_1, \xi') = 0$ has m_i roots on Z_i ($i = 1, 2$).

We may assume P of Weierstrass type in D_1 of order m, and moreover we may write:

$$P = P_2 P_1 = S_1 S_2$$

where P_i and S_i are Weierstrass type of order m_i, and are invertible on Z_j for $i, j \in \{1, 2\}, j \neq i$. This follows immediately from the preparation theorem.

Set $\mathscr{M}_1 = \mathscr{E}_X / \mathscr{E}_X \cdot S_1 = \mathscr{E}_X u_1$, $\mathscr{M}_2 = \mathscr{E}_X / \mathscr{E}_X \cdot P_2 = \mathscr{E}_X u_2$.   Then $\mathscr{M} = \mathscr{M}_1 \oplus \mathscr{M}_2$, $u = S_2^{-1} u_1 \oplus P_1^{-1} u_2$, $S_2 u = u_1$, $P_1 u = u_2$, since $S_2 P_1^{-1} u_2 = S_1^{-1} P_2 u_2 = 0$, $P_1 S_2^{-1} u_1 = P_2^{-1} S_1 u_1 = 0$.

Any $w_j \in \mathscr{M}_{jY} (j=1,2)$ is generated by $\{1_{Y\to X} \otimes D_1^k u_j; \ 0 \leqslant k \leqslant m_j - 1\}$ Since $u_1 = S_2 u$, $u_2 = P_1 u$, the sections of $\mathscr{M}_Y$ are generated by $\{1_{Y\to X} \otimes D_1^k u; \ 0 \leqslant k \leqslant m-1\}$. It remains to prove that these sections are free in $\mathscr{M}_Y$.

Let $\mathscr{N}_1 = \mathscr{E}_X / \mathscr{E}_X P_1 = \mathscr{E}_X v_1$. Then:

$$\mathscr{M} = \mathscr{N}_1 \oplus \mathscr{M}_2, \quad u = v_1 \oplus P_1^{-1} u_2.$$

Assume $1_{Y\to X} \otimes Q u = 0$, where $Q = \sum_{j=0}^{m-1} Q_j(x, D') D_1^j$. Dividing Q by P_1 we find $Q = G(x,D) + H P_1$, where G is of Weierstrass type of order less than m_1 in D_1, and H is of order less than m_2 in D_1. Then $1_{Y\to X} \otimes Q v_1 = 1_{Y\to X} \otimes G v_1 = 0$ and by the induction hypothesis, $G \in x_1 \mathscr{E}_X$. Similarly,

$$1_{Y\to X} \otimes Q P_1^{-1} u_2 - 1_{Y\to X} \otimes H u_2 = 0,$$

and $H \in x_1 \mathscr{E}_X$. Thus $Q \in x_1 \mathscr{E}_X$.   □

The following result will allow us to formulate the Cauchy problem in the next chapter.

Let f be a holomorphic map from Y to X.

Proposition 3.5.2. *Let $\mathscr{M}$ and $\mathscr{N}$ be two left coherent $\mathscr{E}_X$-modules on an open set $U \subset T^* X$. Assume f non characteristic for $\mathscr{M}$ and for $\mathscr{N}$. Then there exists for each j a canonical morphism on $\overline{\omega}^{-1}(U)$:*

(3.5.1)
$$\overline{\omega}^{-1} \mathscr{E}\mathit{xt}^j_{\mathscr{E}_X}(\mathscr{M}, \mathscr{N}) \to \mathscr{E}\mathit{xt}^j_{\rho^{-1}\mathscr{E}_Y}(\mathscr{E}_{Y\to X} \otimes_{\overline{\omega}^{-1}\mathscr{E}_X} \overline{\omega}^{-1}\mathscr{M}, \mathscr{E}_{Y\to X} \otimes_{\overline{\omega}^{-1}\mathscr{E}_X}\overline{\omega}^{-1}\mathscr{N}).$$

Proof. Let $\mathscr{M}^{\boldsymbol{\cdot}}$ (resp. $\mathscr{N}^{\boldsymbol{\cdot}}$) be a complex of finite length of finite free $\mathscr{E}_X$-modules quasi-isomorphic to $\mathscr{M}$ (resp. $\mathscr{N}$).

Let:

$$\mathscr{M}_Y^{\boldsymbol{\cdot}} = \mathscr{E}_{Y\to X} \otimes_{\overline{\omega}^{-1}\mathscr{E}_X}\overline{\omega}^{-1}\mathscr{M}^{\boldsymbol{\cdot}},$$

$$\mathscr{N}_Y^{\boldsymbol{\cdot}} = \mathscr{E}_{Y\to X} \otimes_{\overline{\omega}^{-1}\mathscr{E}_X}\overline{\omega}^{-1}\mathscr{N}^{\boldsymbol{\cdot}}.$$

We have a canonical homomorphism:

$$\overline{\omega}^{-1}\mathscr{H}\mathit{om}_{\mathscr{E}_X}(\mathscr{M}^{\boldsymbol{\cdot}}, \mathscr{N}^{\boldsymbol{\cdot}}) \to \mathscr{H}\mathit{om}_{\mathscr{E}_Y}(\mathscr{M}_Y^{\boldsymbol{\cdot}}, \mathscr{N}_Y^{\boldsymbol{\cdot}}).$$

Now let $\mathscr{L}_Y^{\boldsymbol{\cdot}}$ be a complex of injective $\rho^{-1}\mathscr{E}_Y$-modules quasi-isomorphic to $\mathscr{N}_Y^{\boldsymbol{\cdot}}$. We get a morphism:

(3.5.2) $$\overline{\omega}^{-1}\mathscr{H}\mathit{om}_{\mathscr{E}_X}(\mathscr{M}^{\boldsymbol{\cdot}}, \mathscr{N}^{\boldsymbol{\cdot}}) \to \mathscr{H}\mathit{om}_{\mathscr{E}_Y}(\mathscr{M}_Y^{\boldsymbol{\cdot}}, \mathscr{L}_Y^{\boldsymbol{\cdot}}).$$

The complex $\mathscr{M}_Y^{\boldsymbol{\cdot}}$ is quasi-isomorphic to $\mathscr{E}_{Y\to X} \otimes_{\overline{\omega}^{-1}\mathscr{E}_X}\overline{\omega}^{-1}\mathscr{M}$ and $\mathscr{L}_Y^{\boldsymbol{\cdot}}$ is quasi-isomorphic to $\mathscr{E}_{Y\to X} \otimes_{\overline{\omega}^{-1}\mathscr{E}_X}\overline{\omega}^{-1}\mathscr{N}$ by Theorem 3.4.2. Thus taking the j-th cohomology of each term in (3.5.2) we get the result.   □

Remark 3.5.3. The language of derived categories permits us to generalize Proposition 3.5.2 by removing the hypotheses on $\mathscr{M}$ and $\mathscr{N}$. In that case it should be written as follows (cf. Kashiwara [5]): there exists a canonical morphism:

(3.5.3)
$$\bar{\omega}^{-1} \mathbb{R} \mathscr{H}om_{\mathscr{E}_X}(\mathscr{M}, \mathscr{N}) \to$$
$$\to \mathbb{R} \mathscr{H}om_{\bar{\omega}^{-1}\mathscr{E}_Y}(\mathscr{E}_{Y\to X} \overset{\mathbb{L}}{\otimes}_{\bar{\omega}^{-1}\mathscr{E}_X} \bar{\omega}^{-1}\mathscr{M}, \mathscr{E}_{Y\to X} \overset{\mathbb{L}}{\otimes}_{\bar{\omega}^{-1}\mathscr{E}_X} \bar{\omega}^{-1}\mathscr{N}).$$

Remark 3.5.4. There exists a nice theory of inverse images of holonomic $\mathscr{D}_X$-modules (Kashiwara [4]), and also of direct images of such Modules, but only under suitable assumptions (Bernstein [2], Kashiwara-Schapira [4]). If $\mathscr{M}$ is a holonomic $\mathscr{D}_X$-module, and f a holomorphic map from Y to X, M. Kashiwara has proved that all the $\mathscr{D}_Y$-modules $\mathscr{T}or_j^{\mathscr{D}_X}(\mathscr{O}_Y, \mathscr{M})$ are holonomic (in particular they are coherent). But the characteristic varieties of those $\mathscr{D}_Y$-modules are in general difficult to calculate, except in the regular case (Kashiwara-Schapira [3]). We shall not explain this theory here, and refer to Kashiwara's paper.

Example 3.5.5. We give the simplest example of restriction of a holonomic Module in the characteristic case.

Let $X = \mathbb{C}$ with a coordinate t, $Y = \{0\}$, $\mathscr{M} = \mathscr{D}_{\mathbb{C}}/\mathscr{D}_{\mathbb{C}}(tD_t - \alpha)$, with $\alpha \notin \mathbb{N}$, $u = 1 \bmod \mathscr{D}_{\mathbb{C}}(tD_t - \alpha)$. Then $\mathscr{M}/t\mathscr{M}$ is generated by the family $(D_t^j u)_{j \in \mathbb{N}}$ over $\mathbb{C}$. But:

$$(tD_t - \alpha + j) D_t^j u = D_t^j (tD_t - \alpha) u = 0,$$

which proves that $D_t^j u \in t\mathscr{M}$ for all j, thus $\mathscr{M}/t\mathscr{M} = 0$.

Similarly let $v \in \mathscr{M}$, with $tv = 0$. Then $v = Ru$, and $tRu = 0$, that is $tR = Q(tD_t - \alpha)$, for some operator Q. But that equation implies $Q = tS$ and $R = S \circ (tD_t - \alpha)$ hence $Ru = 0$, and t is injective on $\mathscr{M}$. In other words we have proved:

$$\mathscr{T}or_j^{\mathscr{D}_X}(\mathscr{O}_Y, \mathscr{M}) = 0 \quad \forall j.$$

To end this section, let us give a result on the structure of the induced systems on a submanifold Y, when we do not assume any more that Y is non characteristic. We shall need this result in Chapter III, § 1.5.

Proposition 3.5.6. *Let $\mathscr{M}$ be a left coherent $\mathscr{D}_X$-module, Y a submanifold of X. Then for each j, the $\mathscr{D}_Y$-module $\mathscr{T}or_j^{\mathscr{D}_X}(\mathscr{D}_{Y\to X}, \mathscr{M})$ is locally a union of an increasing sequence of coherent $\mathscr{D}_Y$-modules.*

Proof. Let us choose local coordinates $(y, t) = (y_1, \ldots, y_q, t_1, \ldots, t_p)$ on X such that $Y = \{(y, t); \ t = 0\}$, and let us denote by $\tilde{\mathscr{D}}_X$ the sub-Ring of $\mathscr{D}_X$ of operators which do not depend on D_t. It is immediately proved, using the results of § 1.4, that $\tilde{\mathscr{D}}_X$ is a noetherian Ring, and any coherent $\mathscr{D}_X$-module is $\tilde{\mathscr{D}}_X$-pseudo-coherent.

We shall calculate $\mathscr{T}\!or_j^{\mathscr{D}_X}(\mathscr{D}_{Y\to X}, \mathscr{M})$ by the Koszul complex associated to the sequence $(t_1, \ldots, t_p)$ (cf. Appendix B.4):

$$\ldots \to \mathscr{M}^{(j-1)} \xrightarrow{d_{j-1}} \mathscr{M}^{(j)} \xrightarrow{d_j} \mathscr{M}^{(j+1)} \to \ldots$$

(where $\mathscr{M}^{(j)} = \mathscr{M} \otimes_{\mathbb{Z}} \overset{j}{\Lambda} \mathbb{Z}^p$).

Let $\mathscr{M}_0$ be a (locally defined) $\mathscr{O}_X$-module of finite type which generates $\mathscr{M}$. Set $\mathscr{M}_k = \mathscr{D}_X(k) \mathscr{M}_0$, $\tilde{\mathscr{M}}_k = \tilde{\mathscr{D}}_X \mathscr{M}_k$, $\mathscr{M}_k^{(j)} = \mathscr{M}_k \otimes_{\mathbb{Z}} \overset{j}{\Lambda} \mathbb{Z}^p$, $\tilde{\mathscr{M}}_k^{(j)} = \tilde{\mathscr{M}}_k \otimes_{\mathbb{Z}} \overset{j}{\Lambda} \mathbb{Z}^p$.

First we remark that the left $\tilde{\mathscr{D}}_X$-modules $\mathrm{Ker}\, d_j \cap \tilde{\mathscr{M}}_k^{(j)}$ and $\mathrm{Im}\, d_{j-1} \cap \tilde{\mathscr{M}}_k^{(j)}$ are coherent. For example we have:

$$\mathrm{Ker}\, d_j \cap \tilde{\mathscr{M}}_k^{(j)} = \bigcup_l \left(\tilde{\mathscr{D}}_X (\mathrm{Ker}\, d_j \cap \mathscr{M}_l^{(j)}) \cap \tilde{\mathscr{M}}_k^{(j)} \right)$$

and the coherency follows since $\tilde{\mathscr{D}}_X$ is noetherian. The proof for $\mathrm{Im}\, d_{j-1} \cap \tilde{\mathscr{M}}_k^{(j)}$ is similar.

Now $\mathscr{T}\!or_j^{\mathscr{D}_X}(\mathscr{D}_{Y\to X}, \mathscr{M})$ is the union of the coherent $\tilde{\mathscr{D}}_X$-modules:

$$\mathscr{L}_k \underset{\mathrm{def}}{=} (\mathrm{Ker}\, d_j \cap \tilde{\mathscr{M}}_k^{(j)})/(\mathrm{Im}\, d_{j-1} \cap \tilde{\mathscr{M}}_k^{(j)})$$

and it remains to make use of the following obvious fact.

Lemma 3.5.7. *Let $\mathscr{L}$ be a coherent $\tilde{\mathscr{D}}_X$-module such that $\mathscr{J}_Y \mathscr{L} = 0$. Then $\mathscr{L}$ is $\mathscr{D}_Y$-coherent.*

Exercises to II.3

Ex. 3.1. Let $(x) = (x_1, \ldots, x_n)$ be coordinates on $\mathbb{C}^n$, $Y = \{x; x_1 = 0\}$, $\mathscr{M} = \mathscr{D}_X/\mathscr{D}_X P$. Calculate $\mathscr{T}\!or_j^{\mathscr{D}_X}(\mathscr{O}_Y, \mathscr{M})$ in the following cases.
  i) $P = 0$,
  ii) $P = x_2 x_1 D_1 + 1$,
  iii) $P = (x_1 D_1)^2 + x_1 D_1 + x_2$,
  iv) $P = x_1 D_1 + D_2$,
  v) $P = x_1 D_1 + x_1 D_2$.
In particular show that in case iii) and v) those Modules are coherent over $\mathscr{D}_Y$, but not in the other cases.

Ex. 3.2. Calculate $\mathscr{T}\!or_j^{\mathscr{O}_X}(\mathscr{O}_X/x_1^2 \mathscr{O}_X, \mathscr{M})$ when $\mathscr{M}$ is one of the systems of Example 3.1.

Ex. 3.3. Assume Y non characteristic for $\mathscr{M}$. Prove that $\mathscr{T}\!or_j^{\mathscr{O}_X}(\mathscr{O}_X/x_1^2 \mathscr{O}_X, \mathscr{M}) = 0$ for $j \neq 0$, and that $\mathscr{M}/x_1^2 \mathscr{M}$ is coherent over $\mathscr{D}_Y$.

Ex. 3.4. Let $W_n = \mathbb{C}[x, D_x]$ be the Weyl algebra on $\mathbb{C}^n$, and let M be a finitely generated right W_n-module. Give a sufficient condition similar to that of Theorem 3.4.4 in order that M/MD_n be finitely generated over W_{n-1}.

Ex. 3.5. Prove that on the Weyl algebra the operations of restriction to a linear subspace and the operation of direct image by a linear projection are equivalent by the Fourier transformation (which interchanges (x, D_x) with $(-D_\xi, \xi)$).

Ex. 3.6. Let t be a coordinate on $\mathbb{C}$, $Y = \{t = 0\}$. Consider the coherent $\mathscr{D}_{\mathbb{C}}$-modules:

$$\mathscr{L} = \mathscr{D}_{\mathbb{C}}/\mathscr{D}_{\mathbb{C}} \cdot D_t, \quad \mathscr{M} = \mathscr{D}_{\mathbb{C}}/\mathscr{D}_{\mathbb{C}} \cdot (D_t t), \quad \mathscr{N} = \mathscr{D}_{\mathbb{C}}/\mathscr{D}_{\mathbb{C}} \cdot t$$

and the exact sequence:

$$0 \to \mathscr{L} \xrightarrow{t} \mathscr{M} \to \mathscr{N} \to 0.$$

Calculate the long exact sequence obtained by applying the functor $\mathscr{O}_Y \otimes_{\mathscr{O}_X} \cdot$ to this exact sequence.

Ex. 3.7. Let $(x) = (x_1, \ldots, x_n)$ be coordinates on $\mathbb{C}^n$, $Y = \{x; x_1 = x_2 = 0\}$, $\mathscr{M} = \mathscr{D}_X/\mathscr{D}_X \cdot P$. Calculate in which cases the Modules $\mathscr{T}\!or_j^{\mathscr{O}_X}(\mathscr{O}_Y, \mathscr{M})$ are $\mathscr{D}_Y$-coherent (cf. Chapter III, § 1):
  i) $P = x_1 D_1 + 2x_2 D_2 + \lambda$,
  ii) $P = x_1 D_1 - x_2 D_2 - 1/2$,
  iii) $P = x_1 D_2 + 1$,
  iv) $P = x_1 D_1 - 2x_2 D_2$,
  v) $P = x_1 D_1 - x_2 D_2$,
  vi) $P = x_1 D_2 + x_2 D_1$.

Ex. 3.8. Let $\mathscr{M}$ be the holonomic Module of Ex. 2.15 and let $Y = \left\{x \in \mathbb{C}^n; \sum_{j=1}^{n} x_j^2 = 0, \ x \neq 0\right\}$. Calculate the induced systems $\mathscr{T}\!or_j^{\mathscr{O}_X}(\mathscr{O}_Y, \mathscr{M})$. Same question with Y replaced by $X' = \{0\}$.

Notes

Many results of § 1 are well known, even if they do not always appear under that form in the literature. We refer to Bourbaki [1], Sato-Kashiwara-Kawaï [1], Kashiwara [5], Malgrange [2], Lejeune [1], Björk [1], Schapira [3], Houzel [2]. For example Proposition 1.3.1 is an immediate generalization of an argument of M. Kashiwara [5] who first defined the characteristic cycle of a coherent $\mathscr{E}_X$-module. We also follow partly J.E. Björk, especially for Propositions 1.2.4 and 1.4.1, and C. Houzel, especially for Proposition 1.1.7 and Remarks 1.1.9, 1.1.10.

The involutivity of the characteristic variety was first conjectured by Guillemin-Quillen-Sternberg [1], then proved by Sato-Kashiwara-Kawaï [1] with the help of quantized contact transformations and infinite order microdifferential equations. Later M. Kashiwara [unpublished] and B. Malgrange [3] removed the use of infinite order operators in the proof, until O. Gabber [1] gave a purely algebraic formulation to it (Theorem 1.3.4). Let us mention that an involutivity theorem for sheaves (including Theorem 2.3.1) has recently been stated by Kashiwara-Schapira [4].

Theorem 2.1.1 is crucial to the application of the algebraic results of § 1, and we follow Sato-Kashiwara-Kawaï [1] for its proof. All the other results of § 2 are essentially due to M. Kashiwara [3], but the fact that projective $\mathscr{E}_X$-modules are stably finite free was first noticed by E. Andronikof [1] [2] (cf. Proposition 2.3.6 b), c)), and the exact sequence (2.5.1) is a construction due to the author. Let us also mention that another proof of Theorem 2.1.1 (and its corollaries) has been proposed by L. Boutet de Monvel [1].

The operations on $\mathscr{E}_X$-modules are defined in Sato-Kashiwara-Kawaï [1], (cf. Bernstein [2] and Kashiwara [1] for the differential case), and most of the results of § 3 belong to those authors, with the exception of the multiplicity formula of Kashiwara [5], in Theorem 3.4.2. However our construction of operations on $\mathscr{E}_X$-modules is rather different than that of Sato-Kashiwara-Kawaï, since we do not use coholomological methods, and proceed naively, starting with $\mathscr{B}_{S|X}$. (Such a method is also used in the book of Pham [1], mainly for differential Modules).

Chapter III. Cauchy Problem and Propagation

Summary

In many cases, the study of the holomorphic solutions of systems of differential equations, that is, of the sheaves $\mathcal{E}xt^j_{\mathcal{D}_X}(\mathcal{M}, \mathcal{O}_X)$, reduces to the study of $\text{char}(\mathcal{M})$, the characteristic variety of $\mathcal{M}$.

However, sometimes we need to manipulate other sheaves of solutions, as for example the sheaves $\mathcal{E}xt^j_{\mathcal{E}_X}(\mathcal{M}, \mathcal{L})$, for a system $\mathcal{L}$ with simple characteristics along an involutive manifold V, or the sheaves $\mathcal{E}xt^j_{\mathcal{D}_X}(\mathcal{M}, \mathcal{B}_{Z|X})$ for Z a submanifold of X, or else the sheaves $\mathcal{T}or^{\mathcal{D}_X}_j(\mathcal{M}, \mathcal{D}_{X \to Y})$ in the study of direct images of (right) $\mathcal{D}_X$-modules. In these cases the characteristic variety of $\mathcal{M}$ is too rough a tool, and various microcharacteristic varieties have to be introduced.

The microcharacteristic variety of $\mathcal{M}$ along V, denoted by $C_V(\mathcal{M})$, is just the "normal cone" to $\text{char}(\mathcal{M})$ along V. It was first introduced by Bony-Schapira [2] and Kashiwara-Schapira [1] in order to study the sheaves $\mathcal{E}xt^j_{\mathcal{E}_X}(\mathcal{M}, \mathcal{L}^\infty)$, where $\mathcal{L}^\infty$ is the Module generated by $\mathcal{L}$ over $\mathcal{E}^\infty_X$, the Ring of infinite order microdifferential operators. Although we did not introduce the Ring $\mathcal{E}^\infty_X$ here, we still make a study of the variety $C_V(\mathcal{M})$. First it appears to be of some importance in many questions, such as in the study of "regular holonomic Modules". Furthermore it serves as an introduction to the study of $C^1_V(\mathcal{M})$, the 1-microcharacteristic variety of $\mathcal{M}$ along V. This last variety takes into account not only the principal symbols of the operators of the system $\mathcal{M}$, but also the lower order terms, as the so-called "Levi condition" does in the case of one operator.

When defining the characteristic variety of a differential Module $\mathcal{M}$, there are two methods. One can either endow $\mathcal{M}$ with a good filtration and take the graded associated Module, or else tensorize $\mathcal{M}$ with the Ring $\mathcal{E}_X$ on T^*X. Along an involutive manifold V the situation is similar. In order to define microcharacteristic varieties in $T_V T^*X$, one can either endow $\mathcal{M}$ with suitable filtrations and take the graded associated Modules, or else construct new Rings on $T_V T^*X$ in which microdifferential operators become invertible in the non microcharacteristic directions. We present here the first method, following T. Monteiro-Fernandes [1], [3] and refer to Y. Laurent [1], [3] for the second one.

We also introduce $\mathrm{char}_f^1(\mathcal{M})$, the "relative 1-characteristic variety" associated to a smooth map, and following Laurent-Schapira [1], we define $\hat{C}_{T_Y X}(\mathcal{M})$, the "formal microcharacteristic variety" associated to a submanifold Y of X, in order to give a bound for the characteristic variety of the systems induced by $\mathcal{M}$ on Y and to give a criterion of coherency of these induced systems.

Next we study the Cauchy problem for systems. If Y is a submanifold of X, or more generally if f is a holomorphic map from Y to X, we want to prove the isomorphisms between the sheaves $\mathcal{E}xt^j_{\mathcal{D}_X}(\mathcal{M}, \mathcal{L})|_Y$ of the solutions of $\mathcal{M}$ in $\mathcal{L}$ (in the neigborhood of Y) and the sheaves $\mathcal{E}xt^j_{\mathcal{D}_Y}(\mathcal{M}_Y, \mathcal{L}_Y)$, where $\mathcal{M}_Y$ (resp. $\mathcal{L}_Y$) is the induced system by $\mathcal{M}$ (resp. by $\mathcal{L}$) on Y. When $\mathcal{L} = \mathcal{O}_X$, this is the Cauchy-Kowalewski theorem in the formulation of M. Kashiwara [1]. When $\mathcal{L}$ corresponds to the sheaf of holomorphic functions with polar or logarithmic singularities along hypersurfaces transversal to Y, and $\mathcal{M}$ is associated to a single operator with simple characteristics, this is Hamada's theorem (Hamada [1]). Here we give a general formulation of these results, following the method of Kashiwara-Schapira [1] with the help of the microdifferential Cauchy-Kowalewski theorem of Chapter I.

By "propagation" we mean the following. Let $\mathcal{F}^{\cdot}$ be a sheaf (rather, a complex of sheaves) on a real manifold X. There is "propagation" in the direction $(x_0, \xi_0) \in T^*X$ if for each (x_1, ξ_1) sufficiently close to (x_0, ξ_0), and for any real C^1-function ϕ such that $\phi(x_1) = 0$, and $d\phi(x_1) = \xi_1$, there are no sections of $\mathcal{F}^{\cdot}$ supported by the closed set $\{x; \phi(x) \geq 0\}$ in a neighborhood of x_1, that is, $(\mathcal{H}^j_{\{\phi \geq 0\}}(\mathcal{F}^{\cdot}))_{x_1} = 0$ for all j. The micro-support of $\mathcal{F}^{\cdot}$, denoted by $SS(\mathcal{F}^{\cdot})$, is the closed conic subset of T^*X where there is no "propagation" (cf. Kashiwara-Schapira, [4]).

We prove that the micro-support of the complex of solutions of $\mathcal{M}$ in $\mathcal{L}|_{\Sigma}$ (where Σ is a bicharacteristic leaf of the involutive manifold V), is contained in the restriction to $T^*\Sigma$ of the variety $C_V^1(\mathcal{M})$. Similarly the micro-support of the complex of solutions of $\mathcal{M}$ in $\mathcal{B}_{Z|X}$ is contained in $T^*Z \cap \hat{C}_{T_Z X}(\mathcal{M})$, and the micro-support of the complex of solutions of $\mathcal{M}$ in $\mathcal{D}_{X \to Y}$ is contained in $\mathrm{char}_f^1(\mathcal{M})$ (assuming $f: X \to Y$ is smooth).

The proofs are all based on the "refined" (microdifferential) Cauchy-Kowalewski theorem, starting with a theorem on extension across non characteristic real hypersurfaces (Zerner's theorem in the case of one operator acting on holomorphic functions) and using non characteristic deformations, as in Bony-Schapira [1].

Then we recall how to obtain "weak constructibility" theorems from the knowledge of the micro-support, using Kashiwara's microlocal interpretation of Whitney's conditions. Since this study only makes use of the real structure of X, we assume X is real analytic and prove that if the micro-support $SS(\mathcal{F}^{\cdot})$ is contained in a closed conic isotropic subanalytic subset of T^*X, then there exists an $\mathbb{R}$-analytic stratification $X = \bigsqcup_{\alpha} X_\alpha$ such that the sheaves $H^j(\mathcal{F}^{\cdot})|_{X_\alpha}$ are locally constant.

When applying this result to the complex of solutions of $\mathscr{M}$ in $\mathscr{O}_X$, we obtain Kashiwara's constructibility theorem for holonomic Modules. We also obtain various generalizations of this result, including a "relative" version (using the "relative 1-characteristic variety").

At this stage, we should mention that it is possible to develop a "microlocal theory of sheaves" (Kashiwara-Schapira [4]), that is, to keep the geometrical aspect of the theory of microdifferential equations, and to ignore everything else. In other words, one may consider sheaves on real manifolds and study them from the point of view of their micro-support. Then it is possible to perform quantized contact transformations, to prove the involutivity of $SS(\mathscr{F})$, to study inverse and direct images, etc ... and many classical results on the solutions of microdifferential systems appear as particular cases of this theory. But let us emphasize that the results so obtained only concern microfunction solutions, or infinite order systems, and by no means sheaves as, for example, the sheaf of meromorphic functions, or more generally, coherent $\mathscr{E}_X$-modules, since these geometrical methods do not take into account the "growth conditions". This is also the main difference between the varieties $C_V(\mathscr{M})$ and $C_V^1(\mathscr{M})$: the first one is purely geometrical and the second one more "algebraic". (However, let us mention that in the case of "regular holonomic Modules" this difference disappears, cf. Kashiwara-Kawaï [1]). Moreover, in this new context, the Ring $\mathscr{E}_X^\infty$ is in fact not so-well suited, and should be replaced by the Ring $\mathscr{E}_X^{\mathbb{R}}$ of Sato-Kashiwara-Kawaï [1], constructed by Sato's microlocalization process. At this point, we would clearly get into another subject (for which we refer to Kashiwara-Schapira [4]), and that is the reason why we did not introduce the Ring $\mathscr{E}_X^\infty$, and have prefered to concentrate on $\mathscr{E}_X$.

§ 1. Microcharacteristic Varieties

1.1. Normal Cones

Let X be a complex manifold, Y a submanifold, $\mathscr{J}_Y$ the defining Ideal of Y in $\mathscr{O}_X$, that is, the subsheaf of sections of $\mathscr{O}_X$ vanishing on Y. We endow the ring $(\mathscr{O}_X)|_Y$ with the filtration generated by $\mathscr{J}_Y$:

(1.1.1)
$$\begin{cases} F_k^Y \mathscr{O}_X = (\mathscr{J}_Y^{-k})|_Y & k \leq 0, \\ F_k^Y \mathscr{O}_X = (\mathscr{O}_X)|_Y & k \geq 0. \end{cases}$$

We denote by $\mathscr{S}_Y$ the graded Ring:

(1.1.2)
$$\mathscr{S}_Y = \bigoplus_{k \leq 0} \mathscr{J}_Y^{-k} / \mathscr{J}_Y^{-k+1}$$

and we denote by $\sigma_Y(\cdot)$ the principal symbol map, from $(\mathscr{O}_X)|_Y$ to $\mathscr{S}_Y$.

Let $T_Y X$ be the normal bundle to Y in X, τ the projection from $T_Y X$ to Y. Let $\mathscr{O}_{[T_Y X]}$ be the sub-Ring of $\mathscr{O}_{T_Y X}$ of sections which are polynomial with respect to the fibers of the bundle. Then there is a natural isomorphism:

(1.1.3) $$\mathscr{S}_Y \cong \tau_* \mathscr{O}_{[T_Y X]}.$$

If we choose a system of local coordinates $(x_1, \ldots, x_n) = (x', x'')$ on X such that $Y = \{x; \, x'' = 0\}$, and if we denote by $(x', \langle \frac{\partial}{\partial x''}, \tilde{x}'' \rangle) = (x', \tilde{x}'')$ the associated coordinates on $T_Y X$, then we can identify $\mathscr{S}_Y$ with the Ring $\mathscr{O}_Y[\tilde{x}'']$ of polynomials in the variables $\tilde{x}''$ with coefficients in $\mathscr{O}_Y$, and if f is a section of $(\mathscr{O}_X)|_Y$, $\sigma_Y(f)$ is the lowest degree term of the Taylor expansion of f along Y:

(1.1.4) $$\begin{cases} f(x) = \sum_{|\alpha| = m} a_\alpha(x') x''^\alpha + 0(|x''|^{m+1}), \\ \sigma_Y(f)(x', \tilde{x}'') = \sum_{|\alpha| = m} a_\alpha(x') \tilde{x}''^\alpha. \end{cases}$$

Example 1.1.1. Take $X = \mathbb{C}^2$ with coordinates (x_1, x_2), $Y = \{x; \, x_2 = 0\}$, $f(x) = x_2 - x_1^2$. Then $\sigma_Y(f) = -x_1^2$.

Let $x \in Y$. The commutative ring $\mathscr{O}_{X,x}$ being noetherian, the filtered ring $F^Y \mathscr{O}_{X,x}$ is noetherian (this is the "Artin-Rees lemma", cf. Bourbaki [1]). Moreover the ring $\mathscr{O}_{X,x}$ being zariskian, the filtered ring $F^Y \mathscr{O}_{X,x}$ is zariskian. The graded Ring $\mathscr{S}_Y$ being locally isomorphic to the polynomial Ring $\mathscr{O}_Y[\tilde{x}'']$, it is graded coherent and noetherian. Thus the hypotheses of Proposition 1.4.1 of Chapter II are satisfied.

Let $\mathscr{M}$ be a coherent $\mathscr{O}_X$-module. We set:

(1.1.5) $$\mathscr{S}_Y(\mathscr{M}) = \bigoplus_{k \leq 0} \mathscr{I}_Y^{-k} \mathscr{M} / \mathscr{I}_Y^{-k+1} \mathscr{M} = \mathscr{S}_Y \otimes_{\mathscr{O}_X} \mathscr{M}.$$

Thus $\mathscr{S}_Y(\mathscr{M})$ is the graded Module associated to the good filtration $(\mathscr{I}_Y^{-k} \mathscr{M})_k$ on $\mathscr{M}$. It is a coherent graded $\mathscr{S}_Y$-module, and the $\mathscr{O}_{T_Y X}$-module obtained by tensorizing its inverse image on $T_Y X$ by $\mathscr{O}_{T_Y X}$ will be coherent.

Definition 1.1.2. Let $\mathscr{M}$ be a coherent $\mathscr{O}_X$-module. The normal cone of $\mathscr{M}$ along Y, denoted $C_Y(\mathscr{M})$, is the analytic set:

$$\text{supp}(\mathscr{O}_{T_Y X} \otimes_{\tau^{-1} \mathscr{S}_Y} \tau^{-1} \mathscr{S}_Y(\mathscr{M})).$$

For S a closed analytic subset of X, $\mathscr{I}_S$ the defining Ideal of S, one also defines the normal cone to S along Y, denoted by $C_Y(S)$, as the set $C_Y(\mathscr{O}_X / \mathscr{I}_S)$.

Proposition 1.1.3. a) *The set $C_Y(\mathscr{M})$ is a closed analytic subset of $T_Y X$, conic for the action of $\mathbb{C}^\times$ on $T_Y X$.*

b) *If* $0 \to \mathscr{L} \to \mathscr{M} \to \mathscr{N} \to 0$ *is an exact sequence of coherent* $\mathscr{O}_X$-*modules, then:*

$$C_Y(\mathscr{M}) = C_Y(\mathscr{L}) \cup C_Y(\mathscr{N}).$$

c) *Let* $\mathscr{M} = \mathscr{O}_X / \mathscr{I}$ *where* $\mathscr{I}$ *is a coherent Ideal of* $\mathscr{O}_X$. *Then:*

$$C_Y(\mathscr{M}) = \{\theta \in T_Y X;\ \sigma_Y(f)(\theta) = 0 \quad \forall f \in \mathscr{I}\}.$$

Proof. Let $\mathsf{F}^Y \mathscr{M}$ be a good filtration on the coherent $\mathscr{O}_X$-module $\mathscr{M}$. Then the graded Module $\mathsf{G}^Y \mathscr{M}$ will be $\mathscr{S}_Y$-coherent (II, Proposition 1.4.1) and the set:

$$\mathrm{supp}(\mathscr{O}_{T_Y X} \otimes_{\tau^{-1} \mathscr{S}_Y} \tau^{-1} \mathsf{G}^Y \mathscr{M})$$

will depend only on $\mathscr{M}$ by (II, Proposition 1.3.1 and Remark 2.6.3). The result now follows again from (II, Proposition 1.3.1 and Remark 2.6.3).   □

Remark 1.1.4. The set $C_Y(\mathscr{M})$ depends only on the support of $\mathscr{M}$. In fact using Proposition 1.1.3, one proves easily the equality $C_Y(\mathscr{M}) = C_Y(\mathrm{supp}\,\mathscr{M})$.

Remark 1.1.5. It is possible to give a purely geometric construction of the normal cone $C_Y(S)$, for S a subset of X, using only the real structure of the manifold X. We shall come back to this construction in § 4.

Now we apply the preceding discussion to the case of a (locally closed) conic involutive manifold V in $T^* X$. Let U be an open set in $T^* X$, V an involutive closed conic submanifold of U.

Remark first that $T_V T^* X$ is endowed with two natural actions of $\mathbb{C}^\times$, one coming from the bundle structure over V, the other from the conic structure of V.

Example 1.1.6. Let $(x) = (x', x'')$ be a system of local coordinates on X, $(x', x'';\ \xi', \xi'')$ the associated coordinates on $T^* X$, V the Lagrangean manifold: $V = \{(x', x'';\ \xi', \xi'');\ x'' = \xi' = 0\}$. Let $(x', \xi'';\ \langle \frac{\partial}{\partial x''}, \tilde{x}'' \rangle + \langle \frac{\partial}{\partial \xi'}, \tilde{\xi}' \rangle)$ $= (x', \xi'', \tilde{x}'', \tilde{\xi}')$ be the associated coordinates on $T_V T^* X$. Then the two actions of $\mathbb{C}^\times$ are described by:

$$(x', \xi'', \tilde{x}'', \tilde{\xi}') \to (x', c_1 \xi'', \tilde{x}'', c_1 \tilde{\xi}'),$$
$$(x', \xi'', \tilde{x}'', \tilde{\xi}') \to (x', \xi'', c_2 \tilde{x}'', c_2 \tilde{\xi}').$$

Let $\mathscr{M}$ be a coherent $\mathscr{E}_X$-module on U. Locally, we may endow $\mathscr{M}$ with a good filtration, and set as in (II, § 2):

$$\overline{\mathrm{gr}}(\mathscr{M}) = \mathscr{O}_{T^* X} \otimes_{\mathrm{gr}(\mathscr{E}_X)} \mathrm{gr}(\mathscr{M}).$$

Applying (II, Proposition 1.3.1), and Proposition 1.1.3, we find that the set $C_V(\overline{\mathrm{gr}}(\mathscr{M}))$ depends only on $\mathscr{M}$.

Definition 1.1.7. One sets:

$$C_V(\mathcal{M}) = C_V(\overline{\mathrm{gr}}(\mathcal{M})).$$

Then $C_V(\mathcal{M})$ is called the microcharacteristic variety of $\mathcal{M}$ along V.

Theorem 1.1.8. a) *Let $\mathcal{M}$ be a coherent $\mathcal{E}_X$-module. Then $C_V(\mathcal{M})$ is a closed analytic subset of $T_V U$, conic for both actions of $\mathbb{C}^\times$ on $T_V T^* X$.*

b) *Let $0 \to \mathcal{L} \to \mathcal{M} \to \mathcal{N} \to 0$ be an exact sequence of coherent $\mathcal{E}_X$-modules on U. Then:*

$$C_V(\mathcal{M}) = C_V(\mathcal{L}) \cup C_V(\mathcal{N}).$$

c) *Let $\mathcal{M} = \mathcal{E}_X / \mathcal{I}$, where $\mathcal{I}$ is a coherent Ideal of $\mathcal{E}_X$. Then:*

$$C_V(\mathcal{M}) = \{\theta \in T_V T^* X;\ \sigma_V(\sigma(P))(\theta) = 0 \quad \forall P \in \mathcal{I}\}.$$

Proof. Since $\mathcal{O}_{T^* X}$ is faithfully flat over the graded Ring $\mathrm{gr}(\mathcal{E}_X)$ (cf. (II, Lemma 2.2.2) for a precise statement), the results follow from the corresponding ones for $\mathcal{O}_{T^* X}$-modules, that is, from Proposition 1.1.3, except for the fact that $C_V(\mathcal{M})$ is conic for both actions of $\mathbb{C}^\times$. But the functor $C_V(\cdot)$ being additive, this follows from assertion c). □

Remark 1.1.9. Applying Remark 1.1.4 we find: $C_V(\mathcal{M}) = C_V(\mathrm{supp}(\mathcal{M}))$.

1.2. 1-Microcharacteristic Variety

As in the preceding section, we denote by V a closed conic involutive manifold in an open subset U of $T^* X$. Following M. Kashiwara and T. Oshima [1] we introduce:

Definition 1.2.1. The sub-Algebra $\mathcal{E}_V$ of $\mathcal{E}_X|_V$ is the Algebra generated over $\mathcal{E}_X(0)|_V$ by the sections P of $\mathcal{E}_X(1)|_V$ such that $\sigma_1(P)|_V = 0$.

We endow $\mathcal{E}_V$ with the induced filtration by $\mathcal{E}_X|_V$ and set:

$$\mathcal{E}_V(k) = (\mathcal{E}_X(k)|_V) \cap \mathcal{E}_V.$$

Remark that $\mathcal{E}_V(k) = \mathcal{E}_X(k)|_V$ for $k \leqslant 0$.

Let $\mathcal{J}_V$ be the defining Ideal of V in $\mathcal{O}_{T^* X}$. We set for $j \in \mathbb{Z}$:

$$\mathcal{J}_V(j) = \mathcal{J}_V \cap \mathcal{O}_{T^* X}(j)$$

and we denote by $\mathcal{J}_V^k(1)$ the k-th power of $\mathcal{J}_V(1)$, with the convention:

$$\mathcal{J}_V^k(1) = \mathcal{O}_{T^* X}(k) \quad \text{for} \quad k \leqslant 0.$$

Then we have a natural isomorphism of graded Rings on V:

$$(1.2.1) \qquad \mathrm{gr}(\mathcal{E}_V) \cong \bigoplus_{k \in \mathbb{Z}} (\mathcal{J}_V^k(1))|_V.$$

For the study of $\mathscr{E}_V$ we shall make the following important restriction which will be valid untill the end of § 1.2:

(1.2.2) $\quad \begin{cases} \text{The manifold } V \text{ is contained in } \dot{T}^*X; \text{ moreover there exists a} \\ \text{(locally closed) submanifold } V' \text{ of } P^*X \text{ such that } V = \gamma^{-1}(V'). \end{cases}$

Remark that this condition is satisfied if V is either regular involutive, or Lagrangean.

Recall that γ is the natural mapping $\dot{T}^*X \to P^*X$.

Proposition 1.2.2. a) *For each $p \in V$, the filtration on $\mathscr{E}_{V,p}$ is right and left zariskian.*

b) *The Ring $\mathscr{E}_V$ is right and left noetherian.*

c) *Let $\mathscr{M}$ be a coherent $\mathscr{E}_X$-module. Then $\mathscr{M}|_V$ is $\mathscr{E}_V$-pseudo-coherent.*

d) *Let $\mathscr{L}$ and $\mathscr{M}$ be two coherent $\mathscr{E}_X$-modules, with $\mathscr{L} \subset \mathscr{M}$, and let $\mathscr{M}_0$ be a coherent $\mathscr{E}_V$-module contained in $\mathscr{M}|_V$. Then $(\mathscr{L}|_V) \cap \mathscr{M}_0$ is $\mathscr{E}_V$-coherent.*

Proof. a), b). By the results of (II, § 1.1, § 1.4, § 2.1) we only have to prove that $\text{gr}(\mathscr{E}_V)$ is a noetherian graded Ring. Since it is generated by the two graded sub-Rings $\text{gr}(\mathscr{E}_X(0))$ and $\text{gr}^+(\mathscr{E}_V) = \bigoplus_{k \geqslant 0} \mathscr{H}_V^k(1)|_V$ and the first one is noetherian, it is enough to prove that $\text{gr}^+(\mathscr{E}_V)$ is noetherian. Let $f_1, \ldots, f_p$ be a set of generators of the coherent $\mathcal{O}_{T^*X}(0)$-module $\mathscr{H}_V(1)$, let $T_1, \ldots, T_p$ be indeterminates, and let ϕ be the surjective homomorphism of graded $\mathcal{O}_{T^*X}(0)$-algebras:

$$\mathcal{O}_{T^*X}(0)[T_1, \ldots, T_p] \xrightarrow[\phi]{} \bigoplus_{k \geqslant 0} \mathscr{H}_V^k(1),$$

which assigns f_i to T_i. We want to prove that $\text{Ker}\,\phi$ is coherent. Let ϕ_m be the restriction of ϕ to the space of polynomials of degree $\leqslant m$, and let $\mathscr{H}_m$ be the Ideal generated by $\text{Ker}\,\phi_m$ in $\mathcal{O}_{T^*X}(0)[T_1, \ldots, T_p]$. Since $\text{Ker}\,\phi_m$ is coherent over $\mathcal{O}_{T^*X}(0)$, $\mathscr{H}_m$ is coherent, and the increasing sequence $\{\mathscr{H}_m\}_m$ whose union is $\text{Ker}\,\phi$, is locally stationary since $\mathcal{O}_{T^*X}(0)[T_1, \ldots, T_p] = \mathcal{O}_{T^*X}(0) \otimes_{\mathbb{Z}} \mathbb{Z}[T_1, \ldots, T_p]$ is a noetherian Ring.

c) By (II, Proposition 1.4.2), it is enough to prove that $\mathscr{E}_V(k)$ is coherent over $\mathscr{E}_X(0)$. But $\mathscr{E}_V(k)$ is locally finitely generated, and is contained in $\mathscr{E}_X(k)$ which is $\mathscr{E}_X(0)$-coherent.

d) Consider the exact sequence

$$0 \to \mathscr{L} \to \mathscr{M} \xrightarrow[\psi]{} \mathscr{N} \to 0,$$

where we put $\mathscr{N} = \mathscr{M}/\mathscr{L}$. Set $\mathscr{L}_0 = \mathscr{M}_0 \cap \mathscr{L}$, $\mathscr{N}_0 = \psi(\mathscr{M}_0)$. Then $\mathscr{N}_0$ is $\mathscr{E}_V$-coherent by c), thus $\mathscr{L}_0$ is $\mathscr{E}_V$-coherent. This completes the proof. □

In Section 1.1 we introduced the graded Algebra:

$$\mathscr{S}_V = \bigoplus_{k \geq 0} \mathscr{J}_V^k / \mathscr{J}_V^{k+1}.$$

Now we shall introduce new Algebras, involving the homogeneity of V. We set:

(1.2.3)
$$\mathscr{S}_V(0) = \bigoplus_{k \geq 0} \mathscr{J}_V^k(0) / \mathscr{J}_V^{k+1}(0),$$

(1.2.4)
$$\mathscr{S}_V(1) = \bigoplus_{k \geq 0} \mathscr{J}_V^k(1) / \mathscr{J}_V(0) \mathscr{J}_V^k(1).$$

Let $\mathscr{O}_V(j)$ be the sheaf of holomorphic functions on V homogeneous of degree $j \in \mathbb{Z}$: $\mathscr{O}_V(j) \cong \mathscr{O}_{P^*X}(j) / \mathscr{J}_V(j)$. It follows from (II, Lemma 2.2.2) that the Ring $\mathscr{O}_V$ is faithfully flat over the graded Ring $\bigoplus_{j \in \mathbb{Z}} \mathscr{O}_V(j)$. Moreover this last Ring is faithfully flat over $\mathscr{O}_V(0)$. In fact, denoting the image of V in P^*X, by V', this Ring is locally isomorphic to $\gamma^{-1} \mathscr{O}_{V'}[T^{-1}, T]$, where T is an indeterminate, and $\mathscr{O}_V(0)$ is locally isomorphic to $\gamma^{-1} \mathscr{O}_{V'}$. Since $\mathscr{S}_V = \mathscr{O}_V \otimes_{\mathscr{O}_V(0)} \mathscr{S}_V(0)$, we find that $\mathscr{S}_V$ is faithfully flat over $\mathscr{S}_V(0)$.

Lemma 1.2.3. *The graded Ring $\mathscr{S}_V$ is faithfully flat over the graded Ring $\mathscr{S}_V(1)$.*

Proof. Let us choose an invertible section of $\mathscr{O}_V(1)$, say T, and define the isomorphism of graded Algebra $\tilde{T}$ on $\mathscr{S}_V$ as follows: for a section f of $\mathscr{J}_V^k / \mathscr{J}_V^{k+1}$ we set:

$$\tilde{T}(f) = T^k f.$$

Then $\tilde{T}$ induces an isomorphism from $\mathscr{S}_V(0)$ to $\mathscr{S}_V(1)$.   □

As a corollary of Lemma 1.2.3, we find that the graded Ring $\mathscr{S}_V(1)$ is coherent.

To a coherent graded $\mathscr{S}_V(1)$-module $\mathscr{F}$ we associate the set:

(1.2.5)
$$C_V^1(\mathscr{F}) \underset{\mathrm{def}}{=} \mathrm{supp}(\mathscr{O}_{T_V T^*X} \otimes_{\tau^{-1}\mathscr{S}_V(1)} \tau^{-1}\mathscr{F})$$

Then $C_V^1(\mathscr{F})$ is a closed analytic subset of $T_V U$, conic for both actions of $\mathbb{C}^\times$ on $T_V T^*X$, and the correspondence $\mathscr{F} \mapsto C_V^1(\mathscr{F})$ is additive.

Now we remark that $\mathscr{S}_V(1)$ may also be defined as:

(1.2.6)
$$\mathscr{S}_V(1) = \mathrm{gr}(\mathscr{E}_V) / \mathscr{O}_{T^*X}(-1) \mathrm{gr}(\mathscr{E}_V),$$

where $\mathscr{O}_{T^*X}(-1)$ acts on the graded Ring $\mathrm{gr}(\mathscr{E}_V)$ by the shift -1.

We denote by $\sigma_V^1(\cdot)$ the map from $\mathrm{gr}(\mathscr{E}_V)$ to $\mathscr{S}_V(1)$ given by (1.2.6):

Definition 1.2.4. Let P be a section of order k of $\mathscr{E}_V$ on a connected open set. Let m be the order of annihilation of $\sigma(P)$ on V. We set:

$$\begin{cases} \sigma_V^1(\sigma(P))=\sigma_V(\sigma(P)) & \text{if } k=m, \\ \sigma_V^1(\sigma(P))=0 & \text{if } k<m. \end{cases}$$

Let τ be the projection $T_V T^* X \to V$. We have now two structures of $\tau^{-1}\mathrm{gr}(\mathscr{E}_V)$-modules on $\mathscr{O}_{T_V T^* X}$, one given by $\sigma_V(\cdot)$, the other by $\sigma_V^1(\cdot)$, but they are different.

Example 1.2.5. Let $V=\{(x,\xi)\in T^*\mathbb{C}^3; \ x_3=\xi_1=\xi_2=0, \ \xi_3\neq 0\}$, and let $P= D_1^2+x_3^2 D_2 D_3$, $Q=D_1^3 D_3^{-1}+x_3 D_2^2$. Then $\sigma_V^1(\sigma(P))=\sigma_V(\sigma(P))=\tilde{\xi}_1^2$, $\sigma_V^1(\sigma(Q)) =0$, $\sigma_V(\sigma(Q))=\xi_3^{-1}\tilde{\xi}_1^3+\tilde{x}_3\tilde{\xi}_2^2$.

Let $\mathscr{M}$ be a coherent $\mathscr{E}_V$-module endowed with a good filtration. Then the set $C_V^1(\mathscr{S}_V(1)\otimes_{\mathrm{gr}(\mathscr{E}_V)}\mathrm{gr}(\mathscr{M}))$ depends only on $\mathscr{M}$, not on the choice of the good filtration. In fact this set is determined by the radical of the annihilator of $\mathscr{S}_V\otimes_{\mathscr{S}_V(1)}\mathscr{S}_V(1)\otimes_{\mathrm{gr}(\mathscr{E}_V)}\mathrm{gr}(\mathscr{M})$ by (II, Remark 2.6.3 c), thus by the radical of the annihilator of $\mathscr{S}_V(1)\otimes_{\mathrm{gr}(\mathscr{E}_V)}\mathrm{gr}(\mathscr{M})$, by Lemma 1.2.3, and the correspondence which associates this Ideal to $\mathrm{gr}(\mathscr{M})$ is additive by (II, Remark 1.3.6), hence this set depends only on $\mathscr{M}$ by (II, Proposition 1.3.1).

Definition 1.2.6. Let $\mathscr{M}$ be a coherent $\mathscr{E}_V$-module. One sets:

$$C_V^1(\mathscr{M})=\mathrm{supp}(\mathscr{O}_{T_V T^* X}\otimes_{\tau^{-1}\mathscr{S}_V(1)}\tau^{-1}(\mathscr{S}_V(1)\otimes_{\mathrm{gr}(\mathscr{E}_V)}\mathrm{gr}(\mathscr{M}))),$$

where $\mathrm{gr}(\mathscr{M})$ is the graded Module associated to a good filtration on $\mathscr{M}$.

It is clear that $C_V^1(\mathscr{M})$ is a closed analytic set of $T_V U$, conic for both action of $\mathbb{C}^\times$, and the correspondence $C_V^1(\cdot)$, which associates $C_V^1(\mathscr{M})$ to a coherent $\mathscr{E}_V$-module $\mathscr{M}$, is additive.

Now we consider a coherent $\mathscr{E}_X$-module $\mathscr{M}$ on U. Locally on U, $\mathscr{M}$ is finitely generated, and there exists a coherent $\mathscr{E}_V$-sub-module $\mathscr{M}_0\subset\mathscr{M}|_V$ which generates $\mathscr{M}|_V$ (Proposition 1.2.2).

Lemma 1.2.7. The set $C_V^1(\mathscr{M}_0)$ depends only on $\mathscr{M}$.

Proof. Let $\mathscr{M}_0'$ be another coherent $\mathscr{E}_V$-module which generates $\mathscr{M}|_V$. Set $\mathscr{M}_k=\mathscr{E}_X(k)\mathscr{M}_0$. Since we have:

(1.2.7) $$\mathscr{E}_X(k)\mathscr{E}_V=\mathscr{E}_V\mathscr{E}_X(k)$$

the $\mathscr{E}_X(0)$-module $\mathscr{M}_k$ is naturally a left $\mathscr{E}_V$-module, and it is locally finitely generated, thus $\mathscr{E}_V$-coherent. Locally there exists an integer $r\geq 0$ such that $\mathscr{M}_{-r}\subset\mathscr{M}_0'\subset\mathscr{M}_r$. By the additivity of $C_V^1(\cdot)$, it is enough to prove:

(1.2.8) $$C_V^1(\mathscr{M}_0)=C_V^1(\mathscr{M}_k).$$

The natural $\mathscr{E}_V$-linear map:

$$\mathscr{E}_X(k)\mathscr{E}_V\otimes_{\mathscr{E}_V}\mathscr{M}_0\to\mathscr{M}_k$$

is an isomorphism, since if we choose an invertible operator T of order k in $\mathscr{E}_X$ (this is always possible locally on $\dot{T}^*X$), any section of $\mathscr{E}_X(k)\mathscr{E}_V$ may be written as $T\circ P$, with P in $\mathscr{E}_V$. Let us endow $\mathscr{M}_0$ with a good $\mathscr{E}_V$-filtration, and $\mathscr{E}_X(k)\mathscr{E}_V$ with the filtration $(\mathscr{E}_X(k)\mathscr{E}_V)_l=T\circ\mathscr{E}_V(l)$. This last filtration is right and left free over $\mathscr{E}_V$, since $T\circ\mathscr{E}_V(l)=\mathscr{E}_V(l)\circ T$, and we have:

$$\mathrm{gr}(\mathscr{E}_X(k)\mathscr{E}_V\otimes_{\mathscr{E}_V}\mathscr{M}_0)=\mathrm{gr}(\mathscr{E}_X(k)\mathscr{E}_V)\otimes_{\mathrm{gr}(\mathscr{E}_V)}\mathrm{gr}(\mathscr{M}_0).$$

Since $\mathrm{gr}(\mathscr{E}_X(k)\mathscr{E}_V)$ is free of rank one over $\mathrm{gr}(\mathscr{E}_V)$ we get the result. $\quad\square$

Definition 1.2.8. Let $\mathscr{M}$ be a coherent $\mathscr{E}_X$-module on $U\subset\dot{T}^*X$. The 1-microcharacteristic variety of $\mathscr{M}$ along V, $C_V^1(\mathscr{M})$, is the set $C_V^1(\mathscr{M}_0)$, where $\mathscr{M}_0$ is a locally defined coherent sub-$\mathscr{E}_V$-module of $\mathscr{M}|_V$ which generates it.

Theorem 1.2.9. a) *Let $\mathscr{M}$ be a coherent $\mathscr{E}_X$-module. Then $C_V^1(\mathscr{M})$ is a closed analytic subset of $T_V U$, conic for both actions of $\mathbb{C}^\times$ on $T_V T^*X$.*

b) *Let $0\to\mathscr{L}\to\mathscr{M}\to\mathscr{N}\to0$ be an exact sequence of coherent $\mathscr{E}_X$-modules. Then:*

$$C_V^1(\mathscr{M})=C_V^1(\mathscr{L})\cup C_V^1(\mathscr{N}).$$

c) *Let $\mathscr{M}=\mathscr{E}_X/\mathscr{I}$, where $\mathscr{I}$ is a coherent Ideal. Then:*

$$C_V^1(\mathscr{M})=\{\theta\in T_V T^*X;\ \sigma_V^1(\sigma(P))(\theta)=0\quad\forall P\in\mathscr{I}\cap\mathscr{E}_V\}.$$

d) *We have the inclusion:*

$$C_V(\mathscr{M})\subset C_V^1(\mathscr{M}).$$

Proof. a) Let $\mathscr{M}_0$ be a coherent $\mathscr{E}_V$-module which generates $\mathscr{M}|_V$. Then $C_V^1(\mathscr{M}_0)$ is analytic by Definition 1.2.6. The fact that $C_V^1(\mathscr{M})$ is conic for both action of $\mathbb{C}^\times$ will follow from b) and c).

b) Let $\mathscr{M}_0$ be a coherent $\mathscr{E}_V$-module which generates $\mathscr{M}|_V$. Replacing $\mathscr{M}_0$ by $\mathscr{E}_X(k)\mathscr{M}_0$ for $k\gg0$, we may assume (locally) that $\mathscr{M}_0\cap\mathscr{L}=\mathscr{L}_0$ generates $\mathscr{L}|_V$. Applying Proposition 1.2.2 we find that $\mathscr{L}_0$ is $\mathscr{E}_V$-coherent. Let $\mathscr{N}_0$ be the image of $\mathscr{M}_0$ in $\mathscr{N}$. Then we have the exact sequence of coherent $\mathscr{E}_V$-modules:

$$0\to\mathscr{L}_0\to\mathscr{M}_0\to\mathscr{N}_0\to0$$

and as already mentioned, we get $C_V^1(\mathscr{M}_0)=C_V^1(\mathscr{L}_0)\cup C_V^1(\mathscr{N}_0)$ by applying (II, Remark 1.3.6).

c) Put $\mathscr{I}_V=\mathscr{E}_V\cap\mathscr{I}$, $\mathscr{M}_0=\mathscr{E}_V/\mathscr{I}_V$. Then $\mathscr{I}_V$ is a coherent $\mathscr{E}_V$-module by Proposition 1.2.2. We endow $\mathscr{M}_0$ with the filtration image of that of $\mathscr{E}_V$. Then $\mathrm{gr}(\mathscr{M}_0)=\mathrm{gr}(\mathscr{E}_V)/\mathrm{gr}(\mathscr{I}_V)$, and we have:

$$\mathscr{S}_V(1)\otimes_{\mathrm{gr}(\mathscr{E}_V)}\mathrm{gr}(\mathscr{M}_0)\cong\mathscr{S}_V(1)/\mathscr{K}_V,$$

where $\mathcal{H}_V$ is the image of $\mathrm{gr}(\mathcal{H}_V)$ in $\mathcal{S}_V(1)$, that is $\mathcal{H}_V$ is generated by $\{\sigma_V^1\sigma(P);\ P\in\mathcal{E}_V\cap\mathcal{I}\}$. Applying Lemma 1.2.3 and (II, Remark 2.6.3), we get the result.

d) Follows from Theorem 1.1.8 c), Theorem 1.2.9 c) and the additivity of $C_V(\cdot)$ and $C_V^1(\cdot)$. □

Remark 1.2.10. It is of course possible to define $\sigma_V^1(\sigma(P))$ as in Definition 1.2.4 without assuming the conic involutive manifold V satisfies hypothesis (1.2.2).

1.3. Characteristic Varieties Associated to a Submersion

Let f be a *smooth* (i.e. *df* is surjective at each point $x\in X$) holomorphic surjective map from X to Y. We identify $X\times_Y T^*Y$ with a sub-bundle of T^*X, and define the relative cotangent bundle $T^*(X/Y)$, and the projection $\tilde{f}$, by the exact sequence of bundles over X (cf. (II, Example 1.5.4)):

$$(1.3.1)\qquad 0\to X\times_Y T^*Y\to T^*X \xrightarrow{\tilde{f}} T^*(X/Y)\to 0.$$

We have already introduced the Ring $\mathcal{D}_{X/Y}$ on X, of relative differential operators, generated over $\mathcal{O}_X$ by the vector fields tangent to the leaves of f. In other words:

$$(1.3.2)\qquad \mathcal{D}_{X/Y}=\{P\in\mathcal{D}_X;\ [P, f^{-1}\mathcal{O}_Y]=0\},$$

We endow $\mathcal{D}_{X/Y}$ with the filtration induced from $\mathcal{D}_X$. Then the graded Ring $\mathrm{gr}(\mathcal{D}_{X/Y})$ may be identified with $\pi_*\mathcal{O}_{[T^*(X/Y)]}$, the sub-Ring of $\pi_*\mathcal{O}_{T^*(X/Y)}$ of sections which are polynomials in the fiber of π. Locally on X, f is isomorphic to a projection: $f: Z\times Y\to Y$, and $\mathcal{D}_{X/Y}$ is then isomorphic to the Ring $\mathcal{O}_X\otimes_{g^{-1}\mathcal{O}_Z}g^{-1}\mathcal{D}_Z$, where g is the projection on Z. Using exactly the same arguments as for $\mathcal{D}_X$ and $\mathcal{O}_X$, we get:

Proposition 1.3.1. a) *$\forall x\in X$, the filtered Ring $\mathcal{D}_{X/Y,x}$ is right and left zariskian.*

b) *The Ring $\mathcal{D}_{X/Y}$ is right and left noetherian (hence coherent).*

c) *Let $\mathcal{M}$ be a coherent $\mathcal{D}_X$-module. Then $\mathcal{M}$ is $\mathcal{D}_{X/Y}$-pseudo-coherent.*

d) *Let $\mathcal{L}$ and $\mathcal{M}$ be two coherent $\mathcal{D}_X$-modules with $\mathcal{L}\subset\mathcal{M}$, and $\mathcal{M}_0\subset\mathcal{M}$ a coherent $\mathcal{D}_{X/Y}$-module. Then $\mathcal{L}\cap\mathcal{M}_0$ is $\mathcal{D}_{X/Y}$-coherent.*

c) *The Ring $\mathcal{D}_X$ is flat over $\mathcal{D}_{X/Y}$.*

Let $\mathcal{M}$ be a coherent $\mathcal{D}_{X/Y}$-module. Proceeding as for $\mathcal{D}_X$-modules, we may locally endow $\mathcal{M}$ with a good filtration and define the set in $T^*(X/Y)$:

$$(1.3.3)\qquad \mathrm{char}_{X/Y}(\mathcal{M})\underset{\mathrm{def}}{=}\mathrm{supp}(\mathcal{O}_{T^*(X/Y)}\otimes_{\pi^{-1}\mathrm{gr}(\mathcal{D}_{X/Y})}\pi^{-1}\mathrm{gr}(\mathcal{M})).$$

This is a closed conic analytic subset of $T^*(X/Y)$, which depends only on $\mathcal{M}$, and the correspondence $\mathcal{M} \mapsto \operatorname{char}_{X/Y}(\mathcal{M})$ is additive with respect to exact sequences of coherent $\mathcal{D}_{X/Y}$-modules.

Proposition 1.3.2. *Let $\mathcal{M}$ be a coherent $\mathcal{D}_{X/Y}$-module. Then:*

$$\operatorname{char}_{X/Y}(\mathcal{M}) = \tilde{f}(\operatorname{char}(\mathcal{D}_X \otimes_{\mathcal{D}_{X/Y}} \mathcal{M})).$$

Proof. Let us endow $\mathcal{M}$ with a good filtration. Then if we use the fact that $\mathcal{O}_{T^*X}$ is faithfully flat over $\tilde{f}^{-1}\mathcal{O}_{T^*(X/Y)}$ we obtain:

$$\tilde{f}^{-1}\operatorname{char}_{X/Y}(\mathcal{M}) = \operatorname{supp}(\mathcal{O}_{T^*X} \otimes_{\pi^{-1}(\mathcal{D}_{X/Y})} \pi^{-1}\operatorname{gr}(\mathcal{M})).$$

But $\operatorname{gr}(\mathcal{D}_X)$ is flat over $\operatorname{gr}(\mathcal{D}_{X/Y})$, and we have (II, Proposition 1.2.4):

$$\operatorname{gr}(\mathcal{D}_X) \otimes_{\operatorname{gr}(\mathcal{D}_{X/Y})} \operatorname{gr}(\mathcal{M}) - \operatorname{gr}(\mathcal{D}_X \otimes_{\mathcal{D}_{X/Y}} \mathcal{M}).$$

Thus:
$$\tilde{f}^{-1}\operatorname{char}_{X/Y}(\mathcal{M}) = \operatorname{supp}(\mathcal{O}_{T^*X} \otimes_{\pi^{-1}\operatorname{gr}(\mathcal{D}_X)} \pi^{-1}\operatorname{gr}(\mathcal{D}_X \otimes_{\mathcal{D}_{X/Y}} \mathcal{M}))$$
$$= \operatorname{char}(\mathcal{D}_X \otimes_{\mathcal{D}_{X/Y}} \mathcal{M}). \quad \square$$

Lemma 1.3.3. *Let $\mathcal{M}$ be a coherent $\mathcal{D}_X$-module, $\mathcal{M}_0 \subset \mathcal{M}$ a coherent $\mathcal{D}_{X/Y}$-sub-module which generates $\mathcal{M}$. Then $\operatorname{char}_{X/Y}(\mathcal{M}_0)$ depends only on $\mathcal{M}$.*

Proof. First we remark that:

(1.3.4) $$\mathcal{D}_X(k) \mathcal{D}_{X/Y} = \mathcal{D}_{X/Y} \mathcal{D}_X(k)$$

and that the Module $\mathcal{D}_X(k) \mathcal{D}_{X/Y}$ is locally free over $\mathcal{D}_{X/Y}$ as a right Module and as a left Module. Moreover for each k the filtration:

(1.3.5) $$(\mathcal{D}_X(k) \mathcal{D}_{X/Y})_j = (\mathcal{D}_X(k) \mathcal{D}_{X/Y}(j)) = (\mathcal{D}_{X/Y}(j) \mathcal{D}_X(k))$$

is both right and left finite free over $\mathcal{D}_{X/Y}$.

Now put $\mathcal{M}_k = \mathcal{D}_X(k) \mathcal{M}_0$. This is a left $\mathcal{D}_{X/Y}$-module by (1.3.4), and as in the proof of Lemma 1.2.7 it is enough to check that $\operatorname{char}_{X/Y}(\mathcal{M}_0) = \operatorname{char}_{X/Y}(\mathcal{M}_k)$.

We endow $\mathcal{M}_0$ we a good filtration. Then:

$$\operatorname{gr}(\mathcal{D}_X(k) \mathcal{D}_{X/Y} \otimes_{\mathcal{D}_{X/Y}} \mathcal{M}_0) = \operatorname{gr}(\mathcal{D}_X(k) \mathcal{D}_{X/Y}) \otimes_{\operatorname{gr}(\mathcal{D}_{X/Y})} \operatorname{gr}(\mathcal{M}_0).$$

Since $\operatorname{gr}(\mathcal{D}_X(k) \mathcal{D}_{X/Y})$ is finite free over $\operatorname{gr}(\mathcal{D}_{X/Y})$, we obtain:

$$\operatorname{char}_{X/Y}(\mathcal{M}_0) = \operatorname{char}_{X/Y}(\mathcal{D}_X(k) \mathcal{D}_{X/Y} \otimes_{\mathcal{D}_{X/Y}} \mathcal{M}_0).$$

Finally we have a surjective $\mathcal{D}_{X/Y}$-linear morphism:

$$\mathcal{D}_X(k) \mathcal{D}_{X/Y} \mathcal{M}_0 \to \mathcal{M}_k$$

and we also have the inclusion:

$$\mathcal{M}_0 \subset \mathcal{M}_k,$$

which completes the proof. $\quad \square$

Definition 1.3.4. Let $\mathcal{M}$ be a coherent $\mathcal{D}_X$-module. The relative 1-characteristic variety of $\mathcal{M}$ over Y, $\mathrm{char}_f^1(\mathcal{M})$, is the variety $\mathrm{char}_{X/Y}(\mathcal{M}_0)$, where $\mathcal{M}_0$ is a locally defined coherent $\mathcal{D}_{X/Y}$-sub-module of $\mathcal{M}$ which generates $\mathcal{M}$.

Using the same arguments as for the proofs of Theorem 1.1.8 or 1.2.9, we get:

Theorem 1.3.5. a) *Let $\mathcal{M}$ be a coherent $\mathcal{D}_X$-module. Then $\mathrm{char}_f^1(\mathcal{M})$ is a closed conic analytic subset of $T^*(X/Y)$.*

b) *if $0 \to \mathcal{L} \to \mathcal{M} \to \mathcal{N} \to 0$ is an exact sequence of coherent $\mathcal{D}_X$-modules, we have:*

$$\mathrm{char}_f^1(\mathcal{M}) = \mathrm{char}_f^1(\mathcal{L}) \cup \mathrm{char}_f^1(\mathcal{N}).$$

c) *Let $\mathcal{M} = \mathcal{D}_X / \mathcal{I}$, where $\mathcal{I}$ is a coherent Ideal of $\mathcal{D}_X$. Then:*

$$\mathrm{char}_f^1(\mathcal{M}) = \{\theta \in T^*(X/Y); \ \sigma(P)(\theta) = 0 \quad \forall P \in \mathcal{I} \cap \mathcal{D}_{X/Y}\}.$$

d) *We have the inclusion:*

$$\tilde{f}(\mathrm{char}(\mathcal{M})) \subset \mathrm{char}_f^1(\mathcal{M}),$$

(The last statement follows immediately from Proposition 1.3.2).

Remark 1.3.6. The set $\tilde{f}(\mathrm{char}(\mathcal{M}))$ is a closed conic analytic subset of $T^*(X/Y)$, since it is the image by a linear map of a closed conic analytic set of T^*X. Moreover the correspondence $\mathcal{M} \mapsto \tilde{f}(\mathrm{char}(\mathcal{M}))$ is additive. We may call this set the "relative characteristic variety of $\mathcal{M}$ along Y", and denote it by $\mathrm{char}_f(\mathcal{M})$. If $\mathcal{M} = \mathcal{D}_X / \mathcal{I}$, we have:

$$\mathrm{char}_f(\mathcal{M}) = \{\theta \in T^*(X/Y); \ g(\theta) = 0 \quad \forall g \in \mathrm{gr}(\mathcal{I}) \cap \mathrm{gr}(\mathcal{D}_{X/Y})\}.$$

Exemples 1.3.7. a) Take $Y = \{p'\}$. Then $T^*(X/Y) = T^*X$, $\mathcal{D}_{X/Y} = \mathcal{D}_X$, and $\mathrm{char}_f^1(\mathcal{M}) = \mathrm{char}_f(\mathcal{M}) = \mathrm{char}(\mathcal{M})$.

b) Take $Y = X$. Then $T^*(X/X) = T_X^* X$, $\mathcal{D}_{X/X} = \mathcal{O}_X$, $\mathrm{char}_f(\mathcal{M}) = \mathrm{char}_f^1(\mathcal{M}) = \mathrm{char}(\mathcal{M}) \cap T_X^* X$.

c) Let $X = \mathbb{C}^2$ with coordinates (y, t), $Y = \mathbb{C}$, $f(y, t) = y$. Let $\mathcal{M}_i = \mathcal{D}_X / \mathcal{D}_X P_i$, $i = 1, 2, 3$, where:

$$P_1 = t D_t^2 + D_y, \quad P_2 = t D_t^2 + y, \quad P_3 = t D_t^2 + D_y^2.$$

Let (y, t, τ) be the coordinates on $T^*(X/Y)$. Then:

$$\mathrm{char}_f(\mathcal{M}_1) = \mathrm{char}_f(\mathcal{M}_2) = \mathrm{char}_f^1(\mathcal{M}_2) = \{(y, t, \tau); \ t\tau = 0\},$$
$$\mathrm{char}_f^1(\mathcal{M}_1) = \mathrm{char}_f(\mathcal{M}_3) = \mathrm{char}_f^1(\mathcal{M}_3) = T^*(X/Y).$$

1.4. Microcharacteristic Varieties Associated to an Embedding

Let $\Lambda \xrightarrow{\tau} Y$ be a complex vector bundle over the manifold Y. We have already introduced the subsheaf $\mathcal{O}_{[\Lambda]}$ of $\mathcal{O}_\Lambda$ of sections which are polynomials in the fiber of Λ.

Now we introduce $\mathcal{D}_{[\Lambda]}$, the sub-Ring of $\mathcal{D}_\Lambda$, of differential operators with coefficients in $\mathcal{O}_{[\Lambda]}$.

If we choose local coordinates $(y, z) = (y_1, \ldots, y_q; z_1, \ldots, z_p)$, where (z) are the linear coordinates on the fiber, then a section P of $\mathcal{D}_{[\Lambda]}$ is written as a finite sum:

$$(1.4.1) \qquad P = \sum_{\alpha, \beta, \gamma} a_{\alpha, \beta, \gamma}(y) z^\alpha D_z^\beta D_y^\gamma,$$

where $\alpha \in \mathbb{N}^p$, $\beta \in \mathbb{N}^p$, $\gamma \in \mathbb{N}^q$ and $a_{\alpha, \beta, \gamma}$ is a section of $\mathcal{O}_Y$.

Let e_Λ be the Euler vector field on Λ. In local coordinates (y, z), e_Λ is written as:

$$(1.4.2) \qquad e_\Lambda = \sum_{i=1}^p z_i \frac{\partial}{\partial z_i},$$

but e_Λ is intrinsically defined by the condition that $e_\Lambda(f) = kf$, for $f \in \mathcal{J}_Y^k / \mathcal{J}_Y^{k+1}$, $\mathcal{J}_Y$ denoting the defining Ideal of Y, the zero section of Λ. We set:

$$(1.4.3) \qquad \mathcal{D}_{\Lambda, k} = \{P \in \mathcal{D}_{[\Lambda]}; \ [e_\Lambda, P] = -kP\}.$$

Then $\mathcal{D}_{\Lambda, k}$ is the subsheaf of $\mathcal{D}_{[\Lambda]}$ of sections which can be written in the coordinates (y, z) as (1.4.1), with $|\beta| - |\alpha| = k$. The Ring $\mathcal{D}_{[\Lambda]}$ is graded by:

$$(1.4.4) \qquad \mathcal{D}_{[\Lambda]} = \bigoplus_{k \in \mathbb{Z}} \mathcal{D}_{\Lambda, k}.$$

Now let Y be a submanifold of codimension p of the manifold X, and let $\Lambda = T_Y X$, be the normal bundle to Y in X.

Let $\mathcal{J}_Y$ be the defining Ideal of Y in $\mathcal{O}_X$. Following M. Kashiwara [6], we have already introduced (II, Example 1.5.5), the filtration $\mathsf{F}^Y \mathcal{D}_X$ on $(\mathcal{D}_X)|_Y$ by setting:

$$(1.4.5) \qquad \mathsf{F}_k^Y \mathcal{D}_X = \{P \in (\mathcal{D}_X)|_Y; \ P\mathcal{J}_Y^l \subset \mathcal{J}_Y^{l-k} \quad \forall l\}.$$

Let $\mathsf{G}^Y \mathcal{D}_X$ be the graded Ring associated to this filtration.

Lemma 1.4.1. *The two graded Rings $\mathsf{G}^Y \mathcal{D}_X$ and $\tau_* \mathcal{D}_{[\Lambda]}$ are naturally isomorphic.*

Proof. Let P be a section of $\mathsf{F}^Y \mathcal{D}_X$, $\hat{\sigma}_Y(P)$ its principal symbol in $\mathsf{G}^Y \mathcal{D}_X$, say of order k. For each l, $\hat{\sigma}_Y(P)$ defines a morphism from $\mathcal{J}_Y^l / \mathcal{J}_Y^{l+1}$ to $\mathcal{J}_Y^{l-k} / \mathcal{J}_Y^{l-k+1}$, and thus a section of $\mathcal{H}om_{\mathbb{C}}(\mathcal{O}_{[\Lambda]}, \mathcal{O}_{[\Lambda]})$. One verifies that the morphism on $\tau_* \mathcal{O}_{[A]}$ so defined is in fact a homogeneous differential operator by choosing local coordinates (y, t) on X, with $Y = \{(y, t); t = 0\}$. At the same time we see that this map from $G^Y \mathcal{D}_X$ to $\tau_* \mathcal{D}_{[A]}$ is an isomorphism. $\square$

Definition 1.4.2. We denote by $\hat{\sigma}_Y(\cdot)$ the principal symbol map from $F^Y \mathcal{D}_X$ to $\tau_* \mathcal{D}_{[A]}$.

Remark that for each $\lambda \in A$, the ring $\mathcal{D}_{[A], \lambda}$ is noetherian, thus graded noetherian (II, Remark 1.1.10): the proof is the same as for $\mathcal{D}_{X,x}$. One sees also that $\mathcal{D}_{[A]}$ is coherent, and $\tau_* \mathcal{D}_{[A]}$ is graded coherent. In fact the graded Ring $\tau_* \mathcal{D}_{[A]}$ is locally isomorphic to $\mathcal{D}_Y \otimes W_p$, where W_p is the Weyl algebra over $\mathbb{C}^p$, the graduation on $\mathcal{D}_Y \otimes W_p$ being defined by the graduation on W_p.

Proposition 1.4.3. a) *Let $x \in Y$. Then the filtered ring $F^Y \mathcal{D}_{X,x}$ is noetherian.*

b) *Let $\mathcal{M}$ be a coherent $\mathcal{D}_X$-module endowed with a good $F^Y \mathcal{D}_X$-filtration. Then $\mathrm{gr}(\mathcal{M})$ is $G^Y \mathcal{D}_X$-coherent.*

Proof. a) Set $\mathcal{A}_k = F_k^Y \mathcal{D}_X$. Applying (II, Proposition 1.1.8), we only have to prove that $\mathcal{A}_{0,x}$ is filtered noetherian. Applying (II, Proposition 1.1.7), we only have to prove that the associated formal graded Ring:

$$S_x = \bigoplus_{k \geq 0} \mathcal{I}_Y^k \mathcal{A}_{0,x}$$

is noetherian.

We may endow S_x with the standard filtration, inherited from the standard filtration on $\mathcal{D}_X$. The graded ring $\mathrm{gr}(S_x)$ is given by:

$$\mathrm{gr}(S_x) = \bigoplus_{k \geq 0} \mathcal{I}_{Y,x}^k \, \mathrm{gr}(\mathcal{A}_{0,x}),$$

where now $\mathcal{A}_{0,x}$ is endowed with the standard filtration induced by $\mathcal{D}_{X,x}$.

Let B be the algebra:

(1.4.6) $$B = \mathbb{Z}[\eta, z, \xi]/J,$$

where $\eta = (\eta_i)_{1 \leq i \leq q}$, $z = (z_i)_{1 \leq i \leq p}$, $\xi = (\xi_{i,j})_{1 \leq i,j \leq p}$, and J is the ideal generated by $(z_i \xi_{jk} - z_j \xi_{ik})$ for $1 \leq i, j, k \leq p$. Then:

(1.4.7) $$\mathrm{gr}(\mathcal{A}_0) \cong \mathcal{O}_Y \otimes_{\mathbb{Z}} B,$$

hence $\mathrm{gr}(\mathcal{A}_{0,x})$ is noetherian, and $\mathrm{gr}(S_x)$ being the formal graded ring associated to the ideal $\mathcal{I}_{Y,x}$ of the commutative noetherian ring $\mathrm{gr}(\mathcal{A}_{0,x})$, is itself noetherian. Finally we see that S_x is noetherian by applying (II, Proposition 1.1.5).

b) We shall apply Corollary 1.4.4 of Chapter II. Thus we only have to check that the hypotheses of Proposition 1.4.3 are satisfied for $\mathcal{A}_k = F_k^Y \mathcal{D}_X$. We have already noticed that $G \mathcal{A} = \tau_* \mathcal{D}_{[A]}$ is graded coherent. Let us prove that $\mathcal{A}_0$ is coherent. For that purpose we endow $\mathcal{A}_0$ with the usual filtration induced by $\mathscr{D}_X$ and apply (II, Proposition 1.4.1). Since this filtration is zariskian, the coherency of $\mathscr{A}_0$ follows from the coherency of $\mathrm{gr}(\mathscr{A}_0) \cong \mathscr{O}_Y \otimes_{\mathbf{Z}} \mathsf{B}$.

Now each $\mathscr{A}_k$ is obviously locally finitely generated over $\mathscr{A}_0$, hence $\mathscr{A}_0/\mathscr{A}_{-l}(l \geqslant 0)$ is coherent. Finally let $l \geqslant 0$, and let us denote by Z the ideal generated by $(z_1, \ldots, z_p)$ in B. We have:

$$\mathrm{gr}(\mathscr{A}_0/\mathscr{A}_{-l}) \cong \mathscr{O}_Y \otimes_{\mathbf{Z}} \mathsf{B}/\mathsf{Z}^l.$$

Let $k \in \mathbf{Z}$, $m \geqslant 0$, $m \leqslant l$. Then we may find a finitely generated B/Z^l-module M such that:

$$\mathrm{gr}(\mathscr{A}_k/\mathscr{A}_{k-m}) \cong \mathscr{O}_Y \otimes_{\mathbf{Z}} \mathsf{M}.$$

Since $\mathscr{O}_Y$ is a noetherian Ring, we obtain that $\mathrm{gr}(\mathscr{A}_k/\mathscr{A}_{k-m})$ is coherent over $\mathrm{gr}(\mathscr{A}_0/\mathscr{A}_{-l})$. Applying (II, Proposition 1.4.1) we find that $\mathscr{A}_k/\mathscr{A}_{k-m}$ is coherent over $\mathscr{A}_0/\mathscr{A}_{-l}$, which completes the proof. $\square$

The Ring $\mathscr{D}_\Lambda$ is flat over $\mathscr{D}_{[\Lambda]}$. Thus the functor which associates to a coherent graded $\mathscr{D}_{[\Lambda]}$-module $\mathscr{F}$ the set $\mathrm{char}(\mathscr{D}_\Lambda \otimes_{\mathscr{D}_{[\Lambda]}} \mathscr{F})$, is additive. Applying (II, Proposition 1.3.1) we find that if $\mathscr{M}$ is a coherent $\mathscr{D}_X$-module endowed with a good $F^Y \mathscr{D}_X$-filtration, for each $x \in Y$, the germ at x of the set $\mathrm{char}(\mathscr{D}_\Lambda \otimes_{\tau^{-1}\mathscr{D}_{[\Lambda]}} \tau^{-1} \mathrm{gr}(\mathscr{M}))$, depends only on $\mathscr{M}$.

Definition 1.4.4. Let $\mathscr{M}$ be a coherent $\mathscr{D}_X$-module. We define the "formal microcharacteristic variety of $\mathscr{M}$ along Y", denoted by $\hat{C}_{T_Y X}(\mathscr{M})$, as the set $\mathrm{char}(\mathscr{D}_\Lambda \otimes_{\tau^{-1}\mathscr{D}_{[\Lambda]}} \tau^{-1} \mathrm{gr}(\mathscr{M}))$ where $\mathrm{gr}(\mathscr{M})$ is the graded Module associated to a (locally defined) good $F^Y \mathscr{D}_X$-filtration.

We summarize:

Theorem 1.4.5. a) *Let $\mathscr{M}$ be a coherent $\mathscr{D}_X$-module. Then $\hat{C}_{T_Y X}(\mathscr{M})$ is a closed analytic subset of $T^* T_Y X$, conic for both actions of $\mathbf{C}^\times$ on this space.*

b) *Let $0 \to \mathscr{L} \to \mathscr{M} \to \mathscr{N} \to 0$ be an exact sequence of coherent $\mathscr{D}_X$-modules. Then:*

$$\hat{C}_{T_Y X}(\mathscr{M}) = \hat{C}_{T_Y X}(\mathscr{L}) \cup \hat{C}_{T_Y X}(\mathscr{N}).$$

c) *Let $\mathscr{M} = \mathscr{D}_X/\mathscr{I}$ for a coherent Ideal $\mathscr{I}$. Then:*

$$\hat{C}_{T_Y X}(\mathscr{M}) = \{\theta \in T^* T_Y X; \ \sigma(\hat{\sigma}_Y(P))(\theta) = 0 \quad \forall P \in \mathscr{I}\}.$$

Remark 1.4.6. The symplectic structure on $T^* X$ defines an isomorphism H from $T^* T^* X$ to $T T^* X$ (cf. Appendix A). Since $T_Y^* X$ is a Lagrangean submanifold of $T^* X$, H induces an isomorphism:

(1.4.8) $$T^* T_Y^* X \simeq T_{T_Y^* X} T^* X.$$

Moreover there exists a natural isomorphism:

$$(1.4.9) \qquad T^* T_Y X \simeq T^* T_Y^* X.$$

If (y, z) are local coordinates on X such that $Y = \{(y, z); z = 0\}$, and if $(y, \langle \tilde{z}, \frac{\partial}{\partial z} \rangle)$ and $(y, \langle \zeta, dz \rangle)$ are the associated coordinates on $T_Y X$ and $T_Y^* X$ respectively, then the isomorphism (1.4.9) is described by:

$$(y, \tilde{z}; \langle \eta, dy \rangle - \langle \zeta, d\tilde{z} \rangle) \to (y, \zeta; \langle \eta, dy \rangle + \langle \tilde{z}, d\zeta \rangle).$$

(We shall not prove here that the isomorphism (1.4.9) is intrinsically defined, cf. Kashiwara-Schapira [4]).

Let $\mathcal{M}$ be a coherent $\mathcal{D}_X$-module. To $\mathcal{M}$ are now naturally associated three analytic sets in $T^* T_Y X$: the formal microcharacteristic variety $\hat{C}_{T_Y X}(\mathcal{M})$, and the microcharacteristic varieties of $\mathcal{M}$ along $T_Y^* X$.

Let us take an example.

Let (y_1, y_2, z) be the coordinates on $X = \mathbb{C}^3$, $Y = \{(y, z); z = 0\}$, $\mathcal{M} = \mathcal{D}_X / \mathcal{D}_X P$ where $P = z D_{y_1}^2 + D_{y_2}$. Then:

$$C_{T_Y^* X}(\mathcal{M}) = \{(y, \zeta, \eta, \tilde{z}); \tilde{z} \eta_1^2 = 0\}, \quad C_{T_Y^* X}^1(\mathcal{M}) = T^* T_Y^* X,$$

$$\hat{C}_{T_Y X}(\mathcal{M}) = \{(y, \zeta, \eta, \tilde{z}); \eta_2 = 0\}.$$

Remark 1.4.7. Let $\mathcal{M}$ be a holonomic $\mathcal{D}_X$-module, and let us (locally) endow $\mathcal{M}$ with a good $F^Y \mathcal{D}_X$-filtration. Then $G^Y \mathcal{M}$ is a holonomic $\mathcal{D}_{T_Y X}$-module. This result can be easily proved using Laurent's second microlocalization (cf. Laurent-Schapira [1]) or else using purely algebraic method (J.E. Björk, unpublished).

1.5. Characteristic Variety of the Systems Induced on a Submanifold

Let $\mathcal{N}$ be a $\mathcal{D}_Y$-module on a manifold Y, $\mathcal{N}$ being locally the union of an increasing sequence of coherent $\mathcal{D}_Y$-modules $\mathcal{N}_j, j \in \mathbb{N}$. By the additivity property of the characteristic variety of coherent $\mathcal{D}_Y$-modules, we find that the set $\bigcup_j \text{char}(\mathcal{N}_j)$ depends only on $\mathcal{N}$. We denote it by $\text{char}(\mathcal{N})$ and call it the characteristic variety of $\mathcal{N}$. Remark that if $0 \to \mathcal{N}_1 \to \mathcal{N}_2 \to \mathcal{N}_3 \to 0$ is an exact sequence of $\mathcal{D}_Y$-modules, each $\mathcal{N}_i$ ($i = 1, 2, 3$) being locally the union of an increasing sequence of coherent $\mathcal{D}_Y$-modules, we have:

$$(1.5.1) \qquad \text{char}(\mathcal{N}_2) = \text{char}(\mathcal{N}_1) \cup \text{char}(\mathcal{N}_3).$$

Now assume Y is a submanifold of a manifold X, and let $\mathcal{M}$ be a coherent $\mathcal{D}_X$-Module. Applying Proposition 3.5.6 of Chapter II, we may consider $\text{char}(\mathcal{T}or_j^{\mathcal{D}_X}(\mathcal{D}_{Y \to X}, \mathcal{M}))$.

Definition 1.5.1. Let $\mathcal{M}$ be a coherent $\mathcal{D}_X$-module. We set:

$$\mathrm{char}(\mathcal{M}_Y^{\cdot}) = \bigcup_{j \in \mathbb{N}} \mathrm{char}(\mathcal{T}\!\mathit{or}_j^{\mathcal{D}_X}(\mathcal{D}_{Y \to X}, \mathcal{M})).$$

The purpose of this section is to give a bound to $\mathrm{char}(\mathcal{M}_Y^{\cdot})$. First we need some geometric preliminaries.

Let $\Lambda = T_Y X$ be the normal bundle to Y in X, τ the projection $\Lambda \to Y$. Since τ is smooth, it defines an immersion:

$$T^*Y \underset{Y}{\times} \Lambda \hookrightarrow T^*\Lambda.$$

Since T^*Y is sent to $T^*Y \underset{Y}{\times} \Lambda$ by the zero section of Λ, we get injections:

(1.5.2) $$T^*Y \hookrightarrow T^*Y \underset{Y}{\times} \Lambda \hookrightarrow T^*\Lambda.$$

Let us choose a system of local coordinates (y, z) on Λ, and let $((y, z); \langle \eta, dy \rangle + \langle \zeta, dz \rangle)$ be the associated coordinates on $T^*\Lambda$. Then:

$$T^*Y \underset{Y}{\times} \Lambda = \{(y, z; \eta, \zeta); \ \zeta = 0\},$$

$$T^*Y = \{(y, z; \eta, \zeta); \ z = \zeta = 0\}.$$

Theorem 1.5.2. *Let $\mathcal{M}$ be a coherent $\mathcal{D}_X$-module. Then:*

$$\mathrm{char}(\mathcal{M}_Y^{\cdot}) \subset T^*Y \cap \hat{C}_{T_Y X}(\mathcal{M}).$$

Proof. First we shall reduce the proof to the case where $\mathcal{M} = \mathcal{D}_X / \mathcal{D}_X \cdot P$, for a differential operator P.

Let $\theta \in T^*Y$, with $\theta \notin \hat{C}_{T_Y X}(\mathcal{M})$.

Let $(u_1, \ldots, u_N)$ be a system of generators of $\mathcal{M}$ in a neighborhood of $\pi(\theta) \in Y$. Let $\mathcal{M}_i = \mathcal{D}_X u_i$ be the sub-$\mathcal{D}_X$-module generated by u_i. Applying Theorem 1.4.5 we see that $\theta \notin \hat{C}_\Lambda(\mathcal{M}_i)$, and that there exists a differential operator P_i such that:

(1.5.3) $$\begin{cases} P_i u_i = 0, \\ \sigma(\hat{\sigma}_Y(P_i))(\theta) \neq 0. \end{cases}$$

(Recall that $\hat{\sigma}_Y(P_i)$ is a differential operator on Λ).

Let ψ be the $\mathcal{D}_X$-linear morphism:

$$\bigoplus_{i=1}^{N} \mathcal{D}_X / \mathcal{D}_X \cdot P_i \underset{\psi}{\longrightarrow} \mathcal{M},$$

which associates u_i to $1 \bmod \mathcal{D}_X \cdot P_i$, and let $\mathcal{K} = \ker \psi$. Again by Theorem 1.4.5. we get that $\theta \notin \hat{C}_\Lambda(\mathcal{K})$.

Let us apply the functor $\mathcal{D}_{Y \to X}$ to the exact sequence:

(1.5.4) $$0 \to \mathcal{K} \to \mathcal{L} \to \mathcal{M} \to 0,$$

where $\mathscr{L} = \bigoplus_{i=1}^{N} \mathscr{D}_X / \mathscr{D}_X \cdot P_i$. We get a long exact sequence:

(1.5.5)
$$\ldots \to \mathscr{T}or_j^{\mathscr{D}_X}(\mathscr{D}_{Y \to X}, \mathscr{K}) \to \mathscr{T}or_j^{\mathscr{D}_X}(\mathscr{D}_{Y \to X}, \mathscr{L})$$
$$\to \mathscr{T}or_j^{\mathscr{D}_X}(\mathscr{D}_{Y \to X}, \mathscr{M}) \to \ldots .$$

Assume we have proved that $\mathrm{char}(\mathscr{T}or_j^{\mathscr{D}_X}(\mathscr{D}_{Y \to X}, \mathscr{N}))$ is contained in $T^* Y \cap \hat{C}_\Lambda(\mathscr{N})$ for all j and all coherent $\mathscr{D}_X$-modules $\mathscr{N}$ of the type $\mathscr{D}_X / \mathscr{D}_X \cdot P$, and also for all coherent Modules $\mathscr{N}$ and all j such that $0 \leqslant j \leqslant j_0$. By (1.5.5) and (1.5.1)) we get that θ does not belong to $\mathrm{char}(\mathscr{T}or_{j_0+1}^{\mathscr{D}_X}(\mathscr{D}_{Y \to X}, \mathscr{M}))$. Thus it is enough to prove the theorem when $\mathscr{M} = \mathscr{D}_X / \mathscr{D}_X \cdot P$. Now we assume $\mathscr{M} = \mathscr{D}_X / \mathscr{D}_X P$, and we assume that for some $\theta \in T^* Y$.

(1.5.6)
$$\sigma(\hat{\sigma}_Y(P))(\theta) \neq 0.$$

Let (y, z) be a system of local coordinates such that $Y = \{(y, z); z = 0\}$. Let $\mathscr{J}_Y$ be the defining Ideal of Y (i.e. the Ideal generated by $z = (z_1, \ldots, z_p)$). Then by the assumption (1.5.6), we may decompose P as:

(1.5.7)
$$\begin{cases} P = Q + R + S, \\ Q \in \mathscr{D}_Y(m), \quad \sigma_m(Q)(\theta) \neq 0, \\ R \in F_0^Y(\mathscr{D}_X) \cap \mathscr{D}_X(m) \cap \mathscr{J}_Y \mathscr{D}_X, \\ S \in F_{-1}^Y(\mathscr{D}_X). \end{cases}$$

The groups $\mathscr{T}or_j^{\mathscr{D}_X}(\mathscr{D}_{Y \to X}, \mathscr{M})$ may be calculated as the cohomology groups of the complex obtained after applying the functor $\mathscr{D}_{Y \to X} \otimes_{\mathscr{D}_X} \cdot$ to the complex $0 \to \mathscr{D}_X \xrightarrow{P} \mathscr{D}_X \to 0$, where P operates on the right. Thus these groups are 0 for $j \neq 1, 0$, and for $j = 1$ (resp. 0) they are the kernel (resp. co-kernel) of P acting on the right on $\mathscr{D}_{Y \to X}$:

(1.5.8)
$$0 \to \mathscr{D}_{Y \to X} \xrightarrow{P} \mathscr{D}_{Y \to X} \to 0.$$

First we shall prove that P is injective.

Let us write a section of $\mathscr{D}_{Y \to X} = \mathscr{D}_X / \mathscr{J}_Y \mathscr{D}_X$ as a finite sum:

(1.5.9)
$$u = \sum_{0 \leqslant |\alpha| \leqslant k} u_\alpha(y, D_y) \delta^{(\alpha)}$$

where $u_\alpha \in \mathscr{D}_Y$, $\delta^{(\alpha)}$ is the class of $D^\alpha \mod \mathscr{J}_Y \mathscr{D}_X$.

Let us assume u of order k for the filtration for which a non zero section of $\mathscr{D}_Y$ is of order 0, and $\delta^{(\alpha)}$ of order $|\alpha|$. Let us denote by $\bar{u}_\alpha$ the principal symbol of a section u_α of $\mathscr{D}_Y$. Then:

$$u \circ P = \sum_{|\alpha| = k} u_\alpha \delta^{(\alpha)} \circ (Q + R) + \sum_{0 \leqslant |\beta| < k + m} v_\beta \delta^{(\beta)}$$
$$= \sum_{|\alpha| = k} v_\alpha \delta^{(\alpha)} + \sum_{0 \leqslant |\beta| < k} w_\beta \delta^{(\beta)}.$$

Since the order of R with respect to D_y is at most $m - 1$, we have:

$$\bar{v}_\alpha = \bar{u}_\alpha \cdot \sigma_m(Q) \quad \forall \alpha, \ |\alpha| = k$$

which proves the injectivity.

Finally let us calculate the characteristic variety of the $\mathscr{D}_Y$-module $\mathscr{D}_X/(\mathscr{J}_Y\mathscr{D}_X+\mathscr{D}_X P)=\mathscr{T}\!or_0^{\mathscr{D}_X}(\mathscr{D}_{Y\to X},\mathscr{M})$.
Let u be the class of $1\bmod\mathscr{D}_X\cdot P$ in $\mathscr{D}_X/\mathscr{D}_X\cdot P$, and let $v_\alpha=1_{Y\to X}\otimes D_z^\alpha u$ that is the image of $D_z^\alpha u$ in $\mathscr{M}/\mathscr{J}_Y\mathscr{M}$. By the definition of u we have:

$$(Q+R+S)u=0,$$

thus:

$$(Q+R+S)(D_z^\alpha u)+[D_z^\alpha,R]u+[D_z^\alpha,S]u=0.$$

We get:

$$(Q+R_\alpha')v_\alpha+\sum_{|\beta|=|\alpha|}R_\beta v_\beta+\sum_{|\gamma|<|\alpha|}S_\gamma v_\gamma=0,$$

where R_α', R_β, S_γ are sections of $\mathscr{D}_Y$, and $\operatorname{ord}(R_\beta)<m$, $\operatorname{ord}(R_\alpha')<m$. For $k\in\mathbb{N}$, define $\mathscr{N}_k$ as the sub-$\mathscr{O}_Y$-module of $\mathscr{M}/\mathscr{J}_Y\mathscr{M}$ generated by the (v_α)'s for $|\alpha|\leqslant k$. We get:

$$Q\mathscr{N}_k\subset\mathscr{D}_Y(m-1)\mathscr{N}_k+\mathscr{D}_Y\mathscr{N}_{k-1}.$$

Let $\tilde{\mathscr{N}}_k=\mathscr{D}_Y\mathscr{N}_k$. Applying (II, Proposition 3.5.6) we see that $\tilde{\mathscr{N}}_k$ is coherent over $\mathscr{D}_Y$. Arguing by induction on k, we only have to prove that $\theta\notin\operatorname{char}(\tilde{\mathscr{N}}_k/\tilde{\mathscr{N}}_{k-1})$. Set $\mathscr{L}_k=\tilde{\mathscr{N}}_k/\tilde{\mathscr{N}}_{k-1}$, and let $\mathscr{L}_k^j$ be the image of $\mathscr{D}_Y(j)\mathscr{N}_k$ in $\mathscr{L}_k$. The family $(\mathscr{L}_k^j)_j$ defines a good filtration on $\mathscr{L}_k$, and we have:

$$Q\mathscr{L}_k^j\subset\mathscr{L}_k^{j+m-1}.$$

Thus:

$$\sigma(Q)\operatorname{gr}(\mathscr{L}_k)=0,$$

which completes the proof.   □

Remark 1.5.3. If $\mathscr{M}$ is a regular holonomic Module in the sense of Kashiwara-Kawaï [1], then we have for Y a submanifold of X:

$$\operatorname{char}(\mathscr{M}_Y'')\subset T^*Y\cap C_{T_Y^*X}(\operatorname{char}(\mathscr{M})),$$

(cf. Kashiwara-Schapira [3], [4]).

1.6. Coherency of the Systems Induced on a Submanifold

Let Y be a submanifold of X of codimension d. Let $\Lambda=T_YX$ be the normal bundle to Y, $\tau:\Lambda\to Y$ its projection. We have already defined in § 1.3 the Ring $\mathscr{D}_{\Lambda/Y}$ of relative differential operators over Y. Now we set:

$$(1.6.1)\qquad\mathscr{D}_{\Lambda,0/Y}=\mathscr{D}_{\Lambda,0}\cap\mathscr{D}_{\Lambda/Y}.$$

Definition 1.6.1. Let $\mathcal{M}$ be a coherent $\mathcal{D}_X$-module. We shall say that $\mathcal{M}$ is elliptic along Y if there exists a coherent $\mathcal{O}_X$-module $\mathcal{M}_0 \subset \mathcal{M}$ which generates $\mathcal{M}$ and a differential operator b on X such that:

$$(1.6.2) \qquad \hat{\sigma}_Y(b) \in \mathcal{D}_{\Lambda,0/Y},$$

$$(1.6.3) \qquad b \mathcal{M}_0 \subset (F^Y_{-1} \mathcal{D}_X) \mathcal{M}_0,$$

(1.6.4) there exists a system of local coordinates (y,t) on the vector bundle Λ, t being linear in the fibers, such that:

$$\sigma(\hat{\sigma}_Y(b))(y,t,\bar{t}) \neq 0 \quad \text{for} \quad t \neq 0,$$

where $\bar{t}$ is the complex conjugate to t.

In other words, if (y,t) are coordinates such that $Y = \{(y,t); \ t=0\}$, then b may be written as:

$$(1.6.5) \qquad b(y,t,D_y,D_t) = \sum_{0 \leq |\alpha| = |\beta| \leq m} a_{\alpha,\beta}(y) t^\alpha D_t^\beta + Q,$$

where $Q \in F^Y_{-1} \mathcal{D}_X$, and

$$(1.6.6) \qquad \sum_{|\alpha| = |\beta| = m} a_{\alpha,\beta}(y) t^\alpha \bar{t}^\beta \neq 0 \quad \text{for} \quad t \neq 0.$$

Lemma 1.6.2. *Let $\mathcal{M}$ be a coherent $\mathcal{D}_X$-module, $F^Y \mathcal{M}$ a good $F^Y \mathcal{D}_X$-filtration on $\mathcal{M}$, $\mathcal{M}_0$ a coherent $\mathcal{D}_{\Lambda,0/Y}$-module which generates $G^Y \mathcal{M}$. Then:*
   i) *the $\mathrm{gr}(\mathcal{D}_{\Lambda,0/Y})$-ideal $\mathrm{Icar}(\mathcal{M}_0)$ depends only on $\mathcal{M}$.*
   ii) *the correspondence which associates the Ideal $\mathrm{Icar}(\mathcal{M}_0)$ to the coherent $\mathcal{D}_X$-module $\mathcal{M}$ is additive.*

Since the proof is similar to that of Lemma 1.3.3 we do not repeat it. In the situation of Lemma 1.6.2 we set:

$$(1.6.7) \qquad \mathrm{Icar}^1_{\Lambda,0}(\mathcal{M}) = \mathrm{Icar}(\mathcal{M}_0).$$

Remark that assertion ii) of the lemma means that if $0 \to \mathcal{L} \to \mathcal{M} \to \mathcal{N} \to 0$ is an exact sequence of coherent $\mathcal{D}_X$-modules, then $\mathrm{Icar}^1_{\Lambda,0}(\mathcal{M}) = \mathrm{Icar}^1_{\Lambda,0}(\mathcal{L}) \cap \mathrm{Icar}^1_{\Lambda,0}(\mathcal{N})$.   □

Lemma 1.6.3. a) *Let $\mathcal{M}_0$ and b satisfying conditions (1.6.2) and (1.6.3). Then:*

$$\sigma(\hat{\sigma}_Y(b)) \in \mathrm{Icar}^1_{\Lambda,0}(\mathcal{M}).$$

   b) *Conversely let $f \in \mathrm{Icar}^1_{\Lambda,0}(\mathcal{M})$.*
   Then there exists $\mathcal{M}_0$ and b satisfying conditions (1.6.2) and (1.6.3) such that for some $N \geq 1$, $\sigma(\hat{\sigma}_Y(b)) = f^N$.

Proof. a) Set $F_k^Y \mathscr{M} = (F_k^Y \mathscr{D}_X) \mathscr{M}_0$. Let $\mathscr{M}_0'$ be the image of $\mathscr{M}_0$ in $G^Y \mathscr{M}$, and let $\mathscr{M}_0'' = \mathscr{D}_{\Lambda,0/Y} \mathscr{M}_0'$. Then $\hat{\sigma}_Y(b) \mathscr{M}_0' = 0$, thus $\sigma(\hat{\sigma}_Y(b))$ belongs to $\mathrm{Icar}(\mathscr{M}_0'')$, that is to $\mathrm{Icar}_{\Lambda,0}^1(\mathscr{M})$.

b) Let $(u_1, \ldots, u_N)$ be a system of generators of $\mathscr{M}$ and let $\mathscr{M}_0$ be the coherent $\mathscr{O}_X$-module generated by these sections. Let $F_k^Y \mathscr{M} = (F_k^Y \mathscr{D}_X) \mathscr{M}_0$, and let $\bar{u}_1, \ldots, \bar{u}_N$ be the images of $u_1, \ldots, u_N$ in $G^Y \mathscr{M}$. By hypothesis, for each $j = 1, \ldots, N$, there exists a section $\bar{b}_j$ of $\mathscr{D}_{\Lambda,0/Y}$ such that $\bar{b}_j \bar{u}_j = 0$ and $\sigma(\bar{b}_j) = f^{N_j}$, for some $N_j \geqslant 0$. Applying Remark 1.3.8 of Chapter II, we may even choose the same operator $\bar{b}$ for all j.

Let $b \in \mathscr{D}_X$ such that $\hat{\sigma}_Y(b) = \bar{b}$. Then $(\mathscr{M}_0, b)$ satisfies (1.6.2) and (1.6.3). $\square$

Remark 1.6.4. Assume $\mathscr{M}$ is elliptic along Y. Let $\mathscr{M}_0$ be a coherent $\mathscr{O}_X$-module which generates $\mathscr{M}$. Then there exists an operator b in $\mathscr{D}_X$, satisfying (1.6.2), (1.6.3), (1.6.4). In fact this follows immediately from the proof of Lemma 1.6.3.

Proposition 1.6.5. *Assume $\mathscr{M}$ is elliptic along Y. Let $\mathscr{N}$ be a coherent sub-Module of $\mathscr{M}$. Then $\mathscr{N}$ is elliptic along Y.*

Proof. By Lemma 1.6.2 we know that $\mathrm{Icar}_{\Lambda,0}^1(\mathscr{M})$ is contained in $\mathrm{Icar}_{\Lambda,0}^1(\mathscr{N})$. Then the Proposition follows from Lemma 1.6.3. $\square$

Theorem 1.6.6. *Let $\mathscr{M}$ be a coherent $\mathscr{D}_X$-module. Assume $\mathscr{M}$ is elliptic along Y. Then the $\mathscr{D}_Y$-modules $\mathscr{T}or_j^{\mathscr{D}_X}(\mathscr{D}_{Y \to X}, \mathscr{M})$ are all coherent.*

Sketch of proof. Using the same method as for example in the proof of Theorem 1.5.2, and using Proposition 1.6.5, we may reduce the proof to the case where $\mathscr{M} = \mathscr{D}_X / \mathscr{D}_X \cdot b$.

Let us fix a system of coordinates (y, t) as in Definition 1.6.1. Then a section u of $\mathscr{D}_{Y \to X}$ may be written as a finite sum:

$$(1.6.8) \qquad u = \sum_\alpha a_\alpha(y, D_y) \delta_t^\alpha,$$

where $a_\alpha(y, D_y)$ is a section of $\mathscr{D}_Y$ and $\delta_t^\alpha = 1_{Y \to X} \cdot D_t^\alpha$. We shall denote by $\mathscr{D}_{Y \to X}(N)$ the subsheaf of $\mathscr{D}_{Y \to X}$ of sections which may be written as in (1.6.8) with $|\alpha| \leqslant N$. We also set:

$$(1.6.9) \qquad \mathscr{S}_N = \mathscr{D}_{Y \to X}(N) / \mathscr{D}_{Y \to X}(N-1).$$

Using Proposition 3.5.6 of Chapter II, we only have to prove that the kernel and the cokernel of b acting on the right on $\mathscr{D}_{Y \to X}$ are finitely generated $\mathscr{D}_Y$-modules. Thus it is enough to prove that, locally on Y, there exists N_0 such that b acting on $\mathscr{S}_N$ is an isomorphism for $N \geqslant N_0$.

Since $F_{-1}^Y \mathscr{D}_X$ acts as zero on $\mathscr{S}_N$, we may assume b is as in (1.6.5) with $Q = 0$.

We order the set of multi-integers $\alpha \in \mathbb{N}^d$ such that $|\alpha| = N$. Then the family $(\delta_t^\alpha)_{|\alpha|=N}$ gives a basis of the free $\mathscr{D}_Y$-module $\mathscr{S}_N$, and b is represented in that basis by a square matrix with entries in $\mathscr{O}_Y$.

It remains to check, using (1.6.6), that this matrix is invertible for N big enough, but for that rather technical part of the proof we refer to Kashiwara-Kawaï-Sjöstrand [1].   □

Exercices to III.1

Ex. 1.1.   Let (x, ξ) be   the   coordinates   on $T^*\mathbb{C}^n$,   and   let $V = \{(x, \xi); \ \xi_1 = 0, \xi_n \neq 0\}$. Let $P = D_1^m + \sum\limits_{0 \leqslant j \leqslant m-1} A_j(x, D')D_1^j$ be a microdifferential operator of order $m > 0$, with A_j of order at most $m - j - 1$, not depending on D_1. Let $\mathscr{M} = \mathscr{E}_X / \mathscr{E}_X P$. Prove that $C_V(\mathscr{M}) = V$ (the zero section of $T_V \dot{T}^* X$), but that $C_V^1(\mathscr{M}) = V$ if A_j is of order at most zero for all j, and $C_V^1(\mathscr{M}) = T_V \dot{T}^* X$ otherwise.

Ex. 1.2. With the notations of § 1.3, consider the $\mathscr{D}_X$-module $\mathscr{B}_{Y|X}$, endowed with its natural filtration (I, § 4). Prove that $(F_k^Y \mathscr{D}_X)\mathscr{B}_{Y|X}(0) = \mathscr{B}_{Y|X}(k)$, and that for this $F^Y \mathscr{D}_X$-filtration on $\mathscr{B}_{Y|X}$ we have $G^Y(\mathscr{B}_{Y|X}) = \mathscr{B}_{Y|\Lambda}$.

Ex. 1.3.   Let $(y, z) = (y_1, \ldots, y_q; z_1, \ldots, z_p)$ be   the   coordinates   on X, $Y = \{(y, z); \ z = 0\}$. Let b be a non zero polynomial, Q a section of $F_{-1}^Y \mathscr{D}_X$, P the operator $b\left(\sum\limits_{j=1}^{p} z_j \dfrac{\partial}{\partial z_j}\right) + Q$, $\mathscr{M}$ the Module $\mathscr{D}_X / \mathscr{D}_X \cdot P$. Assume that the polynomial b has no roots $\alpha \in \mathbb{N} \cup \{0\}$. Then prove that $\mathrm{Tor}_j^{\mathscr{D}_X}(\mathscr{D}_{Y \to X}, \mathscr{M}) = 0 \quad \forall j$.

Ex. 1.4. Let Y be a hypersurface of X, non characteristic for $\mathscr{M}$. Prove that $\mathscr{M}$ is elliptic along Y.

Ex. 1.5. Let (y, t) be the coordinates on $X = \mathbb{C}^2$, and let $Y = \{(y, t); \ t = 0\}$. Let $b = (t D_t)^2 + \alpha(y)t D_t + \beta(y)$. Calculate $\mathscr{T}\!o\!r_j^{\mathscr{D}_X}(\mathscr{O}_Y, \mathscr{D}_X / \mathscr{D}_X \cdot b)$.

Ex. 1.6. Let $Y = \{(y, z) \in \mathbb{C}^2; \ z = 0\}$ and let $\mathscr{M}$ be the holonomic $\mathscr{D}_X$-module $\mathscr{M} = \mathscr{D} / \mathscr{D}_{\mathbb{C}^2}(z - y^2, 2y D_z + D_y) = \mathscr{D}_{\mathbb{C}^2} u$. Find a non zero polynomial b and an operator $Q \in F_{-1}^Y \mathscr{D}_X$, such that $(b(z D_z) + Q)u = 0$.

Ex. 1.7. By applying Theorem 1.5.2 give a bound to the characteristic varieties of the induced systems for the Modules of (II. Ex. 3.1 and Ex. 3.7).

Ex. 1.8. We keep the notations of Ex. 1.6. Calculate $C_{T_Y X}(\mathscr{M})$ and calculate the induced systems by $\mathscr{M}$ on Y.

Ex. 1.9. Let (t, y_1, y_2) be the coordinates on $X = \mathbb{C}^3$, f the projection $(t, y_1, y_2) \mapsto (y_1, y_2)$ on $Y = \mathbb{C}^2$. Let Z be the hypersurface: $Z = \{(t, y_1, y_2) \in X; \ y_1 - t y_2 = 0\}$.

   i) Calculate the set $T_Z^* X$.

   ii) Let ϕ be the projection $T^* X \to T^*(X/Y)$, and let Σ_y denote the leaf $f^{-1}(y)$, for $y \in Y$. Calculate $(T_Z^* X) \underset{X}{\times} \Sigma_y$.

Ex. 1.10. Let $b(y, t, D_y, D_t)$ be as in (1.6.5), (1.6.6). Prove that the kernel and the cokernel of b acting on the sheaf $\mathcal{O}_Y[[t]]$ of formal series with respect to t with coefficients in $\mathcal{O}_Y$, are $\mathcal{O}_Y$-coherent.

§ 2. The Cauchy Problem

2.1. The Cauchy Problem for a System

Let Y and X be two complex manifolds, f a holomorphic map from Y to X. Let $\overline{\omega}$ and ρ be the natural maps associated to f:

$$T^* Y \underset{\rho}{\leftarrow} Y \underset{X}{\times} T^* X \underset{\overline{\omega}}{\to} T^* X.$$

In this section U will denote an open set in $\dot{T}^* X$, and V a conic involutive manifold, closed in U, and satisfying condition (1.2.2).

Definition 2.1.1. Let $\mathcal{M}$ be a coherent $\mathcal{E}_X$-module on U. We say that f is non 1-microcharacteristic for $\mathcal{M}$ along V at $p \in V$, if for any function ϕ defined in a neighborhood of p and satisfying $\phi \circ \overline{\omega} = 0$, $d\phi(p) \neq 0$, we have: $H_\phi(p) \notin C_V^1(\mathcal{M})$.

Here we have kept the notation H_ϕ for denoting the image of the Hamiltonian vector field of ϕ (which is a vector field on $T^* X$) into $T_V T^* X$. When f is an immersion, we also say under the same hypothesis that Y is non 1-microcharacteristic for $\mathcal{M}$ along V.

Remark that if f is non 1-microcharacteristic, then for any function ϕ as in the definition, the vector $H_\phi(p)$ does not belong to $C_V(\mathcal{M})$, and therefore does not belong to $C_{\{p\}}(\text{char}(\mathcal{M}))$. This last condition implies that f is non characteristic for $\mathcal{M}$ in a neighborhood of p: there exists a neighborhood W of p such that ρ is finite over $\overline{\omega}^{-1}(\text{char}(\mathcal{M}) \cap W)$ (cf. II, Definition 3.2.1).

If p belongs to $\text{char}(\mathcal{M})$, and if moreover f is non 1-microcharacteristic for $\mathcal{M}$ along V, then there exists a neighborhood W of p such that ρ is finite over $\overline{\omega}^{-1}(V \cap W)$. Thus we are in a situation where we may apply Proposition 3.5.2 of Chapter II.

The aim of this section is to prove:

Theorem 2.1.2. *Let $\mathscr{M}$ be a coherent $\mathscr{E}_X$-module on U and let $\mathscr{L}$ be a system with simple characteristics along V (cf. I, Definition 6.2.2). Assume f is non 1-microcharacteristic for $\mathscr{M}$ along V at each $p \in V$. Then the natural morphisms*

$$\overline{\omega}^{-1}\mathscr{E}xt^j_{\mathscr{E}_X}(\mathscr{M}, \mathscr{L}) \to \mathscr{E}xt^j_{\rho^{-1}\mathscr{E}_Y}(\mathscr{E}_{Y \to X} \otimes_{\overline{\omega}^{-1}\mathscr{E}_X} \overline{\omega}^{-1} \mathscr{M}, \mathscr{E}_{Y \to X} \otimes_{\overline{\omega}^{-1}\mathscr{E}_X} \overline{\omega}^{-1} \mathscr{L})$$

are isomorphisms for all j.

Proof. For the sake of simplicity we shall not always write the symbols ρ and $\overline{\omega}$.

By decomposing f into an embedding and a projection it is sufficient to check separately both cases. In the case where f is the projection $Z \times X \to X$, the theorem states the isomorphism

$$f^{-1}\mathscr{E}xt^j_{\mathscr{E}_X}(\mathscr{M}, \mathscr{L}) \cong \mathscr{E}xt^j_{\mathscr{E}_{Z \times X}}(\mathscr{O}_Z \hat{\otimes} \mathscr{M}, \mathscr{O}_Z \hat{\otimes} \mathscr{L})$$

which is valid for any couple $(\mathscr{M}, \mathscr{L})$ of coherent $\mathscr{E}_X$-modules.

Now assume Y is a submanifold of X. We want to prove that for any $p \in Y \underset{X}{\times} T^*X \cap \mathrm{char}(\mathscr{M}) \cap V$ and any j, we have the isomorphisms:

$$(2.1.1) \quad \mathscr{E}xt^j_{\mathscr{E}_X}(\mathscr{M}, \mathscr{L})_p \xrightarrow{\sim} \mathscr{E}xt^j_{\rho^{-1}(\mathscr{E}_Y)}(\mathscr{E}_{Y \to X} \otimes_{\mathscr{E}_X} \mathscr{M}, \mathscr{E}_{Y \to X} \otimes_{\mathscr{E}_X} \mathscr{L})_p.$$

First we reduce the proof to the case where both manifolds V and $Y \underset{X}{\times} T^*X$ are regular involutive in a neighborhood of p.

We use the trick of the dummy variable (II, § 2.5) and introduce new coordinates, t on $\mathbb{C}$, $(t; \tau)$ on $T^*\mathbb{C}$, and define the map ψ from $T^*X \times \dot{T}^*\mathbb{C}$ to T^*X by setting $\psi((x, t; \xi, \tau)) = (x; \xi/\tau)$. Then there exists an exact sequence of left $\psi^{-1}(\mathscr{E}_X)$-modules:

$$(2.1.2) \qquad 0 \to \psi^{-1}\mathscr{L} \to \mathscr{L} \hat{\otimes} \mathscr{E}_{\mathbb{C}} \to (\mathscr{L} \hat{\otimes} \mathscr{E}_{\mathbb{C}})^2 \to \mathscr{L} \hat{\otimes} \mathscr{E}_{\mathbb{C}} \to 0$$

and it is enough to prove the analogous of (2.1.1) with $\mathscr{E}_X$, $\mathscr{M}$, $\mathscr{L}$ replaced by $\mathscr{E}_{X \times \mathbb{C}}$, $\mathscr{L} \hat{\otimes} \mathscr{E}_{\mathbb{C}}$, $\mathscr{M} \hat{\otimes} \mathscr{E}_{\mathbb{C}}$ respectively, at $\tilde{p} = (p; (0; dt))$.

Thus we may assume V and $Y \underset{X}{\times} T^*X$ regular involutive from the beginning. We put $W = Y \underset{X}{\times} T^*X$, $\mathscr{K} = \mathscr{E}_{Y \to X}$. Then $\mathscr{K}$ is a right $\mathscr{E}_X$-module with simple characteristics along the regular involutive manifold W, and we have:

$$(2.1.3) \qquad \rho^{-1}(\mathscr{E}_Y) \cong \mathscr{E}nd_{\mathscr{E}_X}(\mathscr{K}).$$

Since a system with simple characteristics along a regular involutive manifold is locally unique, up to isomorphism (I, Corollary 6.2.3), it is enough to prove the isomorphisms:

(2.1.4) $\mathcal{E}xt^j_{\mathcal{E}_X}(\mathcal{M}, \mathcal{L})_p \simeq \mathcal{E}xt^j_{\mathcal{E}nd(\mathcal{K})}(\mathcal{K} \otimes_{\mathcal{E}_X} \mathcal{M}, \mathcal{K} \otimes_{\mathcal{E}_X} \mathcal{L})_p$

under the assumption that $\mathcal{L}$ is a system with simple characteristics along V, $\mathcal{K}$ a system with simple characteristics along W, and V and W are regular involutive manifolds such that for any function ϕ such that $\phi|_W = 0$, $d\phi \neq 0$, the vector H_ϕ is not tangent to V.

Then the hypotheses of Corollary A.4.5 are satisfied, and we may find a contact transformation such that in the new coordinates $(x; \xi)$ on $T^* X$, we have:

(2.1.5) $\begin{cases} p = (0; dx_n), \\ V = \{(x, \xi); \; \xi_1 = \ldots = \xi_r = 0, r < n\}, \\ W = \{(x, \xi); \; x_1 = \ldots = x_d = 0, d \leqslant r\}. \end{cases}$

By quantizing this contact transformation (I, Theorem 5.3.4) we may assume from the start that $Y = \{x \in X; \; x_1 = \ldots = x_d = 0\}$ and p and V are given by (2.1.5).

Let $\mathcal{M}_0$ be a coherent $\mathcal{E}_V$-module which generates $\mathcal{M}|_V$. We see by the proof of Lemma 1.2.7 that the natural morphism:

$$\mathcal{E}_X \otimes_{\mathcal{E}_V} \mathcal{M}_0 \to \mathcal{M}$$

is an isomorphism. Moreover $\mathcal{E}_X = \varinjlim_k \mathcal{E}_X(k)\mathcal{E}_V$ is flat over $\mathcal{E}_V$ since $\mathcal{E}_X(k)\mathcal{E}_V$ is locally free over $\mathcal{E}_V$. Thus:

$$\mathcal{E}xt^j_{\mathcal{E}_X}(\mathcal{M}, \mathcal{L}) \simeq \mathcal{E}xt^j_{\mathcal{E}_V}(\mathcal{M}_0, \mathcal{L}).$$

Let us denote by V' the image of $Y \underset{X}{\times} V$ by ρ, that is:

$$V' = \{(x_{d+1}, \ldots, x_n; \xi_{d+1}, \ldots, \xi_n) \in T^* Y; \xi_{d+1} = \ldots = \xi_r = 0\}.$$

We define the $(\rho^{-1}\mathcal{E}_{V'}, \mathcal{E}_V)$-bimodule $\mathcal{E}_{V' \to V}$ by:

$$\mathcal{E}_{V' \to V} = \mathcal{E}_V/(x_1, \ldots, x_d)\mathcal{E}_V$$

and we set:

$$\mathcal{M}_{0Y} = \rho_*(\mathcal{E}_{V' \to V} \otimes_{\mathcal{E}_V} \mathcal{M}_0).$$

Remark that $\mathcal{E}_{V' \to V}$ is naturally contained in $\mathcal{E}_{Y \to X}$, and that the natural map:

$$\rho^{-1}\mathcal{E}_Y \otimes_{\rho^{-1}\mathcal{E}_{V'}} \mathcal{E}_{V' \to V} \to \mathcal{E}_{Y \to X}$$

is an isomorphism. Hence:

$$\rho^{-1}\mathcal{E}_Y \otimes_{\rho^{-1}\mathcal{E}_{V'}} \mathcal{E}_{V' \to V} \otimes_{\mathcal{E}_V} \mathcal{M}_0 \simeq \mathcal{E}_{Y \to X} \otimes_{\mathcal{E}_V} \mathcal{M}_0$$

$$\simeq \mathcal{E}_{Y \to X} \otimes_{\mathcal{E}_X} \mathcal{M}$$

Then (2.1.1) reduces to the isomorphisms:

(2.1.6) $\mathcal{E}xt^j_{\mathcal{E}_V}(\mathcal{M}_0, \mathcal{L})_p \simeq \mathcal{E}xt^j_{\mathcal{E}_{V'}}(\mathcal{M}_{0Y}, \mathcal{L}_Y)_{\rho(p)}$

Lemma 2.1.3. a) *The Module $\mathscr{M}_{0Y}$ is coherent over $\mathscr{E}_{V'}$.*

b) $\mathcal{T}\!or_j^{\mathscr{E}_V}(\mathscr{E}_{V'\rightarrow V}, \mathscr{M}_0) = 0$ *for* $j \neq 0$.

c) $C_{V'}^1(\mathscr{M}_{0Y}) \subset \tilde{\rho}(C_V^1(\mathscr{M}_0))$, *where* $\tilde{\rho}$ *denotes* *the* *projection* $Y \underset{X}{\times} T_V T^* X \rightarrow T_{V'} T^* Y$.

The proof of this lemma follows exactly the same lines as the proof of Theorem 3.4.2 of Chapter II, and we shall not repeat it.

By this Lemma we may argue by induction on $d = \operatorname{codim} Y$, since if Y' is the hypersurface with equation $\{x \in X; x_1 = 0\}$, we have:

$$\mathscr{M}_{0Y} = (\mathscr{M}_{0Y'})_Y, \quad \mathscr{L}_Y = (\mathscr{L}_{Y'})_Y.$$

(Remark that we only need the lemma when Y is a hypersurface). Now we assume $d = 1$ in (2.1.5).

Let $(u_1, \ldots, u_N)$ be a system of generators of $\mathscr{M}$ in a neighborhood of p. By Theorem 1.2.9, for each $j = 1, \ldots, N$, there exists a section P_j of $\mathscr{E}_V$ such that $P_j u_j = 0$ and $\sigma_V^1 \sigma(P_j)(H_{x_1}) \neq 0$. Put $\mathscr{M}' = \bigoplus_{j=1}^N \mathscr{E}_X / \mathscr{E}_X P_j$. Let ψ be the morphism from $\mathscr{M}'$ to $\mathscr{M}$ which associates u_j to the class $1 \bmod \mathscr{E}_X P_j$ of $\mathscr{M}'$, and let $\mathscr{M}'' = \operatorname{Ker} \psi$. We have an exact sequence of left $\mathscr{E}_X$-modules:

$$(2.1.7) \qquad 0 \rightarrow \mathscr{M}'' \rightarrow \mathscr{M}' \rightarrow \mathscr{M} \rightarrow 0.$$

Since Y is non characteristic for each of these Modules, the sequence obtained by tensorizing (2.1.7) by $\mathscr{E}_{Y\rightarrow X}$ over $\mathscr{E}_X$ remains exact (II, Theorem 3.4.2):

$$(2.1.8) \qquad 0 \rightarrow \mathscr{E}_{Y\rightarrow X} \otimes_{\mathscr{E}_X} \mathscr{M}'' \rightarrow \mathscr{E}_{Y\rightarrow X} \otimes_{\mathscr{E}_X} \mathscr{M}' \rightarrow \mathscr{E}_{Y\rightarrow X} \otimes_{\mathscr{E}_X} \mathscr{M} \rightarrow 0.$$

Let us apply the functor $\mathscr{H}om_{\mathscr{E}_X}(\cdot, \mathscr{L})_p$ to the sequence (2.1.7) and the functor $\mathscr{H}om_{\rho^{-1}\mathscr{E}_Y}(\cdot, \mathscr{E}_{Y\rightarrow X} \otimes_{\mathscr{E}_X} \mathscr{L})_p$ to the sequence (2.1.8). We get two long exact sequences, with a natural morphism of restriction from the first one to the second (II, Proposition 3.5.2):

$$
\begin{array}{ccccccccc}
\ldots & \rightarrow & A_j & \rightarrow & A_j' & \rightarrow & A_j'' & \rightarrow & A_{j+1} & \rightarrow & \ldots \\
 & & \downarrow {\scriptstyle r_j} & & \downarrow {\scriptstyle r_j'} & & \downarrow {\scriptstyle r_j''} & & \downarrow {\scriptstyle r_{j+1}} & & \\
\ldots & \rightarrow & B_j & \rightarrow & B_j' & \rightarrow & B_j'' & \rightarrow & B_{j+1} & \rightarrow & \ldots,
\end{array}
$$

where $A_j = \mathscr{E}\!xt_{\mathscr{E}_X}^j(\mathscr{M}, \mathscr{L})_p$, $B_j = \mathscr{E}\!xt_{\rho^{-1}\mathscr{E}_Y}^j(\mathscr{E}_{Y\rightarrow X} \otimes_{\mathscr{E}_X} \mathscr{M}, \mathscr{E}_{Y\rightarrow X} \otimes_{\mathscr{E}_X} \mathscr{L})_p$, and likewise for A_j', B_j' (resp. A_j'', B_j'') replacing $\mathscr{M}$ by $\mathscr{M}'$ (resp. $\mathscr{M}''$). First we prove that all the r_j''s are isomorphisms.

Let P be a section of $\mathscr{E}_V$, say of order m, such that $\sigma_V^1 \sigma(P)(H_{x_1})$ is not zero. Replacing P by AP, for an invertible microdifferential operator A of order zero, which does not change the $\mathscr{E}_X$-module $\mathscr{E}_X / \mathscr{E}_X P$, we may assume that P is as in (I, (3.1.3)). We may also assume $\mathscr{L} = \mathscr{E}_X / \mathscr{E}_X (D_1, \ldots, D_r)$. Let:

$$\mathscr{L}(k) = \mathscr{E}_X(k) / \mathscr{E}_X(k-1)(D_1, \ldots, D_r).$$

Then P acts naturally on each $\mathscr{L}(k)$, and we know by Theorem 3.1.1. of Chapter I that P is surjective on $\mathscr{L}(0)$, and the kernel of P acting on $\mathscr{L}(0)|_{Y \underset{X}{\times} T^*X}$ is isomorphic to $(\mathscr{L}(0)/x_1\mathscr{L}(0))^m$. The same result holds with $\mathscr{L}(0)$ replaced by $\mathscr{L}(k)$: in fact it can be obtained by a slight modifications of the proof of Theorem 3.1.1 of Chapter I, or else it can be deduced from this theorem by adding a dummy variable t and using the fact that P commutes with D_t and D_t^{-1}. Thus we get:

$$(2.1.9) \quad \begin{cases} \mathscr{H}om_{\mathscr{E}_X}(\mathscr{E}_X/\mathscr{E}_X P, \mathscr{L})_p \simeq (\mathscr{E}_{Y \to X} \otimes_{\mathscr{E}_X} \mathscr{L})_p^m, \\ \mathscr{E}xt^1_{\mathscr{E}_X}(\mathscr{E}_X/\mathscr{E}_X P, \mathscr{L})_p = 0 \end{cases}$$

and since $(\mathscr{E}_{Y \to X} \otimes_{\mathscr{E}_X} \mathscr{E}_X/\mathscr{E}_X P)_p \simeq \mathscr{E}^m_{Y, \rho(p)}$ by Proposition 3.5.1 of Chapter II, we obtain the result in that case.

Now assume that we have proved that r_j, r_j', r_j'' are isomorphism for $j < j_0$ (which is the case for $j_0 \leqslant 0$). Since all the r_j' are isomorphisms, we see that r_{j_0} is injective. Since $\mathscr{M}''$ satisfies the same hypotheses as $\mathscr{M}, r_{j_0}''$ is also injective, thus r_{j_0} is bijective and so is r_{j_0}''. This completes the proof of the theorem.   □

We have proved Theorem 2.1.2 when V is an involutive manifold in $\dot{T}^*X$, but is clear by the proof that the same arguments hold if we take for V the zero section T_X^*X (then $\mathscr{M}$ is a coherent $\mathscr{D}_X$-module) and for $\mathscr{L}$ the sheaf $\mathcal{O}_X$.

We get the "Cauchy-Kowalewski-Kashiwara" theorem:

Theorem 2.1.4. Let $\mathscr{M}$ be a coherent $\mathscr{D}_X$-module. Assume f is non characteristic for $\mathscr{M}$. Then for all j, the natural morphisms:

$$f^{-1} \mathscr{E}xt^j_{\mathscr{D}_X}(\mathscr{M}, \mathcal{O}_X) \to \mathscr{E}xt^j_{\mathscr{D}_Y}(\mathscr{M}_Y, \mathcal{O}_Y)$$

are isomorphisms.

Remark 2.1.5. It is clear by the proof of Theorem 2.1.2 that we could have stated a "filtered version" of Theorem 2.1.2. Let $\mathscr{M}_0$ be a coherent $\mathscr{E}_V$-module which generates $\mathscr{M}$, and let u be a simple generator of $\mathscr{L}$. Set $\mathscr{L}(k) = \mathscr{E}_X(k)u$. Then $\mathscr{L}(k)$ is an $\mathscr{E}_V$-module (cf. Lemma 3.1.1 below) and the same proof as for Theorem 2.1.2 gives the isomorphisms for all j, all k:

$$\begin{aligned} \varpi^{-1} \mathscr{E}xt^j_{\mathscr{E}_V}(\mathscr{M}_0, \mathscr{L}(k))_p \\ \simeq \mathscr{E}xt^j_{\rho^{-1}\mathscr{E}_{V'}}(\mathscr{E}_{Y \to X}(0) \otimes_{\mathscr{E}_X(0)} \mathscr{M}_0, \mathscr{E}_{Y \to X}(0) \otimes_{\mathscr{E}_X(0)} \mathscr{L}(k))_p, \end{aligned}$$

where $V' = \rho(Y \underset{X}{\times} V)$.

We may partly extend Theorem 2.1.4 by replacing $\mathcal{O}_X$ by $\mathscr{B}_{Z|X}$, and using the variety $\hat{C}_{T_Z^*X}(\mathscr{M})$ introduced in Section 1.4.

Proposition 2.1.6. *Let Y be a hypersurface of X, Z a submanifold of X, with Y and Z transversal. Let P be a differential operator on X and let m be the order of $\hat{\sigma}_Z(P)$ (recall that $\hat{\sigma}_Z(P)$ is a differential operator on $T_Z X$). Assume that, $\{\phi=0\}$ denoting an equation of Y in X, we have:*

$$\sigma(\hat{\sigma}_Z(P))(H_\phi)\neq 0 \quad \text{on} \quad Y.$$

Let $\mathcal{M}=\mathcal{D}_X/\mathcal{D}_X\cdot P$. Then:

$$\mathcal{H}om_{\mathcal{D}_X}(\mathcal{M},\mathcal{B}_{Z|X})|_Y \simeq \mathcal{B}^m_{Z\cap Y|Y},$$
$$\mathcal{E}xt^j_{\mathcal{D}_X}(\mathcal{M},\mathcal{B}_{Z|X})|_Y=0 \quad \text{for} \quad j>0.$$

Proof. We choose coordinates $(z,t)=(z_1,\dots,z_l,t_1,\dots,t_d)$ on X such that $Z=\{(z,t);\ t=0\}$ and $Y=\{(z,t);\ z_1=0\}$. We put $z=(z_1,z')$. Then P may be written as:

$$(2.1.10) \qquad \begin{cases} P=P_0+Q, \\ Q\in F^Z_{-1}\mathcal{D}_X, \\ P_0=\displaystyle\sum_{\substack{|\beta|+|\gamma|\leqslant m \\ |\alpha|=|\beta|}} a_{\alpha,\beta,\gamma}(z)\,t^\alpha\,D^\beta_t\,D^\gamma_z \end{cases}$$

and the coefficient of $D^m_{z_1}$ in P_0 does not vanish.

For a section u of $\mathcal{B}_{Z|X}$ defined on a neighborhood of Y, let us denote by $\gamma_Y(u)$ the set of the m first traces of u on Y:

$$(2.1.11) \qquad \gamma_Y(u)=(u|_Y,\dots,(\frac{\partial^{m-1}}{\partial z_1^{m-1}}u)|_Y).$$

We have to solve the Cauchy problem in $\mathcal{B}_{Z|X}$:

$$(2.1.12) \qquad \begin{cases} Pu=v, \quad \gamma_Y(u)=(w), \\ v\in\mathcal{B}_{Z|X}|_Y, \quad (w)\in\mathcal{B}^m_{Z\cap Y|Y}. \end{cases}$$

Let us endow $\mathcal{B}_{Z|X}$ with its natural filtration (cf. I, Definition 4.1.2), and let us denote by $\hat{v}$ the image in $\mathrm{gr}(\mathcal{B}_{Z|X})$ of a section v of $\mathcal{B}_{Z|X}$ (that is, $\hat{v}$ is the principal symbol of v), and similarly for sections of $\mathcal{B}_{Z\cap Y|Y}$.

We may identify P_0 with the principal symbol of P for the filtration $F^Z\mathcal{D}_X$ on $\mathcal{D}_X$, and it is easily seen that for solving (2.1.12), it is enough to solve the Cauchy problem in $\mathrm{gr}(\mathcal{B}_{Z|X})$:

$$(2.1.13) \qquad \begin{cases} P_0\hat{u}=\hat{v}, \quad \gamma_Y(\hat{u})=(\hat{w}), \\ \hat{v}\in\mathrm{gr}(\mathcal{B}_{Z|X})|_Y, \quad \hat{w}\in\mathrm{gr}(\mathcal{B}^m_{Z\cap Y|Y}), \end{cases}$$

we may assume that $\hat{v}$ and $\hat{w}$ are homogeneous sections of degree k. Denoting the canonical generator of $\mathcal{B}_{Z|X}$, as well as that of $\mathcal{B}_{Z\cap Y|Y}$, by δ_t, we have:

$$\hat{v} = \sum_{|\lambda|=k} \hat{v}_\lambda(z)\delta_t^\lambda, \qquad \hat{w} = \sum_{|\lambda|=k} \hat{w}_\lambda(z')\delta_t^\lambda,$$

where $\delta_t^\lambda = D_t^\lambda \delta_t$, and $\hat{v}_\lambda(z)$ (resp. $\hat{w}_\lambda(z')$) is a section of $\mathscr{O}_Z$ (resp. $\mathscr{O}_{Z\cap Y}^m$).

We order the set of multi-integers $\lambda \in \mathbb{N}^d$ such that $|\lambda|=k$, and represent $\hat{v}$ as a column $\vec{v} = {}^t(v_1(z), \ldots, v_{N_k}(z))$, where $N_k = \#\{\alpha \in \mathbb{N}^d, |\alpha|=k\}$. Then for $|\alpha|=|\beta|$, the action of $t^\alpha D_t^\beta$ on $\hat{v}$ is represented by a matrix with entries in $\mathbb{Z}$ acting on $\vec{v}$. Thus we may formulate (2.1.13) as a Cauchy problem:

$$(2.1.14) \qquad \tilde{P}_{0,k}\vec{u} = \vec{v}, \qquad \vec{u}|_Y = (\vec{w}).$$

Now $\tilde{P}_{0,k}$ is a $(N_k \times N_k)$-matrix whose entries are differential operators on Z of order at most m, the terms on the diagonal are exactly of order m, and non characteristic with respect to Y, the terms off the diagonal are of order strictly less than m, and $\vec{v}$ (resp. $(\vec{w})$) is a column of holomorphic functions on Z (resp. $Z \cap Y$). Therefore (2.1.14) is solved by using the classical Cauchy-Kowalewski theorem for such square matrices, (cf. I, Remark 3.3.2). $\square$

2.2 Application: The Cauchy Problem with Data Ramified along Hypersurfaces

Proposition 2.1.6 is not so useful in practice, since if P is a differential operator, Y a non characteristic hypersurface, Z' a hypersurface of Y, then in general there exists no hypersurface Z in X such that $Z \cap Y = Z'$, and such that the hypotheses of the proposition are satisfied for Y, Z, P.

For example take $X = \mathbb{C}^n$, with coordinates $(x_1, \ldots, x_n)$, $Y = \{x \in X;\ x_1 = 0\}$ $Z' = \{x \in Y;\ x_1 = x_2 = 0\}$, $P = \sum_{j=1}^n D_j^2$. We want to solve the Cauchy problem:

$$Pf = 0, \qquad f|_Y = 0, \qquad \frac{\partial f}{\partial x_1}\bigg|_Y = 1/x_2.$$

There are no hypersurfaces $Z \subset X$ transversal to Y, with $Z \cap Y - Z'$, and such that $\sigma(\hat{\sigma}_Z(P))(H_{x_1}) \neq 0$ (recall that $\hat{\sigma}_Z(P)$ has been defined in § 1.4).

Let $S = \{(x, \xi) \in T^*X;\ \sum_{j=1}^n \xi_j^2 = 0\}$ be the characteristic variety of P. It is natural to look for a solution f whose singularities will be the projection on X of the bicharacteristic curves of S issuing from the singularities of $\left(f|_Y, \frac{\partial f}{\partial x_1}\big|_Y\right)$, that is, issuing from $\rho^{-1}(T_{Z'}^*Y) \cap S$. Write $(x, \xi) = (x_1, x_2, x''; \xi_1, \xi_2, \xi'')$. Then:

$$\rho^{-1}(T_{Z'}^*Y) \cap S = \{(x, \xi);\ x_1 = x_2 = 0,\ \xi_1^2 + \xi_2^2 = 0,\ \xi'' = 0\}.$$

The union of the bicharacteristic curves issuing from this set consists of two Lagrangean manifolds:

$$\Lambda^{\pm}=\{(x,\xi);\ x_1\pm ix_2=0,\xi_1\mp i\xi_2=0,\xi''=0\}.$$

In fact

$$H_{\sigma(P)}=2\left(\sum_j \xi_j \frac{\partial}{\partial x_j}\right)$$

$$=2\left(\xi_1 \frac{\partial}{\partial x_1}+\xi_2 \frac{\partial}{\partial x_2}\right)\quad \text{on}\quad \Lambda^+\cup\Lambda^-$$

$$=2\xi_1\left(\frac{\partial}{\partial x_1}\pm i\frac{\partial}{\partial x_2}\right)\quad \text{on}\quad \Lambda^{\mp}.$$

Thus we have found two hypersurfaces, $Z^{\pm}=\{x\in X;\ x_1\pm ix_2=0\}$ which are candidates for carrying the singularities of f.

In order to formulate the theorem let us recall the construction of some $\mathcal{D}_X$-modules naturally associated to a hypersurface Z (cf. I, § 4.2). First recall that $\dot{\pi}_*\mathscr{C}_{Z|X}$ is a coherent $\mathcal{D}_X$-module, endowed with a canonical generator denoted Y_Z. If $\phi=0$ is an equation of Z, we also write $Y(\phi)$ instead of Y_Z, and $\mathcal{D}_X Y(\phi)$ instead of $\dot{\pi}_*\mathscr{C}_{Z|X}$. We have also introduced in (I, § 4.2) the left $\mathcal{D}_X$-module $\mathcal{D}_X \operatorname{Log}\phi$ (after a choice of local coordinates on X), and we have considered the exact sequence:

(2.1.1) $$0\to\mathcal{O}_X\to\mathcal{D}_X\log\phi\to\mathcal{D}_X Y(\phi)\to 0.$$

We set:

(2.2.2) $$\mathcal{O}^1_{Z|X}=(\mathcal{D}_X\log\phi)|_Z.$$

Locally on Z, this $\mathcal{D}_X$-module only depends on Z, up to isomorphism. Now let $Z_1,\ldots,Z_r$ be r hypersurfaces of X, and let $Z=\bigcap_{i=1}^r Z_i$. We define the left $\mathcal{D}_X$-module $\sum_{i=1}^r \mathcal{O}^1_{Z_i|X}$ on Z by the exact sequence:

(2.2.3) $$0\to(\mathcal{O}^{r-1}_X)|_Z \xrightarrow{\lambda} \bigoplus_{i=1}^r (\mathcal{O}^1_{Z_i|X})|_Z \to \sum_{i=1}^r \mathcal{O}^1_{Z_i|X}\to 0,$$

where $\lambda(f_1,\ldots,f_{r-1})=(f_1,f_2-f_1,\ldots,-f_{r-1})$.

We obtain the following diagram of exact sequences on Z:

$$
\begin{array}{ccc}
0 & & 0 \\
\downarrow & & \downarrow \\
0\to(\mathcal{O}_X^{r-1})|_Z \to (\mathcal{O}_X^{r-1})|_Z \to 0 \\
\downarrow{\scriptstyle\lambda} & & \downarrow{\scriptstyle\lambda} \\
0\to(\mathcal{O}_X^r)|_Z \to \left(\overset{r}{\underset{i=1}{\bigoplus}}\ \mathcal{O}_{Z_i|X}^1\right)|_Z \to \left(\overset{r}{\underset{i=1}{\bigoplus}}\ \dot\pi_*\mathscr{C}_{Z_i|X}\right)|_Z \to 0 \\
\downarrow & & \downarrow \\
0\to(\mathcal{O}_X)|_Z \to \overset{r}{\underset{i=1}{\sum}}\ \mathcal{O}_{Z_i|X}^1 \to \left(\overset{r}{\underset{i=1}{\bigoplus}}\ \dot\pi_*\mathscr{C}_{Z_i|X}\right)|_Z \to 0. \\
\downarrow & & \downarrow \\
0 & & 0
\end{array}
$$

Theorem 2.2.1. *Let* Y *be a submanifold of* X, Z *a hypersurface of* Y, $Z_i (i=1,\ldots,r)$ *hypersurfaces of* X *pairwise transversal and transversal to* Y, *with* $Z_i \cap Y = Z$ *for any* i. *Let* $\mathscr{M}$ *be a coherent* $\mathscr{D}_X$-*module such that:*

a) $\mathrm{char}(\mathscr{M}) \cap \rho^{-1}(T_Z^* Y) \subset \overset{r}{\underset{i=1}{\bigcup}}\ T_{Z_i}^* X.$

b) *For any* $i=1,\ldots,r$, Y *is non* 1-*microcharacteristic for* $\mathscr{M}$ *along* $\dot T_{Z_i}^* X$ *on* $\dot T^* X$.
Then for all j *the natural morphisms on* Z:

$$
\mathscr{E}xt_{\mathscr{D}_X}^j\left(\mathscr{M}, \overset{r}{\underset{i=1}{\sum}}\ \mathcal{O}_{Z_i|X}^1\right) \to \mathscr{E}xt_{\mathscr{D}_Y}^j(\mathscr{M}_Y, \mathcal{O}_{Z|Y}^1)
$$

are isomorphisms.

Proof. Let $p \in \dot T_Z^* Y$, $z = \pi(p)$, and let $p_i = \rho^{-1}(p) \cap T_{Z_i}^* X$. Applying Theorem 2.1.2 we get the isomorphisms:

$$
\mathscr{E}xt_{\mathscr{D}_X}^j(\mathscr{M}, \mathscr{C}_{Z_i|X})_{p_i} \simeq \mathscr{E}xt_{\mathscr{E}_{Y,p}}^j(\mathscr{E}_{Y\to X} \otimes_{\mathscr{E}_X} \mathscr{M}_{p_i}, \mathscr{C}_{Z|Y,p}).
$$

By hypothesis a) we have:

$$
(\mathscr{E}_Y \otimes_{\mathscr{D}_Y} \mathscr{M}_Y)_p = \overset{r}{\underset{i=1}{\bigoplus}}\ (\mathscr{E}_{Y\to X} \otimes_{\mathscr{E}_X} \mathscr{M}_{p_i}).
$$

Thus we get:

(2.2.4) $\qquad \mathscr{E}xt_{\mathscr{D}_X}^j\left(\mathscr{M}, \overset{r}{\underset{i=1}{\bigoplus}}\ \dot\pi_*\mathscr{C}_{Z_i|X}\right)|_Z \simeq \mathscr{E}xt_{\mathscr{D}_Y}^j(\mathscr{M}_Y, \dot\pi_*\mathscr{C}_{Z|Y}).$

Now we apply the functor $\mathscr{H}om_{\mathscr{D}_X}(\mathscr{M}, \cdot)$ to the exact sequence (2.2.3) and the functor $\mathscr{H}om_{\mathscr{D}_Y}(\mathscr{M}_Y, \cdot)$ to the exact sequence $0 \to (\mathcal{O}_Y)|_Z \to \mathcal{O}_{Z|X}^1 \to \dot\pi_*\mathscr{C}_{Z|Y} \to 0$. Then by (2.2.4) and Theorem 2.1.4 we obtain the result. $\quad\square$

Finally we shall explain how Theorem 2.2.1 applies to operators with simple characteristics (cf. Leray [1], Hamada [1]).

Proposition 2.2.2. *Let Y be a hypersurface of X, Z a hypersurface of Y, P a differential operator of order m on X. We assume that Y is non characteristic for P and P has "simple characteristics transversal to $Y \underset{X}{\times} T^*X$ at $\rho^{-1}(T^*_{\dot{Z}} Y)$",*
that is:

– if $\phi = 0$ is a local equation of Y, then the Poisson bracket $\{\phi, \sigma(P)\}$ does not vanish at any point of $\rho^{-1}(T^*_{\dot{Z}} Y) \cap \sigma(P)^{-1}(0)$. Then there exists m hypersurfaces $Z_1, \ldots, Z_m$, pairwise transversal, defined in a neighborhood of Z in X, such that:

a) *each Z_i is transversal to Y, and $Z_i \cap Y = Z$, $i = 1, \ldots, m$.*
b) *$T^*_{\dot{Z_i}} X \subset \sigma(P)^{-1}(0)$, $i = 1, \ldots, m$.*

c) *$\rho^{-1}(T^*_{\dot{Z}} Y) \cap \sigma(P)^{-1}(0) \subset \bigcup_{i=1}^{m} T^*_{\dot{Z_i}} X$.*

d) *Y is non 1-microcharacteristic for P (i.e.: for the Module $\mathscr{D}_X / \mathscr{D}_X \cdot P$) on each $\dot{T}^*_{Z_i} X$, $i = 1, \ldots, m$.*

In other words the assumptions of Theorem 2.2.1 are satisfied for $\mathscr{M} = \mathscr{D}_X / \mathscr{D}_X \cdot P$.

Proof. Put $\Lambda' = \rho^{-1}(T^*_{\dot{Z}} Y) \cap \sigma(P)^{-1}(0)$. By the hypothesis, Λ' is the disjoint union of m isotropic manifolds $\Lambda'_1, \ldots, \Lambda'_m$. In fact by choosing local coordinates $(x_1, \ldots, x_n) = (x_1, x_2, x'')$ on X such that $Y = \{x; \ x_1 = 0\}$, $Z = \{(x_2, x''); \ x_2 = 0\}$, we see that $\sigma(P)(x; 1, 0, \ldots, 0)$ is different from zero for $x \in Y$, and the equation $\sigma(P)(x; \tau, 1, 0, \ldots, 0) = 0$ has m distinct roots $\tau_1(x), \ldots, \tau_m(x)$ for $x = (0, 0, x'') \in Z$, since $\dfrac{\partial \sigma(P)}{\partial \xi_1} = -\{x_1, \sigma(P)\}$ is not zero at those points.

Now the Hamiltonian vector field $H_{\sigma(P)}$ is transversal to $Y \underset{X}{\times} T^*X$ at Λ'_i. Therefore the union of the integral curves of this vector field issuing from Λ'_i is a Lagrangean manifold Λ_i, and $\sigma(P)^{-1}(0)$ being involutive, it contains Λ_i (cf. Appendix A).

Let $Z_i = \pi(\Lambda_i)$. Since the projection on X of the vector field $(H_{\sigma(P)})|_{\Lambda_i}$ is a vector field transversal to Y, Z_i is a hypersurface transversal to Y, and Λ_i being Lagrangean, $\Lambda_i = \dot{T}^*_{Z_i} X$.

It remains to prove that Y is non 1-microcharacteristic for P on each Λ_i. Let E_i be an invertible microdifferential operator of order $m-1$ defined in a neighborhood of $Y \underset{X}{\times} \Lambda_i$, and let $P_i = E_i^{-1} \circ P$. Then P_i is of order one, and vanishes on Λ_i. Thus P_i belongs to $\mathscr{E}_{\Lambda_i}$ and $\sigma^1_{\Lambda_i}(\sigma(P))(H_{x_1}) = \sigma_{\Lambda_i}(\sigma(P))(H_{x_1}) \neq 0$. $\square$

Remark 2.2.3. When $\mathscr{M} = \mathscr{D}_X / \mathscr{D}_X \cdot P$ and P satisfies the assumptions of Proposition 2.2.2 we find the Hamada theorem, (Hamada [1]). In fact $\mathscr{O}_{Z|X}^1$ is isomorphic to the sheaf on Z of holomorphic functions having polar or logarithmic singularities along Z.

Exercises to III.2

Ex. 2.1. Let $X = \mathbb{C}^n$ with coordinates $(x_1, \ldots, x_n)$, $n \geqslant 3$, let $Y = \{x \in X; \ x_1 = 0\}$, $Z = \{x \in X; \ x_2 = x_3 = 0\}$. Solve the following Cauchy problems in $\mathscr{B}_{Z|X}$:

$$Pf = 0, \quad f|_Y = 0, \quad \frac{\partial f}{\partial x_1}\Big|_Y = h,$$

where h is the section $\dfrac{1}{x_2^3 x_3^2}$ of $\mathscr{B}_{Z \cap Y|Y}$, and P is one of the operators:

   i) $P_1 = D_1^2 + (x_2 D_2)^2 + (x_3 D_3)^2$,

   ii) $P_2 = D_1^2 + (x_2 D_3)^2$,

   iii) $P_3 = D_1^2 + (x_2 D_3)^2 + x_3^2 D_2$.

Ex. 2.2. State and prove an analogous to Proposition 2.1.6 with the sheaf $\mathscr{B}_{Z|X}$ replaced by the sheaf $\hat{\mathscr{C}}_{Z|X} = \hat{\mathscr{E}}_X \otimes_{\mathscr{E}_X} \mathscr{C}_{Z|X}$.

Ex. 2.3. Let Z be a submanifold of X, Q an operator of $\mathsf{F}_{-1}^Z \mathscr{D}_X$. Prove that $1 + Q$ defines an isomorphism on $\mathscr{B}_{Z|X}$.

Ex. 2.4. Let $X = \mathbb{C}^n$ with coordinates $(x_1, \ldots, x_n) = (x_1, x')$, $Y = \{x \in X; x_1 = 0\}$, $Z = \{x' \in Y; \ x_2 = 0\}$.

Describe the singularities of the holomorphic solutions of the Cauchy problem:

$$Pf = 0, \quad f|_Y = 0, \quad \frac{\partial f}{\partial x_1}\Big|_Y = \frac{1}{x_2},$$

when $P = (D_1 - 2x_1 D_2)^2 + (x_2 + x_1^2) Q_1 + Q_0$, where Q_1 is a differential operator of order one, and Q_0 is a holomorphic function on X.

Ex. 2.5. Let X, Y, Z be as in Exercise 2.4. By solving formally the Cauchy problem:

$$(D_1^2 + D_2)f = 0, \quad f|_Y = 0, \quad \frac{\partial f}{\partial x_1}\Big|_Y = \frac{1}{x_2},$$

prove that there exists no solution of this problem in $\mathscr{B}_{Z|X}$.

Prove that in a neighborhood of Y there exists a unique solution f holomorphic in $X \backslash Z$ (Hint: look at f as a Laurent series, $f = \sum\limits_{n \geqslant 0} a_n(x_1) x_2^{-n}$, cf. Sato-Kashiwara-Kawaï [1 Chap. II, § 5.2, p. 446]).

Ex. 2.6. Let $X, Y, Z, P, Z_1, \ldots, Z_m$ be as in Proposition 2.2.2. Let $p \leqslant m$, $f \in \sum_{i=1}^{m-p} \mathcal{O}^1_{Z_i|X}$. Assume that $Pf = 0$, and the p first traces of f (which belong to $\mathcal{O}^1_{Z_i|Y}$) are zero. Prove that $f = 0$ (cf. Pallu de la Barrière-Schapira [1]).

§ 3. Propagation

3.1. Propagation at the Boundary for one Operator

We denote by U an open set in T^*X, and by V a conic involutive manifold, closed in U.

Let Σ be a bicharacteristic leaf of V (cf. Appendix A), and let p belong to Σ. Since the tangent space $T_p\Sigma$ to Σ at p is the orthogonal space to the tangent space to V, and the dual to $(T_p V)^\perp$ is $(T_V T^*X)_p$, we find:

$$(3.1.1) \qquad\qquad T^*\Sigma \simeq T_V T^*X \underset{V}{\times} \Sigma.$$

If ϕ is a function on Σ, we may locally extend ϕ as a function $\tilde\phi$ on T^*X and the image of $d\phi$ in $T_V T^*X \underset{V}{\times} \Sigma$ by the isomorphism (3.1.1) will be the image in $T_V T^*X \underset{V}{\times} \Sigma$ of the Hamiltonian vector field $H_{\tilde\phi}$. Therefore we denote this vector field by H_ϕ.

We shall also make use of $X^{\mathbb{R}}$, the real underlying manifold to X. (cf. § 4.1 below for more details). There exists a natural isomorphism $T^*X^{\mathbb{R}} \simeq (T^*X)^{\mathbb{R}}$, and for a real function ϕ of class C^1 on $X^{\mathbb{R}}$, this isomorphism associates the vector $\partial\phi(x) \in T^*_x X$ to the vector $d\phi(x) \in T^*_x X^{\mathbb{R}}$ ($d\phi$ is the real differential and $\partial\phi$ the complex one).

In this section we shall often identify X and $X^{\mathbb{R}}$ or T^*X and $T^*X^{\mathbb{R}}$, without risk of confusion.

If ϕ is a real C^0-function on X, we denote by $\{\phi \geqslant 0\}$ the closed subset $\{x \in X; \phi(x) \geqslant 0\}$.

Now let $\mathscr{L}$ be a system with simple characteristics along V, u_0 a simple generator of $\mathscr{L}$. We set:

$$(3.1.2) \qquad\qquad \mathscr{L}(k) = \mathscr{E}_X(k)u_0.$$

Lemma 3.1.1. *For each $k \in \mathbb{Z}$, $\mathscr{L}(k)$ is naturally endowed with a structure of a left $\mathscr{E}_V$-module.*

Proof. Let $\mathscr{I}$ be the annihilator Ideal of u_0. By the hypothesis, $\mathrm{gr}(\mathscr{I})$ is reduced, and coincides with the defining Ideal of V.

Let $P \in \mathscr{E}_V(1)$. Then $\sigma_1(P)$ belongs to $\mathrm{gr}(\mathscr{I})$ and P may be decomposed as $P = Q + R$, with $Q \in \mathscr{I}$, $R \in \mathscr{E}_X(0)$. Thus:

$$(3.1.3) \qquad\qquad P\mathscr{E}_X(k) \subset \mathscr{E}_X(k) + \mathscr{I} \quad \forall k.$$

Since $\mathscr{E}_V$ is generated by $\mathscr{E}_V(1)$, (3.1.3) holds for any $P \in \mathscr{E}_V$, which proves the Lemma. $\square$

Theorem 3.1.2. *Let V be a conic involutive manifold, Σ a bicharacteristic leaf of V, $p \in \Sigma$, ϕ a real C^1-function on Σ. let $\mathscr{L}$ be a system with simple characteristics along V, u_0 a simple generator of $\mathscr{L}$, $\mathscr{L}(k) = \mathscr{E}_X(k) u_0$, and let P be a section of $\mathscr{E}_V$. Assume $\sigma_V^1(\sigma(P))(H_\phi(p)) \neq 0$. Then for all $j \geqslant 0$, all $k \in \mathbb{Z}$, P induces an isomorphism on each of the cohomology groups $(\mathscr{H}^j_{\{\phi \geqslant 0\}}(\mathscr{L}(k)|_\Sigma))_p$ and $(\mathscr{H}^j_{\{\phi \geqslant 0\}}(\mathscr{L}|_\Sigma))_p$.*

(For the definition of $\sigma_V^1(\sigma(P))$ cf. Remark 1.2.10).

Proof. First we shall reduce the problem to the case where V is regular involutive, using the same trick as in the proof of Theorem 2.1.2.

Let t be a coordinate on $\mathbb{C}$, (t, τ) the coordinates on $T^*\mathbb{C}$, ψ the map from $T^*X \times \dot{T}^*\mathbb{C}$ to T^*X defined by $\psi(x, t; \xi, \tau) = (x; \xi/\tau)$. Let $\tilde{V} = \psi^{-1}(V)$, $\tilde{p} = (p; (0, dt))$, and let $\tilde{\Sigma}$ be the leaf of $\tilde{V}$ through $\tilde{p}$. Then ψ is a diffeormorphism from $\tilde{\Sigma}$ to Σ, and we get:

$$(3.1.4) \qquad (\mathscr{H}^j_{\{\phi \circ \psi \geqslant 0\}}(\psi^{-1}\mathscr{L}|_{\tilde{\Sigma}}))_{\tilde{p}} \simeq (\mathscr{H}^j_{\{\phi \geqslant 0\}}(\mathscr{L}|_\Sigma))_p$$

and similarly with $\mathscr{L}$ replaced by $\mathscr{L}(k)$.

Thus by applying the functor $\Gamma_{\{\phi \circ \psi \geqslant 0\}}(\cdot|_{\tilde{\Sigma}})_{\tilde{p}}$ to the sequence (2.1.2) (of Section 2), we find that it is enough to prove the theorem with $\mathscr{L}$ replaced by $\mathscr{L} \hat{\otimes} \mathscr{E}_{\mathbb{C}}$, and similarly for $\mathscr{L}(k)$. Therefore we may assume from the beginning that V is regular involutive.

By a quantized contact transformation, we may even assume that $X = \mathbb{C}^r \times \mathbb{C}^{n-r}$ with coordinates $z = (x, y)$, $x = (x_1, \ldots, x_r)$, $y = (y_{r+1}, \ldots, y_n)$ and:

$$(3.1.5) \qquad \begin{cases} p = (0; dy_n) \\ V \text{ is an open subset of } \{(x, y; \xi, \eta) \in T^*X; \xi = 0\}, \\ \Sigma = \{(x, y; 0, \eta) \in V; y = 0, \eta = (0, \ldots, 0, 1)\}, \\ \mathscr{L} = \mathscr{E}_X/\mathscr{E}_X(D_1, \ldots, D_r), \text{ and } u_0 = 1 \bmod \mathscr{E}_X(D_1, \ldots, D_r). \end{cases}$$

At this stage of the proof, we need to study the cohomology of the sheaves $\mathscr{L}$ and $\mathscr{L}(k)$.

Lemma 3.1.3. a) *The sheaf $\mathscr{L}|_\Sigma$ on Σ satisfies the "analytic extension principle": if ω and Ω are two open subsets of Σ, $\omega \subset \Omega$, Ω connected, ω non empty, and if u is a section of $\mathscr{L}|_\Sigma$ on Ω which is zero on ω, then $u = 0$.*

b) *Let K be a compact subset of Σ. Then for $j \geqslant 0$:*

$$H^j(K, \mathscr{S}|_\Sigma) = \varinjlim_k H^j(K, \mathscr{S}(k)|_\Sigma).$$

c) *Let Ω be a domain of holomorphy in Σ. Then:*

$$H^j(\Omega, \mathscr{S}(k)|_\Sigma) = 0 \qquad \forall k, \ \forall j > 0.$$

Proof of Lemma 3.1.3. Let $Y = \mathbb{C}^{n-r}$, $q = (0; 0, \ldots, 0, 1) \in T^* Y$. Then $\Sigma \simeq \mathbb{C}^r \times \{q\}$, and the sheaf $\mathscr{S}|_\Sigma$ may be identified to the sheaf $\mathcal{O}_{\mathbb{C}^r} \hat{\otimes} \mathscr{E}_{Y,q}$ on $\mathbb{C}^r$ of microdifferential operators on $\mathbb{C}^r \times Y$, defined in a neighborhood of $\mathbb{C}^r \times \{q\}$ and independent of $D_1, \ldots, D_r$. Therefore a) is clear.

As already noticed in the course of Chapter II, there exists a local isomorphism of filtered sheaves between $\mathscr{E}_Y$ and $\mathscr{C}_{Y|Y \times Y}$ (where Y is identified to the diagonal of $Y \times Y$), and $\mathscr{C}_{Y|Y \times Y}$ is isomorphic, by a quantized Legendre transformation, to $\mathscr{C}_{Z|Y \times Y}$, for Z a hypersurface of $Y \times Y$.

Since one can extend all these isomorphisms by adding holomorphic parameters, we find that $\mathscr{S}|_\Sigma$ is isomorphic as a filtered sheaf, to $(\mathscr{C}_{\mathbb{C}^r \times Z|\mathbb{C}^r \times Y \times Y})|_{\mathbb{C}^r \times \{q'\}}$, where $q' \in \dot{T}^*_Z(Y \times Y)$.

Now we change our notations, and replace $\mathbb{C}^r \times Y \times Y$ by Y, and $\mathbb{C}^r \times Z$ by Z. We choose coordinates $(x_1, \ldots, x_N)$ on Y such that:

$$Z = \{x; \ x_N = 0\},$$

$$\Sigma = \{(x, \xi); \ x_{r+1} = \ldots = x_N = \xi_1 = \ldots = \xi_{N-1} = 0, \ \xi_N = 1\}.$$

Let $S = \{x \in Y; \ x_{r+1} = \ldots = x_N = 0\}$ be the projection of Σ on Y. When identifying S and Σ, we have isomorphisms (cf. I, § 4):

$$(3.1.6) \qquad \begin{cases} (\mathscr{C}_{Z|Y})|_\Sigma \simeq (\mathcal{O}_Y)|_S \oplus (\mathscr{B}_{Z|Y})|_S, \\ (\mathscr{C}_{Z|Y}(k))|_\Sigma \simeq (\mathcal{O}_Y)|_S \oplus (\mathscr{B}_{Z|Y}(k))|_S \quad \text{for } k \geqslant 0. \end{cases}$$

Let T be an indeterminate. There exists an isomorphism of filtered sheaves:

$$\mathscr{B}_{Z|Y} = \mathcal{O}_Z[T] \underset{\text{def}}{=} \mathcal{O}_Z \otimes_\mathbb{Z} \mathbb{Z}[T].$$

Let $\mathcal{O}_Z[T]_k$ be the subsheaf of $\mathcal{O}_Z[T]$ of sections of degree less than k. Let K be a compact subset of Z. We have:

$$(3.1.7) \qquad H^j(K, \mathcal{O}_Z[T]) = \varinjlim_k H^j(K, \mathcal{O}_Z[T]_k).$$

In fact let $\mathcal{O}_Z^\bullet$ be a flabby resolution of the sheaf $\mathcal{O}_Z$. Then:

$$\Gamma(K, \mathcal{O}_Z^\bullet[T]) = \varinjlim_k \Gamma(K, \mathcal{O}_Z^\bullet[T]_k).$$

To prove c), we may assume $k = 0$, since all the sheaves $\mathscr{C}_{Z|Y}(k)|_\Sigma$ are isomorphic. Then the result follows from a theorem of Siu [1] which asserts that $H^j(\Omega, (\mathcal{O}_Y)|_S) = 0$ for $j > 0$, Ω a domain of holomorphy in $S \subset Y$.

Proof of Theorem 3.1.2. Continued. Adding a dummy variable t, so that t and D_t commute with P, we see that it is equivalent to prove the theorem for $\mathscr{L}(0)|_\Sigma$ or for $\mathscr{L}(k)|_\Sigma$. For the sake of simplicity in the notation, we set:

$$\mathscr{L}_\Sigma^0 \underset{\text{def}}{=} \mathscr{L}(0)|_\Sigma.$$

First we prove the injectivity of P on $\mathscr{H}^1_{[\phi \geqslant 0]}(\mathscr{L}_\Sigma^0)_p$, following a classical argument due to M. Zerner [1] (cf. also Bony-Schapira [1] and Hörmander [4 th. 9.4.7.]).

Lemma 3.1.4. *In the situation of the theorem let* $\Omega = \{x \in \Sigma; \; \phi(x) < 0\}$ *and let* u *be a section of* $\mathscr{L}_\Sigma^0$ *on* Ω. *Assume that* Pu *extends as a section of* $\mathscr{L}_\Sigma^0$ *to a neighborhood of* p. *Then* u *itself extends to a section of* $\mathscr{L}_\Sigma^0$ *in a neighborhood of* p.

Proof of Lemma 3.1.4. If $d\phi(p) = 0$, then P is invertible in a neighborhood of P, and the conclusion follows. Now assume $d\phi(p) \neq 0$. We choose coordinates $(x_1, \ldots, x_r)$ on Σ, such that $x = 0$ at p and $\phi(x) = \text{Re}\, x_1 - \psi(\text{Im}\, x_1, x_2, \ldots, x_r)$ with $d\psi(0) = 0$.

By dividing P by an invertible operator of order zero, we may assume that P is as in (I, (3.1.3)).

Let $H_\varepsilon = \{x \in \Sigma; \; x_1 = -\varepsilon\}$, and let

$$\gamma_\varepsilon(u) = \left\{ u|_{H_\varepsilon}, \ldots, \frac{\partial^{m-1} u}{\partial x_1^{m-1}}\bigg|_{H_\varepsilon} \right\}$$

be the set of m first traces of u on H_ε, m being the order of P. Consider the Cauchy problem:

$$Pu_\varepsilon = Pu, \qquad \gamma_\varepsilon(u_\varepsilon) = \gamma_\varepsilon(u).$$

Applying Theorem 3.1.1 of Chapter I, we find that $u_\varepsilon = u$ is defined in a ball $B_\varepsilon(\delta R)$ of center $(-\varepsilon, 0, \ldots, 0)$ and radius δR, where δ is a constant depending only on P. Moreover R is defined by:

$$-\varepsilon < \psi(0, x') \quad \text{for} \quad |x'| < R.$$

Since $d\psi(0) = 0$, $\psi(0, x') = 0(|x'|)$, and the ball $B_\varepsilon(\delta R)$ will be a neighborhood of 0 for ε small enough. $\square$

We need a global version of Lemma 3.1.4. The following lemma is very close to a result of L. Hörmander concerning global unicity of the Cauchy problem (cf. Hörmander [1, Th. 5.3.3] and Bony-Schapira [1]).

We endow $\mathbb{C}^r$ with the Hermitian norm, $|x|^2 = \langle x, \bar{x} \rangle$. For a subset Ω of $\Sigma \subset \mathbb{C}^r$ we define "1-microchar(Ω)" as the closure in $\mathbb{C}^r$ of the set of codirections θ such that there exists $x \in \Omega$, with $\sigma^1_V(\sigma(P))(x, \theta) = 0$.

Lemma 3.1.5. *Let* ω *and* Ω *be two convex sets in* Σ, $\omega \subset \Omega$, Ω *open,* ω *locally closed. We assume that any real hyperplane whose conormal belongs to 1-mi-*

crochar(Ω) and which intersects Ω, intersects ω. Let u be a section of $\mathscr{L}_\Sigma^0$ on ω such that Pu extends to Ω. Then u itself extends to Ω.

Proof of Lemma 3.1.5. Let $x_1 \in \omega$. For any $x_2 \in \Omega$ we will find an extension of u to a neighborhood of the segment $[x_1, x_2]$. As the sections of $\mathscr{L}_\Sigma^0$ satisfy the analytic extension principle, we get an extension of u to the open set Ω.

Let $\eta > 0$ be such that the closed ball $\overline{B}(x_2, \eta)$ with center x_2 and radius η, is contained in Ω. The set 1-microchar(Ω) being closed and conic, there exists a compact set K in ω such that any real hyperplane whose conormal belongs to 1-microchar(Ω) which intersects $\overline{B}(x_2, \eta)$ will intersect K. Of course we may assume K convex, and $x_2 \in K$. Let $0 < \varepsilon < \eta$, such that u is defined in $K_{2\varepsilon}$, the set of points at distance at most 2ε from K. For $0 \leqslant t \leqslant 1$ set $x_t = x_1 + t(x_2 - x_1)$, and let M_t be the convex hull of $K_\varepsilon \cup B(x_t, \varepsilon)$. The boundary of M_t is C^1, and at any point of this boundary not belonging to K_ε, the conormal to M_t does not belong to 1-microchar(Ω). Let t_0 be the biggest t such that u extends as a section of $\mathscr{L}_\Sigma^0$ on M_t. By the preceding lemma we cannot have $t < 1$. $\quad\square$

Proof of Theorem 3.1.2. Continued. As in Lemma 3.1.4, we set $\Omega = \{x \in \Sigma; \ \phi(x) < 0\}$, and we assume $x = 0$ at $p \in \partial\Omega$, and $d\phi(0) = (1, 0, \ldots, 0)$. We denote this vector by θ. We write $x = (x_1, x')$, and $\xi = (\xi_1, \xi')$. If Γ is a convex cone with vertex at 0 in $\mathbb{C}^r$, the polar cone Γ^0 is the closed convex cone defined by:

(3.1.8) $$\Gamma^0 = \{\xi \in \mathbb{C}^r; \ \mathrm{Re}\langle x, \xi\rangle \geqslant 0 \quad \forall x \in \Gamma\}.$$

Now we choose $\varepsilon_0 > 0$ and $\alpha_0 > 0$ such that:

(3.1.9) $$|x| \leqslant \varepsilon_0, \quad \xi \neq 0, \quad |\xi'| \leqslant \alpha_0 |\xi_1| \Rightarrow \sigma_V^1(\sigma(P))(x, \xi) \neq 0.$$

Let Γ be the open convex cone with vertex at zero given by:

(3.1.10) $$\xi \in \Gamma^0 \Leftrightarrow (|\mathrm{Im}\,\xi_1| + |\xi'|) \leqslant \alpha_0 \mathrm{Re}\,\xi_1.$$

We put, for $x \in \Sigma, \varepsilon \geqslant 0$:

$$\begin{aligned}
\Gamma_x &= x + \Gamma, \\
K_{x,\varepsilon} &= \Gamma_x \cap \{y \in \Sigma; \ \mathrm{Re}\langle y - x, \theta\rangle = -\varepsilon\}, \\
H_{x,\varepsilon} &= \Gamma_x \cap \{y \in \Sigma; \ \langle y - x, \theta\rangle = -\varepsilon\}, \\
\Gamma_{x,\varepsilon}^+ &= \Gamma_x \cap \{y \in \Sigma; \ \mathrm{Re}\langle y - x, \theta\rangle > -\varepsilon\}.
\end{aligned}$$

First we remark that if a vector $\zeta = (\zeta_1, \zeta')$ satisfies $|\zeta'| > \alpha_0 |\zeta_1|$, then the real hyperplane $L_\zeta = \{x; \ \mathrm{Re}\langle x, \zeta\rangle = 0\}$ intersects $H_{0,\varepsilon}$ for all $\varepsilon > 0$. In fact assuming otherwise, we find that the equations:

(3.1.11) $$\begin{cases} \mathrm{Re}\langle x, \zeta\rangle = 0, \\ \mathrm{Im}\langle x, \theta\rangle = 0, \end{cases}$$

have no common solution $x \in \Gamma$ other than 0. Therefore there exist two positive numbers λ, μ, such that:

(3.1.12)
$$\lambda \zeta + i \mu \theta \in \Gamma^0.$$

By (3.1.10) and (3.1.12) we find $\lambda \neq 0$, thus $\zeta + i \mu / \lambda \theta \in \Gamma^0$ and $|\zeta'| \leqslant \alpha_0 \operatorname{Re} \zeta_1$. This is a contradiction.

Now $\varepsilon_0, \alpha_0, \Gamma$ being given as in (3.1.9) and (3.1.10), let $\varepsilon_1 > 0$ be such that the set $\Gamma_{x,\varepsilon}^+$ is contained in the ball $\{x; \ |x| < \varepsilon_0\}$ for $|x| \leqslant \varepsilon_1, 0 < \varepsilon \leqslant \varepsilon_1$.

Lemma 3.1.6. *Let $x \in \Sigma$ with $|x| \leqslant \varepsilon_1$, and $\varepsilon > 0$ with $\varepsilon \leqslant \varepsilon_1$. Let ω be a convex open set in $\Gamma_{x,\varepsilon}^+$ whose closure contains $K_{x,\varepsilon}$. Then P is an isomorphism from $\mathscr{L}_\Sigma^0(\omega) / \mathscr{L}_\Sigma^0(\Gamma_{x,\varepsilon}^+)$ onto itself.*

Proof of Lemma 3.1.6. The injectivity of P follows immediately from Lemma 3.1.5. Now choose $\varepsilon' > \varepsilon$, $\varepsilon' - \varepsilon$ small enough, and let:

$$\omega' = \omega \cup (\Gamma_{x,\varepsilon'}^+ \setminus \Gamma_{x,\varepsilon}^+).$$

Then:

$$\begin{cases} \Gamma_{x,\varepsilon}^+ \cup \omega' = \Gamma_{x,\varepsilon'}^+, \\ \Gamma_{x,\varepsilon}^+ \cap \omega' = \omega. \end{cases}$$

Since convex open sets are acyclic for the sheaf $\mathscr{L}_\Sigma^0$ by Lemma 3.1.3, we can solve the Cousin problem (cf. Theorem C.3.2 of the Appendix) and decompose a section $u \in \mathscr{L}_\Sigma^0(\omega)$ as $u = u_1 + u_2$, $u_1 \in \mathscr{L}_\Sigma^0(\Gamma_{x,\varepsilon}^+)$, $u_2 \in \mathscr{L}_\Sigma^0(\omega')$. To prove the surjectivity of P we may replace u by u_2 when solving the equation $Pv = u$, and thus we may assume that u is defined in a convex neighborhood of $K_{x,\varepsilon}$. Then we solve the equation $Pv = u$ in a neighborhood of $H_{x,\varepsilon}$, using (I, Theorem 3.1.1), and extend v to a neighborhood of $K_{x,\varepsilon}$, using Lemma 3.1.5 and the preceding remark that any 1-microcharacteristic real hyperplane trough x which intersects $K_{x,\varepsilon}$, intersects $H_{x,\varepsilon}$. □

End the proof of Theorem 3.1.2. We may find $c_2 > 0$ and $\varepsilon_3 > 0$ such that for $|x_1| < \varepsilon_2$, $\phi(x_1) > 0$, the interior of the convex hull of $K_{x_1,\varepsilon_3} \cup \{x\}$ for $x \in \partial \Omega \cap \Gamma_{x_1}$, is contained in Ω. Let ω_x denote the interior of the convex hull of $K_{x_1,\varepsilon_3} \cup \{x\}$, and let $\omega' = \Gamma_{x_1,\varepsilon_3}^+$.

Let us calculate the cohomology groups $H^j(\Omega \cap \omega', \mathscr{L}_\Sigma^0)$ and $H^j(\omega', \mathscr{L}_\Sigma^0)$ using Cech cohomology (cf. Appendix C.3). We consider the open covering $\tilde{\Omega} = \{\omega_x; \ x \in \partial \Omega \cap \omega'\}$ of $\Omega \cap \omega'$, and the covering $\tilde{\omega}' = \{\omega_x' \equiv \omega'; \ x \in \partial \Omega \cap \omega'\}$. Let $\check{C}^\cdot(\tilde{\Omega}, \mathscr{L}_\Sigma^0)$ and $\check{C}^\cdot(\tilde{\omega}', \mathscr{L}_\Sigma^0)$ be the complexes constructed by these coverings. Since any finite intersection of convex open sets is a convex open set, these coverings are acyclic by Lemma 3.1.3. In terms of their cohomology we can then calculate the groups $H^\cdot(\Omega \cap \omega', \mathscr{L}_\Sigma^0)$ and $H^\cdot(\omega', \mathscr{L}_\Sigma^0)$ respectively (Theorem C.3.2). Let I be a finite subset of $\partial \Omega \cap \omega'$. Let

$\omega_I = \bigcap_{x \in I} \omega_x$ and $\omega'_I = \bigcap_{x \in I} \omega'_x = \Gamma^+_{x_1, \varepsilon_3}$. Applying Lemma 3.1.6 we see that P induces an isomorphism on $\mathscr{L}^0_{\Sigma}(\omega_I)/\mathscr{L}^0_{\Sigma}(\omega'_I)$. Let $\mathscr{S}^0$ be the mapping cone of the morphism $0 \to \mathscr{L}^0_{\Sigma} \xrightarrow{P} \mathscr{L}^0_{\Sigma} \to 0$. Then we have a quasi-isomorphism from $\check{C}^{\bullet}(\tilde{\omega}', \mathscr{S}^0)$ to $\check{C}^{\bullet}(\tilde{\Omega}, \mathscr{S}^0)$, that is, for each j an isomorphism:

$$(3.1.13) \qquad H^j(\omega', \mathscr{S}^0) \cong H^j(\omega' \cap \Omega, \mathscr{S}^0).$$

Taking the inductive limit when ω' runs over a fundamental system of neighborhoods of 0, we get:

$$\mathscr{H}^j_{\{\phi \geqslant 0\}}(\mathscr{S}^0)_0 = 0.$$

By the long exact sequence:

$$\dots \to \mathscr{H}^j_{\{\phi \geqslant 0\}}(\mathscr{L}^0_{\Sigma}) \xrightarrow{P} \mathscr{H}^j_{\{\phi \geqslant 0\}}(\mathscr{L}^0_{\Sigma}) \to \mathscr{H}^j_{\{\phi \geqslant 0\}}(\mathscr{S}^0) \to \dots$$

we get the result for $\mathscr{L}(0)|_{\Sigma}$, thus for the $\mathscr{L}(k)|_{\Sigma}$'s.

To prove the theorem for the sheaf $\mathscr{L}|_{\Sigma}$, let us consider two increasing sequences of compact sets, $\{K_n\}$ and $\{L_n\}$, such that $K_n \subset L_n$, $\bigcup_n K_n = \omega' \cap \Omega$, $\bigcup_n L_n = \omega'$, and such that the restriction morphisms $H^j(L_n, \mathscr{S}^0) \to H^j(K_n, \mathscr{S}^0)$ are isomorphisms. In order to construct such sequences, we may take for L_n the closure of $\Gamma^+_{x_n, \varepsilon_n}$ (with the preceding notations), for $x_n \in \omega'$, ε_n small enough, $x_n \xrightarrow{n} x$, and for K_n the set $L_n \cap \{x; \phi \leqslant -\varepsilon'_n\}$, for $0 < \varepsilon'_n \ll \varepsilon_n$, and apply (3.1.13) when ω' (resp. $\omega' \cap \Omega$) runs over a fundamental system of neighborhoods of L_n (resp. K_n). Let $\mathscr{S}^k$ be the mapping cone of $0 \to \mathscr{L}(k)|_{\Sigma} \xrightarrow{P} \mathscr{L}(k)|_{\Sigma} \to 0$, and $\mathscr{S}$ the mapping cone of $0 \to \mathscr{L}|_{\Sigma} \to \mathscr{L}|_{\Sigma} \to 0$. Replacing $\mathscr{S}^0$ by $\mathscr{S}^k$ and taking the inductive limit with respect to k, we find isomorphisms for each j:

$$H^j(L_n, \mathscr{S}) \cong H^j(K_n, \mathscr{S}).$$

Let $\mathscr{S}^{\bullet}$ be a flabby resolution of $\mathscr{S}$. Applying Corollary B.5.2 of the Appendix to the complexes $\Gamma(L_n, \mathscr{S}^{\bullet})$ and $\Gamma(K_n, \mathscr{S}^{\bullet})$, we get the isomorphisms:

$$(3.1.14) \qquad H^j(\omega', \mathscr{S}) \cong H^j(\omega' \cap \Omega, \mathscr{S}).$$

Then the proof goes as for the case of $\mathscr{L}^0_{\Sigma}$. $\square$

When V is the zero section $T^*_X X$ of $T^* X$, and $\mathscr{L} = \mathcal{O}_X$, we find:

Corollary 3.1.7. *Let ϕ be a real C^1-function on X, $x \in X$ and let P be a differential operator on X. Assume that $\sigma(P)(d\phi(x)) \neq 0$. Then for all j, P induces an isomorphism on each of the cohomology groups $(\mathscr{H}^j_{\{\phi \geqslant 0\}}(\mathcal{O}_X))_x$.*

This corollary will be generalized in the next section, replacing $\mathcal{O}_X$ by the sheaf $\mathscr{B}_{Z|X}$ associated to a submanifold Z.

3.2. Propagation for Systems

We shall extend the results of Section 3.1 to Modules. As the theory of the Ring $\mathscr{E}_V$ has only been developed under the assumption (1.2.2), we need this hypothesis in the following statement.

Theorem 3.2.1. *Let U be an open subset of $\dot{T}^*X$, V a conic involutive manifold, closed in U and satisfying condition (1.2.2). Let Σ be a bicharacteristic leaf of V, $p \in \Sigma$, ϕ a real C^1-function on Σ. Let $\mathscr{L}$ be a system with simple characteristics along V, u_0 a simple generator of $\mathscr{L}$, $\mathscr{L}(k) = \mathscr{E}_X(k)u_0$. Let $\mathscr{M}$ be a coherent $\mathscr{E}_X$-module on U, $\mathscr{M}_0$ a coherent $\mathscr{E}_V$-sub-module of $\mathscr{M}|_V$. Assume that: $H_\phi(p) \notin C_V^1(\mathscr{M})$. Then:*

$$\mathscr{E}\kern-0.1em xt^j_{\mathscr{E}_V}(\mathscr{M}_0, \mathscr{H}^i_{\{\phi \geq 0\}}(\mathscr{L}(k)|_\Sigma))_p = 0 \qquad \forall i \geq 0,\ j \geq 0,\ k \in \mathbb{Z},$$

$$\mathscr{E}\kern-0.1em xt^j_{\mathscr{E}_X}(\mathscr{M}, \mathscr{H}^i_{\{\phi \geq 0\}}(\mathscr{L}|_\Sigma))_p = 0 \qquad \forall i \geq 0,\ j \geq 0.$$

Proof. Let $v_1, \ldots, v_N$ be a system of generators of $\mathscr{M}_0$ in a neighborhood of p. By hypothesis there exists $P_l \in \mathscr{E}_V$ $(l = 1, \ldots, N)$ such that $P_l v_l = 0$ and $\sigma^1_V(\sigma(P_l))(H_\phi(p)) \neq 0$ (apply Theorem 1.2.9.). Let $\mathscr{M}_0' = \bigoplus_{l=1}^{N} \mathscr{E}_V/\mathscr{E}_V \cdot P_l$ and let $\mathscr{M}_0''$ be the kernel of the morphism from $\mathscr{M}_0'$ to $\mathscr{M}_0$ which associates v_l to $1 \bmod \mathscr{E}_V P_l$:

$$(3.2.1) \qquad\qquad 0 \to \mathscr{M}_0'' \to \mathscr{M}_0' \to \mathscr{M}_0 \to 0.$$

Applying the functor $\mathscr{H}om_{\mathscr{E}_V}(\cdot, \mathscr{H}^i_{\{\phi \geq 0\}}(\mathscr{L}(k)|_\Sigma))_p$ to the exact sequence (3.2.1) we obtain a long exact sequence:

$$(3.2.2) \qquad\qquad \ldots \to A_j \to A_j' \to A_j'' \to A_{j+1} \to \ldots,$$

where $A_j = \mathscr{E}\kern-0.1em xt^j_{\mathscr{E}_V}(\mathscr{M}_0, \mathscr{H}^i_{\{\phi \geq 0\}}(\mathscr{L}(k)|_\Sigma))$ and similarly for A_j' (resp. A_j'') with $\mathscr{M}_0$ replaced by $\mathscr{M}_0'$ (resp. $\mathscr{M}_0''$). The groups A_j' are 0 for $j \neq 0, 1$ since $\mathscr{E}_V/\mathscr{E}_V P_l$ admits a free presentation of length one. Moreover $A_j' = 0$ for $j = 0, 1$ by Theorem 3.1.2. Thus $A_0 = 0$, and arguing by induction on j, as at the end of the proof of Theorem 2.1.2 we find $A_j = 0\ \forall k$.

The same result holds with $\mathscr{L}(k)$ replaced by $\mathscr{L}$. Now we assume that $\mathscr{M}_0$ generates $\mathscr{M}|_V$. Then:

$$\mathscr{E}\kern-0.1em xt^j_{\mathscr{E}_V}(\mathscr{M}_0, \mathscr{K}) = \mathscr{E}\kern-0.1em xt^j_{\mathscr{E}_X}(\mathscr{M}, \mathscr{K})$$

for any left $\mathscr{E}_V$-module $\mathscr{K}$, since $\mathscr{E}_X \otimes_{\mathscr{E}_V} \mathscr{M}_0 \simeq \mathscr{M}$ and $\mathscr{E}_X$ is flat over $\mathscr{E}_V$, as already observed. $\square$

Replacing Theorem 3.1.2 by Corollary 3.1.7 in the proof, we get:

Theorem 3.2.2. *Let ϕ be a real C^1-function on X, $x \in X$. Let $\mathscr{M}$ be a coherent $\mathscr{D}_X$-module. Assume that $d\phi(x) \notin \mathrm{char}(\mathscr{M})$. Then:*

$$\mathcal{E}xt^{j}_{\mathscr{D}_X}(\mathcal{M}, \mathcal{H}^{i}_{\{\phi \geqslant 0\}}(\mathcal{O}_X))_x = 0 \qquad \forall i \geqslant 0,\ j \geqslant 0.$$

Now we may extend this result replacing $\mathcal{O}_X$ by $\mathscr{B}_{Z|X}$, using Definition 1.4.4.

Theorem 3.2.3. *Let Z be a submanifold of X, ϕ a real C^1-function on Z, $x \in Z$. Let $\mathcal{M}$ be a coherent $\mathscr{D}_X$-module. Assume $d\phi(x) \notin \hat{C}_{T_Z X}(\mathcal{M})$. Then:*

$$\mathcal{E}xt^{j}_{\mathscr{D}_X}(\mathcal{M}, \mathcal{H}^{i}_{\{\phi \geqslant 0\}}(\mathscr{B}_{Z|X}))_x = 0 \qquad \forall i \geqslant 0,\ j \geqslant 0.$$

(Here the sheaf $\mathscr{B}_{Z|X}$ on X, which is supported by Z, is identified to $(\mathscr{B}_{Z|X})|_Z$, its restriction to Z).

Proof. As for the proof of Theorem 3.2.1 we reduce the proof to the case where $\mathcal{M} = \mathscr{D}_X / \mathscr{D}_X P$, using Theorem 1.4.5. Then $\sigma(\hat{\sigma}_Z(P))(H_\phi(x)) \neq 0$, and we may decompose P as $P_0 + Q$, where $P_0 \in F_0^Z \mathscr{D}_X$, $Q \in F_{-1}^Z \mathscr{D}_X$ (cf. § 1.4) and $\sigma(P_0)(d\phi(x)) \neq 0$.

For each k, P operates on $\mathscr{B}_{Z|X}(k)$, and by the same method as for Theorem 3.1.2, it is enough to prove the theorem for each $\mathscr{B}_{Z|X}(k)$ (Lemma 3.1.3 holds for the sheaf $(\mathscr{B}_{Z|X})|_Z$ instead of the sheaf $\mathscr{S}|_\Sigma$).

Consider the exact sequence of sheaves on Z:

$$(3.2.3)_k \quad 0 \to \mathscr{B}_{Z|X}(k-1) \to \mathscr{B}_{Z|X}(k) \to \mathscr{B}_{Z|X}(k)/\mathscr{B}_{Z|X}(k-1) \to 0.$$

Applying the functor $\Gamma_{\{\phi \geqslant 0\}}(\cdot)_x$ to this exact sequence, we find a long exact sequence:

$$(3.2.4) \qquad \dots A_j(k-1) \to A_j(k) \to B_j(k) \to A_{j+1}(k-1) \to \dots,$$

where $A_j(k) = \mathcal{H}^{j}_{\{\phi \geqslant 0\}}(\mathscr{B}_{Z|X}(k))_x$ and $B_j(k) = \mathcal{H}^{j}_{\{\phi \geqslant 0\}}(\mathscr{B}_{Z|X}(k)/\mathscr{B}_{Z|X}(k-1))_x$. In order to prove that P induces an isomorphism on each of the $A_j(k)$'s, we may argue by induction on k, since $A_j(k) = 0$ for $k < 0$, and we only have to prove that P induces an isomorphism on each of the $B_j(k)$ ($j \geqslant 0$, $k \geqslant 0$). Now P operates as P_0 on $\mathscr{B}_{Z|X}(k)/\mathscr{B}_{Z|X}(k-1)$, and this last sheaf is isomorphic to $\mathcal{O}_Z^{N_k}$ (where $N_k = \#\{\alpha \in \mathbb{N}^d; |\alpha| = k, d = \mathrm{codim}\, Z\}$), and the action of P_0 on this sheaf is represented by an $(N_k \times N_k)$-matrix $P_{0,k}$ (cf. the proof of Proposition 2.1.6). The entries of $P_{0,k}$ are sections of $\mathscr{D}_Z$ of order at most m, where m is the order of P_0. All the terms on the diagonal are of order m and non characteristic for $d\phi(x)$ and the terms off the diagonal are of order strictly less than m. If we set $\mathcal{M}_k = \mathscr{D}_Z^{N_k}/\mathscr{D}_Z^{N_k} \cdot P_{0,k}$ then $d\phi(x) \notin \mathrm{char}(\mathcal{M}_k)$ (II, Remark 1.3.7) and it remains to apply Theorem 3.2.2 to complete the proof. $\square$

We have formulated the prolongation theorems for the functor $\mathcal{H}om(\cdot,\cdot)$ but we may also formulate such theorems for the functor $\cdot \otimes \cdot$.

Let us give an example of such a result in a situation which appears naturally in the study of direct images of coherent $\mathscr{D}_X$-modules (cf. Houzel-Schapira [1].

Theorem 3.2.4. *Let f be a smooth map from X to Y, $x \in X$, $\Sigma = f^{-1}f(x)$, ϕ a real C^1-function on Σ. Let $\mathscr{M}$ be a right coherent $\mathscr{D}_X$-module, $\mathscr{M}_0$ a coherent $\mathscr{D}_{X/Y}$-sub-module. Assume that $d\phi(x) \notin \mathrm{char}_f^1(\mathscr{M})$. Then:*

$$\mathscr{T}\!or_j^{\mathscr{D}_{X/Y}}(\mathscr{M}_0, \mathscr{H}^i_{\{\phi \geq 0\}}(\mathscr{D}_{X \to Y}(k)|_\Sigma))_x = 0 \quad \forall j \geq 0, \; i \geq 0, \; k \geq 0,$$

$$\mathscr{T}\!or_j^{\mathscr{D}_X}(\mathscr{M}, \mathscr{H}^i_{\{\phi \geq 0\}}(\mathscr{D}_{X \to Y}|_\Sigma))_x = 0 \quad \forall j \geq 0, \; i \geq 0.$$

Recall that $\mathrm{char}_f^1(\mathscr{M})$ is defined in § 1.3.

The proof may be obtained using the same line of reasoning as for Theorem 3.2.1 or else by adding a dummy variable in order to work outside the zero section, and applying Theorem 3.2.1.

Exercises to III.3

Ex. 3.1. Let $X = \mathbb{C}^n$ with coordinates $(x_1, \ldots, x_n)$, let P be the differential operator $\sum\limits_{j=1}^{n} D_j^2$, and let $\theta = (1, i, 0, \ldots, 0)$. Find a real C^1-function ϕ on X such that $d\phi(0) = \theta$ and P is not injective on $(\mathscr{H}^1_{\{\phi \geq 0\}}(\mathscr{O}_X))_0$.

(Hint: solve a Cauchy problem with singular data at 0, using Theorem 2.2.1)

Ex. 3.2. Let $X = \mathbb{C}^2$ with coordinates (t, z), $Z = \{(t, z); \, t = 0\}$, $P = D_z + t D_t + \alpha$, $\alpha \in \mathbb{C}$, $\mathscr{M} = \mathscr{D}_X / \mathscr{D}_X \cdot P$.

Prove that the sheaf $(\mathscr{E}\!xt^j_{\mathscr{D}_X}(\mathscr{M}, \mathscr{B}_{Z|X}))|_Z$ is zero for $j \neq 0$, and that for $j = 0$ it is a constant sheaf on Z.

Ex. 3.3. Let X and Z be as in Ex. 3.2 and let $P = t D_t + 1/2$, $\mathscr{M} = \mathscr{D}_X / \mathscr{D}_X \cdot P$.

Prove that $T^*Z \cap \hat{C}_{T_Z X}(\mathscr{M}) = T^*Z$, but that P induces an isomorphism on $\mathscr{B}_{Z|X}$, thus on all the cohomology groups $\mathscr{H}^i_{\{\phi \geq 0\}}(\mathscr{B}_{Z|X})$ (where ϕ is any real C^1-function on Z).

Ex. 3.4. Let Y and Z be two submanifolds of X, Y and Z being transversal. Prove that the sheaf $\mathscr{E}\!xt^j_{\mathscr{D}_X}(\mathscr{B}_{Y|X}, \mathscr{B}_{Z|X})$ is zero except for only one j, and that for that integer it is isomorphic to the sheaf $\mathbb{C}_{Y \cap Z}$ (the constant sheaf $\mathbb{C}$ on $Y \cap Z$, 0 on $X \backslash (Y \cap Z)$, cf. (Appendix, Example C.1.2)).

§ 4. Constructibility

4.1. Real and Complex Stratifications

Let X be a real manifold of class C^r ($1 \leqslant r \leqslant \infty$, or $r = \omega$, which means that X is real analytic).

We shall denote by $\{x_n\}$, or $\{x_n\}_n$, a sequence in X ($n \in \mathbb{N}$), and $x_n \xrightarrow{n} x$ will mean that x is the limit of this sequence.

Let S and V be two subsets of X.

Definition 4.1.1. The normal cone $C_x(S, V)$ of S along V at x is the cone of $T_x X$ defined, in a local chart in a neighborhood of x, by: $\theta \in C_x(S, V) \Leftrightarrow$ there exists a sequence $\{(s_n, v_n, c_n)\}$ in $S \times V \times \mathbb{R}^+$ such that $s_n \xrightarrow{n} x, v_n \xrightarrow{n} x, c_n(s_n - v_n) \xrightarrow{n} \theta$.

This definition is independent of the choice of the local chart. One sets:

$$C(S, V) = \bigcup_{x \in X} C_x(S, V).$$

If X is open in a vector space, then $\theta \in T_x X$ does not belong to $C_x(S, V)$ if and only if there exists an open cone Γ with $\theta \in \Gamma$ and a neighborhood U of x such that:

$$((U \cap V) + \Gamma) \cap U \cap S = \emptyset.$$

Remark that for two subsets S_1 and S_2 we have:

$$(4.1.1) \qquad C(S_1 \cup S_2, V) = C(S_1, V) \cup C(S_2, V).$$

If V is smooth, then $C(S, V)$ is invariant by TV. In this case one denotes by $C_V(S)$ the image of $C(S, V)$ in the normal bundle $T_V X = T X \underset{X}{\times} V / TV$.

When X is a complex manifold, V a complex submanifold, S a locally closed analytic subset, we have already introduced in § 1 a notion of complex normal cone $C_V(S)$ in the complex normal bundle $T_V X$.

Let $X^{\mathbb{R}}$ be the real underlying manifold of X. After identifying $(TX)^{\mathbb{R}}$ and $T(X^{\mathbb{R}})$, (cf. the end of this section for a detailed study of this isomorphism), it can be proved that the two definitions of normal cones (Definition 1.1.2 and Definition 4.1.1) are equivalent, (cf. Whitney [2]). Thus the notation $C_V(S)$ is unambiguous.

Now we assume again X is real, of class C^r, and take $r \geqslant 2$. A subset of $T^* X$ will be said to be conic if it is locally homogeneous for the action of $\mathbb{R}^+$.

Definition 4.1.2. Let Λ be a conic subset of T^*X. A locally closed subset Y of X is flat at $y \in Y$ with respect to Λ if for any $p \in \pi^{-1}(Y)$ we have:

$$C_p(\Lambda, \pi^{-1}(Y)) \subset \{v \in T_p T^*X; \ \langle v, \omega(p) \rangle \leqslant 0\}.$$

One says that Y is flat with respect to Λ, if Y is flat at each $y \in Y$. Remark that if Y is flat with respect to Λ then:

(4.1.2) $$\pi^{-1}(Y) \cap \Lambda \subset T_Y^* X.$$

In fact let $p \in \Lambda \cap \pi^{-1}(Y)$. Then $C(\Lambda, \pi^{-1}(Y))$ contains $T_p(\pi^{-1}(Y))$. Hence $\omega(p) = 0$ on $T_p(\pi^{-1}(Y))$ and this last condition implies $p \in T_Y^* X$.

Proposition 4.1.3. *Suppose X is open in $\mathbb{R}^N$, and that a submanifold Y is flat with respect to a conic set Λ in T^*X at a point x_0. Then there exists $\varepsilon > 0$ such that $(x; x - y)$ does not belong to Λ for $x \in X$, $y \in Y$ satisfying $|x - x_0| < \varepsilon$, $|y - x_0| < \varepsilon$, $x \neq y$.*

Proof. We argue by contradiction.

Assume there exist sequences $\{x_n\}$ in X, $\{y_n\}$ in Y such that $x_n \xrightarrow{n} x_0$, $y_n \xrightarrow{n} x_0$, $x_n \neq y_n$, $(x_n; x_n - y_n) \in \Lambda$. Let $\{c_n\}$ be a sequence in $\mathbb{R}^+$ such that a subsequence of $\{c_n(x_n - y_n)\}$ converges to $v \neq 0$. Then $\{(x_n; c_n(x_n - y_n))\}$ is a sequence in Λ which converges to $p = (x_0; v)$ and $\{(y_n; c_n(x_n - y_n))\}$ is a sequence in $\pi^{-1}(Y)$ which converges to p.

Since $c_n((x_n; c_n(x_n - y_n)) - (y_n; c_n(x_n - y_n)))$ converges to $(v, 0)$, $(v, 0)$ belongs to $C(\Lambda, \pi^{-1}(Y))$. Thus:

$$\langle (v, 0), \omega(p) \rangle = \langle v, v \rangle \leqslant 0.$$

This is a contradiction. $\square$

Now let Y and Z be two submanifolds of X, with $Z \cap Y = \emptyset$. One says that Y dominates Z, and writes $Z \prec Y$, if Z is contained in $\overline{Y}$.

Definition 4.1.4. Assume $Z \prec Y$.

i) The pair (Z, Y) satisfies the condition a) of Whitney at $z \in Z$ if, in a local chart in a neighborhood of z, for any sequence $\{y_n\}$ in Y such that $y_n \xrightarrow{n} z$ and such that the sequence of tangent spaces $\{T_{y_n} Y\}$ has a limit $\tau \subset T_z X$, one has: $\tau \supset T_z Z$.

ii) The pair (Z, Y) satisfies the condition b) of Whitney at $z \in Z$ if, in a local chart in a neighborhood of z, for any sequence $\{(z_n, y_n, c_n)\}$ in $Z \times Y \times \mathbb{R}^+$ such that $z_n \xrightarrow{n} z$, $y_n \xrightarrow{n} z$, $c_n(z_n - y_n) \xrightarrow{n} v$ and such that the sequence of tangent spaces $\{T_{y_n} Y\}$ has a limit $\tau \subset T_z X$, one has: $v \in \tau$.

One says that (Z, Y) satisfies the Whitney conditions (resp. at $z \in Z$) if it satisfies conditions a) and b) at each $z \in Z$ (resp. at $z \in Z$). In fact condition b) implies condition a), (cf. Whitney [2]).

Proposition 4.1.5. *Assume* (Z, Y) *satisfies the Whitney conditions at* $z \in Z$. *Then:*

    i) $T^*_Y X \cap \pi^{-1}(z) \subset T^*_z X$.

    ii) *Z is flat with respect to* $T^*_Y X$ *at* z.

Proof. i) Let $\{(y_n, \eta_n)\}$ be a sequence in $T^*_Y X$ with $y_n \underset{n}{\to} z$, $\eta_n \underset{n}{\to} \xi \in T^*_z X$. If the sequence $\{T_{y_n} Y\}$ converges to $\tau \subset T_z X$, then $\tau \supset T_z Z$ by the Whitney conditions. Therefore the sequence $\{(T^*_Y X)_{y_n}\}$ converges to the orthogonal space $\tau^\perp$, and $\tau^\perp \subset (T^*_z X)_z$.

    ii) Let $p = (z, \zeta) \in \pi^{-1}(z)$, and let $q \in C_p(T^*_Y X, \pi^{-1}(Z))$. There exist sequences $\{(y_n, \eta_n)\}$ in $T^*_Y X$, $\{(z_n, \zeta_n)\}$ in $\pi^{-1}(Z)$, $\{c_n\}$ in $\mathbb{R}^+$, such that $(y_n, \eta_n) \underset{n}{\to} p$, $(z_n, \zeta_n) \underset{n}{\to} p$, $c_n((y_n, \eta_n) - (z_n, \zeta_n)) \underset{n}{\to} q = (v, w)$. By extracting a subsequence, we may assume that $\{T_{y_n} Y\}$ converges to a subspace $\tau \subset T_z X$. By the Whitney conditions, $v \in \tau$ and $T_z Z \subset \tau$. Since $\zeta \in T^*_Z X$, $\zeta \in \tau^\perp$. Then:

$$\langle v, \zeta \rangle = \langle q, \omega(p) \rangle = 0. \quad \square$$

Definition 4.1.6. Let X be a real C^r-manifold, $(2 \leqslant r \leqslant \infty$ or $r = \omega)$. A stratification $(X_\alpha)_\alpha$ of X is a partition $X = \bigsqcup_\alpha X_\alpha$ such that:

    i) the family $(X_\alpha)_\alpha$ is locally finite,

    ii) each X_α is a locally closed smooth C^r-manifold,

    iii) for each pair (α, β) such that $X_\alpha \cap \overline{X}_\beta$ is non empty, X_α is contained in $\overline{X}_\beta$ (i.e.: $X_\alpha \prec X_\beta$).

    iv) If moreover, for all pairs (α, β) such that $X_\alpha \prec X_\beta$, the pair (X_α, X_β) satisfies the Whitney conditions, one says that the stratification is a Whitney stratification.

When X is complex analytic one also defines a (complex) stratification of X by requiring the X_α's to be complex manifolds.

A stratification $X = \bigsqcup_\beta X_\beta$, $\beta \in B$ is finer than the stratification $X = \bigsqcup_\alpha X_\alpha$, $\alpha \in A$, if there exists a map $\tau : B \to A$ such that $X_\beta \subset X_{\tau(\beta)}$, $\forall \beta \in B$.

As a corollary of Proposition 4.1.5 we have:

Proposition 4.1.7. *Let* $X = \bigsqcup_\alpha X_\alpha$ *be a Whitney stratification. Set* $\Lambda = \bigsqcup_\alpha T^*_{X_\alpha} X$. *Then:*

    a) Λ *is closed,*

    b) *each* X_α *is flat with respect to* Λ.

From now on we assume X to be a real analytic manifold.

We shall make use of the theory of subanalytic sets of H. Hironaka [1]. For a locally closed subset of X, the property of being subanalytic is local on X, and the family of subanalytic subsets of X is characterized by the following properties:

– the union and the intersection of a locally finite family of subanalytic subsets is subanalytic,

– the complement of a subanalytic subset is subanalytic,

– the closure, the boundary, and the interior of a subanalytic set are subanalytic,

– the image by a proper analytic map of a subanalytic set is subanalytic, and a closed subanalytic set is an image of a real analytic manifold by a proper map.

Remark that semi-analytic sets (i.e. sets which are locally defined by real analytic inequalities) are subanalytic.

We say that a stratification $X = \bigsqcup_\alpha X_\alpha$ is $\mathbb{R}$-analytic if the X_α's are real analytic manifolds and are subanalytic in X. Then:

– a subanalytic set S admits an $\mathbb{R}$-analytic stratification (i.e.: there exists an $\mathbb{R}$-analytic stratification of X, finer than $X = (X \setminus S) \sqcup S$).

Moreover we have (cf. Whitney [1] for the complex case, Hironaka [1], Verdier [1]):

– Let $X = \bigsqcup_\alpha X_\alpha$ be an $\mathbb{R}$-analytic stratification. Then there exists a Whitney $\mathbb{R}$-analytic stratification $X = \bigsqcup_\beta X_\beta$, finer than the one above.

If $X = \bigsqcup_\alpha X_\alpha$ is a Whitney $\mathbb{R}$-analytic stratification, then $\Lambda = \bigsqcup_\alpha T^*_{X_\alpha} X$ is a closed conic subanalytic set in T^*X, and Λ is Lagrangean, in the sense that there exists an open dense smooth submanifold $\Lambda' \subset \Lambda$ which is Lagrangean.

Conversely, we shall study the stratifications associated to Lagrangean subanalytic sets, or rather, to isotropic subanalytic sets.

Definition 4.1.8. Let $\Lambda \subset T^*X$ be a conic subanalytic subset. One says that Λ is isotropic if there exists a dense open smooth manifold $\Lambda' \subset \Lambda$ such that $\omega|_{\Lambda'} = 0$.

Remark that if Λ is as in Definition 4.1.8. and V is a conic subanalytic subset of Λ, then V is isotropic. In fact we may find a Whitney $\mathbb{R}$-analytic stratification $\Lambda = \bigsqcup_\alpha \Lambda_\alpha$ such that V is a union of strata. Thus it is enough to prove that each Λ_α is isotropic. For that purpose we may argue by induction on the codimension of the strata, since $\Lambda_\alpha \prec \Lambda_\beta$ implies $\overline{T^*_{\Lambda_\beta} T^*X} \underset{T^*X}{\times} \Lambda_\alpha \subset T^*_{\Lambda_\alpha} T^*X$ by the Whitney conditions.

Proposition 4.1.9. *Let Λ be a closed conic subanalytic isotropic set in T^*X. Then there exists a Whitney $\mathbb{R}$-analytic stratification $X = \bigsqcup_\alpha X_\alpha$ such that $\Lambda \subset \bigsqcup_\alpha T^*_{X_\alpha} X$.*

Proof. Set $S = \pi(\Lambda)$. Then S is subanalytic since Λ is conic. There exist Whitney $\mathbb{R}$-analytic stratifications $\Lambda = \bigsqcup_{\alpha \in A} \Lambda_\alpha$, $S = \bigsqcup_{\beta \in B} S_\beta$ and a map $\tau : A \to B$ such that $\pi(\Lambda_\alpha) \subset S_{\tau(\alpha)}$ and the projection $\pi : \Lambda_\alpha \to S_{\tau(\alpha)}$ is smooth. Let us prove the inclusion $\Lambda_\alpha \subset T^*_{S_{\tau(\alpha)}} X$. We choose local coordinates $x = (x_1, \ldots, x_n) = (x', x'')$ where $x' = (x_1, \ldots, x_p)$ such that $S_{\tau(\alpha)} = \{x \in X; \, x' = 0\}$. Then $\omega|_{\pi^{-1}(S_{\tau(\alpha)})} = \xi'' dx''$, and this vanishes on Λ_α. Since the linear forms $dx_{p+1}, \ldots, dx_n$ are linearly independent on $S_{\tau(\alpha)}$ and $\pi|_{\Lambda_\alpha}$ is smooth, they are linearly independent on Λ_α. Thus $\xi'' = 0$ on Λ_α, which completes the proof. $\square$

Now we take for X a complex manifold and denote by $X^{\mathbb{R}}$ the underlying real analytic manifold. Let $\overline{X}$ be the complex manifold defined by $(\overline{X})^{\mathbb{R}} = X^{\mathbb{R}}$, and such that the holomorphic functions on $\overline{X}$ are the antiholomorphic functions on X. We identify $X^{\mathbb{R}}$ to the diagonal of $X \times \overline{X}$. Then $X \times \overline{X}$ defines a complexification of $X^{\mathbb{R}}$ (cf. Kashiwara [5]), and we have for $x \in X$:

(4.1.3)
$$T_x X^{\mathbb{R}} \otimes_{\mathbb{R}} \mathbb{C} \simeq T_x X \oplus T_x \overline{X}.$$

The composite application:

$$T X^{\mathbb{R}} \to T X^{\mathbb{R}} \otimes_{\mathbb{R}} \mathbb{C} \simeq T X \underset{X}{\times} T \overline{X} \to T X$$

defines the isomorphism:

(4.1.4)
$$T X^{\mathbb{R}} \simeq (T X)^{\mathbb{R}}.$$

Passing to the dual space, we get the isomorphism:

(4.1.5)
$$T^* X^{\mathbb{R}} \simeq (T^* X)^{\mathbb{R}}.$$

For a real C^1-function ϕ on X, this last isomorphism associates the vector $\partial \phi(x) \in T^*_x X$ to the vector $d\phi(x) \in T^*_x X^{\mathbb{R}}$. Here $d\phi$ is the real total differential and $\partial \phi$ the complex one.

Let ω_X be the canonical 1-form on the complex manifold $T^* X$. Then $\omega_{\overline{X}} = \overline{\omega}_X$, and $\omega_{X \times \overline{X}} = \omega_X + \overline{\omega}_X$. Thus:

(4.1.6)
$$\omega_{X^{\mathbb{R}}} = 2 \operatorname{Re} \omega_X$$

If we choose a system of local complex coordinates $z = (z_1, \ldots, z_n)$ on X, with $z = x + iy$, and denote by (z, ζ) the associated coordinates on $T^* X$, $\zeta = (\zeta_1, \ldots, \zeta_n)$, so that $\omega_X = \sum_{j=1}^n \zeta_j dz_j$, then writing $\zeta = \xi - i\eta$, we get $\omega_{X^{\mathbb{R}}} = 2 \sum_{j=1}^n (\xi_j dx_j + \eta_j dy_j)$.

We shall say for short that a subset of X is $\mathbb{C}$-analytic if it is a (locally closed) complex analytic subspace of X. Similarly we shall say "$\mathbb{C}$-analytic stratification" for "complex analytic stratification". A subset of $T^* X$ is $\mathbb{C}^\times$-conic, (or simply conic if there is no risk of confusion) if it is locally homogeneous for the action of $\mathbb{C}^\times$ on T^*X. Remark that a $\mathbb{C}$-analytic subset of T^*X is $\mathbb{C}^\times$-conic if and only if it is $\mathbb{R}^+$-conic, i.e., conic in $T^*X^\mathbb{R}$. A conic $\mathbb{C}$-analytic subset Λ of T^*X is said to be isotropic if there exists an open dense smooth submanifold $\Lambda' \subset \Lambda$ such that $\omega_X|_{\Lambda'} = 0$. One immediatly sees that a conic $\mathbb{C}$-analytic subset Λ of T^*X is isotropic if and only if it is isotropic in $T^*X^\mathbb{R}$ (i.e.: $\operatorname{Re}\omega_X|_{\Lambda'} = 0$).

By the same proof as for Proposition 4.1.9 one gets:

Proposition 4.1.10. *Let Λ be a closed conic isotropic analytic subset of T^*X. Then there exists a Whitney $\mathbb{C}$-analytic stratification $X = \bigsqcup_\alpha X_\alpha$ such that $\Lambda \subset \bigsqcup_\alpha T^*_{X_\alpha} X$.*

4.2. Micro-support of Sheaves

We shall use the notations of § C.4.

Definition 4.2.1. Let X be a real C^1-manifold, $\mathscr{F}^\cdot$ a complex bounded from below, of sheaves of abelian groups on X. The micro-support of $\mathscr{F}^\cdot$, denoted by $SS(\mathscr{F}^\cdot)$, is the closed conic subset of T^*X defined by the condition:

$p \notin SS(\mathscr{F}^\cdot) \Leftrightarrow$ there exists a conic open neighborhood U of p in T^*X such that for any $x_1 \in X$, and any real function ϕ of class C^1 on X such that $\phi(x_1) = 0$, $d\phi(x_1) \in U$ one has $(\mathscr{H}^j_{\{\phi \geq 0\}}(\mathscr{F}^\cdot))_{x_1} = 0$ for all j.

Remark that:

(4.2.1) $$SS(\mathscr{F}^\cdot) \cap T^*_X X = \operatorname{supp}(\mathscr{F}^\cdot),$$

and that if we have an exact sequence $0 \to \mathscr{F}_1^\cdot \to \mathscr{F}_2^\cdot \to \mathscr{F}_3^\cdot \to 0$ then:

(4.2.2) $$SS(\mathscr{F}_i^\cdot) \subset SS(\mathscr{F}_j^\cdot) \cup SS(\mathscr{F}_k^\cdot)$$

for any i, j, k with $1 \leq i, j, k \leq 3$.

Example 4.2.2. Assume X is a vector space, and let G be a closed convex cone with vertex at 0. Then (cf. Kashiwara-Schapira [4]):

(4.2.3) $$SS(\mathbb{C}_G) \cap \pi^{-1}(0) = G^0,$$

where $G^0 = \{\xi \in X^*; \langle x, \xi \rangle \geq 0 \quad \forall x \in G\}$. (The sheaf $\mathbb{C}_G$ is defined in Appendix C.1.2).

If X is a manifold, Y a submanifold, one finds:

(4.2.4) $$SS(\mathbb{C}_Y) = T^*_Y X.$$

The micro-support is a tool for studying "propagation".

Proposition 4.2.3. *Assume X of class C^r, $r \geqslant 2$. Let $\mathscr{F}^{\boldsymbol{\cdot}}$ be a complex, bounded from below, of sheaves on X. Let Y be a submanifold, and assume Y is flat with respect to $SS(\mathscr{F}^{\boldsymbol{\cdot}})$. Then for all j, the sheaves $\mathscr{H}^j(\mathscr{F}^{\boldsymbol{\cdot}})|_Y$ are locally constant on Y.*

Proof. Let $y_0 \in Y$. We may assume X is open in some Euclidian space $\mathbb{R}^N$, and Y is linear. By Proposition 4.1.3 there exists $\varepsilon > 0$ such that:

$$(4.2.5) \qquad \begin{cases} (x; x-y) \notin SS(\mathscr{F}^{\boldsymbol{\cdot}}) & \text{for } x \in X, \\ y \in Y, \ |x-y_0| < \varepsilon, \ |y-y_0| < \varepsilon, \ x \neq y. \end{cases}$$

Let $U_r(y) = \{x \in X; |x-y| < r\}$.

It is enough to show the isomorphisms:

$$(4.2.6) \qquad H^j(U_\varepsilon(y_0), \mathscr{F}^{\boldsymbol{\cdot}}) \simeq H^j(U_\rho(y), \mathscr{F}^{\boldsymbol{\cdot}})$$

for $y \in Y, \rho > 0, |y-y_0| + \rho < \varepsilon$.

In fact (4.2.6) implies:

$$H^j(U_\varepsilon(y_0), \mathscr{F}^{\boldsymbol{\cdot}}) \simeq \varinjlim_{\rho > 0} H^j(U_\rho(y), \mathscr{F}^{\boldsymbol{\cdot}}) = \mathscr{H}^j(\mathscr{F}^{\boldsymbol{\cdot}})_y .$$

for $y \in U_\varepsilon(y_0) \cap Y$.

Set $\Omega_t = U_{t\varepsilon + (1-t)\rho}(ty_0 + (1-t)y)$. Then $\Omega_1 = U_\varepsilon(y_0)$, $\Omega_0 = U_\rho(y)$, and $\{\Omega_t\}_{0 \leqslant t \leqslant 1}$ is an increasing family of open subsets such that $\bigcup_{t < t_0} \Omega_t = \Omega_{t_0}$, $t_0 > 0$, and $\bigcap_{t > t_0} \Omega_t = \overline{\Omega}_{t_0}$, $t_0 < 1$. Moreover:

$$(\mathscr{H}^j_{X \setminus \Omega_t}(\mathscr{F}^{\boldsymbol{\cdot}}))|_{\partial \Omega_t} = 0$$

by (4.2.5) and the definition of $SS(\mathscr{F}^{\boldsymbol{\cdot}})$. Then (4.2.6) follows from Theorem C.4.3. $\square$

Definition 4.2.4. Let X be a real (resp. complex) analytic manifold, $\mathscr{F}^{\boldsymbol{\cdot}}$ a complex, bounded from below, of sheaves on X. One says that $\mathscr{F}^{\boldsymbol{\cdot}}$ is weakly $\mathbb{R}$-constructible (resp. weakly $\mathbb{C}$-constructible) if there exists an $\mathbb{R}$-analytic (resp. $\mathbb{C}$-analytic) stratification $X = \bigsqcup_\alpha X_\alpha$ such that for all j and all α, the sheaves $\mathscr{H}^j(\mathscr{F}^{\boldsymbol{\cdot}}|_{X_\alpha})$ are locally constant on X_α.

If moreover $\mathscr{F}^{\boldsymbol{\cdot}}$ is a bounded complex of sheaves of vector spaces over $\mathbb{R}$ (resp. $\mathbb{C}$) and if for all $x \in X$ and all j, the spaces $\mathscr{H}^j(\mathscr{F}^{\boldsymbol{\cdot}})_x$ are finite dimensional, then one says that $\mathscr{F}^{\boldsymbol{\cdot}}$ is $\mathbb{R}$-constructible (resp. $\mathbb{C}$-constructible).

As a consequence of Propositions 4.1.9 (resp. 4.1.10), 4.1.7, 4.2.3, one gets:

Theorem 4.2.5. *Let X be a real (resp. complex) analytic manifold, $\mathscr{F}^{\boldsymbol{\cdot}}$ a complex, bounded from below, of sheaves on X. Assume $SS(\mathscr{F}^{\boldsymbol{\cdot}})$ is contained in a closed conic subanalytic (resp. $\mathbb{C}$-analytic) isotropic subset of T^*X. Then $\mathscr{F}^{\boldsymbol{\cdot}}$ is weakly $\mathbb{R}$-constructible (resp. weakly $\mathbb{C}$-constructible). More precisely, there* *exists a Whitney* $\mathbb{R}$*-analytic (resp.* $\mathbb{C}$*-analytic) stratification* $X = \bigsqcup_{\alpha} X_{\alpha}$ *such that* $SS(\mathscr{F}^{\cdot})$ *is contained in* $\bigsqcup_{\alpha} T^{*}_{X_{\alpha}} X$*, and for such a stratification, the sheaves* $\mathscr{H}^{j}(\mathscr{F}^{\cdot})|_{X_{\alpha}}$ *are all locally constant.*

Remark 4.2.6. It has been proved by Kashiwara-Schapira [4] that the micro-support of a complex of sheaves $\mathscr{F}^{\cdot}$ is always an involutive subset of $T^{*}X$ (assuming X of class C^{r}, $r \geqslant 2$). That is to say, if ϕ is a real C^{1}-function on an open set $U \subset T^{*}X$, and $\phi = 0$ on $SS(\mathscr{F}^{\cdot})$, then $SS(\mathscr{F}^{\cdot})$ is invariant by the Hamiltonian flow H_{ϕ}. As a consequence of the involutivity of $SS(\mathscr{F}^{\cdot})$ one can get a converse to Theorem 4.2.5: $\mathscr{F}^{\cdot}$ is weakly $\mathbb{R}$-constructible (resp. weakly $\mathbb{C}$-constructible) if and only if $SS(\mathscr{F}^{\cdot})$ is a closed conic subanalytic (resp. $\mathbb{C}$-analytic) Lagrangean subset of $T^{*}X$.

4.3. Micro-supports and Microcharacteristic Varieties

We may translate the results of § 3 in terms of the language of the micro-support.

Let X be a complex manifold, $\mathscr{M}$ a coherent $\mathscr{E}_{X}$-module on an open subset U of $T^{*}X$.

Since we have not introduced the notion of derived category, and have always worked with complexes of sheaves, we shall assume that $\mathscr{M}$ admits a finite free resolution on U, (recall that this hypothesis is always satisfied locally, cf. Chapter II).

Thus we assume:

(4.3.1) $\mathscr{M}$ is quasi-isomorphic to a bounded complex $\mathscr{M}^{\cdot}$ of free $\mathscr{E}_{X}$-modules of finite rank on U:

$$\mathscr{M}^{\cdot} : 0 \to \mathscr{E}_{X}^{N_{r}} \longrightarrow \cdots \xrightarrow[P_{0}]{} \mathscr{E}_{X}^{N_{0}} \to 0.$$

Theorem 4.3.1. *Let* U*,* V*,* L*,* Σ *be as in Theorem 3.2.1 and let* $\mathscr{M}$ *be a coherent* $\mathscr{E}_{X}$*-module on* U *satisfying (4.3.1). Then:*

$$SS(\mathscr{H}om_{\mathscr{E}_{X}}(\mathscr{M}^{\cdot}, \mathscr{L})|_{\Sigma}) \subset C^{1}_{V}(\mathscr{M}) \underset{V}{\times} \Sigma.$$

(Recall that $T_{V} T^{*}X \underset{V}{\times} \Sigma$ is identified to $T^{*}\Sigma$ by the symplectic isomorphism).

Proof. Let $\mathscr{L}_{\Sigma}^{\cdot}$ be a complex, bounded from below, of flabby sheaves on Σ of $\mathscr{E}_{X}$-modules, $\mathscr{L}_{\Sigma}^{\cdot}$ being quasi-isomorphic to $\mathscr{L}|_{\Sigma}$ (such a complex always exists). Take Godement's resolution of $\mathscr{L}|_{\Sigma}$; cf. Appendix C and Godement [1]).

Let ϕ be a real function of class C^1 on Σ. Then $\mathcal{H}^j_{|\phi \geq 0|}(\mathcal{H}om_{\mathcal{E}_X}(\mathcal{M}^{\cdot}\mathcal{L})|_\Sigma)$ is isomorphic to the j-th cohomology group of the simple complex associated to the double complex:

$$(4.3.2) \qquad 0 \to \Gamma_{|\phi \geq 0|}(\mathcal{L}_\Sigma^{\cdot})^{N_0} \xrightarrow[P_0]{} \ldots \to \Gamma_{|\phi \geq 0|}(\mathcal{L}_\Sigma^{\cdot})^{N_r} \to 0 \,.$$

The j-th cohomology group of the complex $\Gamma_{|\phi \geq 0|}(\mathcal{L}_\Sigma^{\cdot})$ is the sheaf $\mathcal{H}^j_{|\phi \geq 0|}(\mathcal{L}|_\Sigma)$. Thus we know by Theorem 3.2.1 that the complexes:

$$(4.3.3) \qquad 0 \to H^j(\Gamma_{|\phi \geq 0|}(\mathcal{L}_\Sigma^{\cdot}))^{N_0} \xrightarrow[P_0]{} \ldots \to H^j(\Gamma_{|\phi \geq 0|}(\mathcal{L}_\Sigma^{\cdot}))^{N_r} \to 0$$

are exact for any j. Then the theorem follows from Proposition B.1.1 of the Appendix. $\square$

Remark 4.3.2. We could have stated a similar theorem with $\mathcal{M}_0$, $\mathcal{L}(k)$ and $\mathcal{E}_V$ (as in Theorem 3.2.1) instead of $\mathcal{M}$, $\mathcal{L}$, $\mathcal{E}_X$.

Similarly we have, by Theorems 3.2.2, 3.2.3.

Theorem 4.3.3. *Let $\mathcal{M}$ be a coherent $\mathcal{D}_X$-module, $\mathcal{M}^{\cdot}$ a bounded complex of free $\mathcal{D}_X$-modules of finite rank, quasi-isomorphic to $\mathcal{M}$. Let Z be a submanifold of X. Then:*

$$SS(\mathcal{H}om_{\mathcal{D}_X}(\mathcal{M}^{\cdot}, \mathcal{B}_{Z|X})) \subset T^*Z \cap \hat{C}_{T_Z X}(\mathcal{M})\,.$$

In particular:

$$SS(\mathcal{H}om_{\mathcal{D}_X}(\mathcal{M}^{\cdot}, \mathcal{O}_X)) \subset \mathrm{char}(\mathcal{M})\,.$$

Remark 4.3.4. In fact we have the equality:

$$(4.3.4) \qquad SS(\mathcal{H}om_{\mathcal{D}_X}(\mathcal{M}^{\cdot}, \mathcal{O}_X)) = \mathrm{char}(\mathcal{M})\,.$$

This has been proved by Kashiwara–Schapira [4]. The corresponding result for $\mathcal{B}_{Z|X}$ and $\hat{C}_{T_Z X}(\mathcal{M})$ is no longer true (cf. Exercises 4.6 and 4.7 below).

Finally we can also state, applying Theorem 3.2.4:

Theorem 4.3.4. *Let f be a smooth surjective map from X to Y, $y \in Y$, $\Sigma = f^{-1}(y)$. Let $\mathcal{M}$ be a right coherent $\mathcal{D}_X$-module and $\mathcal{M}^{\cdot}$ a bounded complex of free $\mathcal{D}_X$-modules of finite rank, quasi-isomorphic to $\mathcal{M}$. Then:*

$$SS((\mathcal{M}^{\cdot} \otimes_{\mathcal{D}_X} \mathcal{D}_{X \to Y})|_\Sigma) \subset \mathrm{char}_f^1(\mathcal{M}) \underset{X}{\times} \Sigma\,.$$

Applying Theorem 4.2.5 we obtain weak constructibility results under the hypothesis that the microcharacteristic variety is isotropic.

Corollary 4.3.5. *Let U, V, L, Σ be a in Theorem 3.2.1 and let $\mathcal{M}$ be a coherent $\mathcal{E}_X$-module on U. Assume the set $C_V^1(\mathcal{M}) \underset{V}{\times} \Sigma$ is isotropic. Then for all j the sheaves $\mathcal{E}xt_{\mathcal{E}_X}^j(\mathcal{M}, \mathcal{L}|_\Sigma)$ are weakly $\mathbb{C}$-constructible.*

Corollary 4.3.6. *Let Z be a submanifold of X, $\mathcal{M}$ a coherent $\mathcal{D}_X$-module. Assume the set $T^*Z \cap \hat{C}_{T_ZX}(\mathcal{M})$ is isotropic in T^*Z. Then for all j the sheaves $\mathcal{E}xt_{\mathcal{D}_X}^j(\mathcal{M}, \mathcal{B}_{Z|X})$ are weakly $\mathbb{C}$-constructible.*

When $\operatorname{char}(\mathcal{M})$ is isotropic (thus Lagrangean, since $\operatorname{char}(\mathcal{M})$ is involutive), we have a better result, Kashiwara's constructibility Theorem:

Theorem 4.3.7. *Let $\mathcal{M}$ be a holonomic $\mathcal{D}_X$-module. Then for all j, the sheaves $\mathcal{E}xt_{\mathcal{D}_X}^j(\mathcal{M}, \mathcal{O}_X)$ are $\mathbb{C}$-constructible.*

Proof. Let $x_0 \in X$. We choose a system of local coordinates $(x_1, \ldots, x_n)$ on X. Let U_ε be the ball $\{x \in X; |x-x_0| < \varepsilon\}$. For $\varepsilon_0 > 0$ small enough, the balls $U_\varepsilon, \varepsilon \leqslant \varepsilon_0$, have a non characteristic boundary. In fact let $X = \underset{\alpha}{\bigsqcup} X_\alpha$ be a Whitney stratification such that $\operatorname{char}(\mathcal{M}) \subset \underset{\alpha}{\bigsqcup} T_{X_\alpha}^* X$. By considering a finer stratification we may assume that $\{x_0\}$ is a stratum, and then we apply Propositions 4.1.3 and 4.1.5. Now let $\mathcal{M}^\bullet$ be a bounded complex of free $\mathcal{D}_X$-modules of finite rank quasi-isomorphic to $\mathcal{M}$ in a neighborhood of x_0 as in (4.3.1), and let $\mathscr{F}^\bullet = \mathcal{H}om_{\mathcal{D}_X}(\mathcal{M}^\bullet, \mathcal{O}_X)$. Let $F_\varepsilon^\bullet$ be the complex of vector spaces and linear maps:

$$(4.3.5) \qquad 0 \to \mathcal{O}_X^{N_0}(U_\varepsilon) \xrightarrow[P_0]{} \ldots \to \mathcal{O}_X^{N_r}(U_\varepsilon) \to 0.$$

Since $H^j(U_\varepsilon, \mathcal{O}_X) = 0$ for $j > 0$, we have:

$$(4.3.6) \qquad H^j(U_\varepsilon, \mathscr{F}^\bullet) \simeq H^j(F_\varepsilon^\bullet).$$

The balls U_ε having a non characteristic boundary, we have already proved in the course of Proposition 4.2.3 that the complexes $F_\varepsilon^\bullet$ are all quasi-isomorphic for $0 < \varepsilon \ll 1$ and $\underset{\varepsilon > 0}{\varinjlim} H^j(F_\varepsilon^\bullet) \simeq \mathcal{E}xt_{\mathcal{D}_X}^j(\mathcal{M}, \mathcal{O}_X)_{x_0}$.

Thus it is enough to prove that the spaces $H^j(F_\varepsilon^\bullet)$ are finite dimensional. The space $\mathcal{O}_X(U_\varepsilon)$ is naturally endowed with a structure of a Fréchet space, and the restriction mappings $\mathcal{O}_X(U_\varepsilon) \to \mathcal{O}_X(U_{\varepsilon'})$, $0 < \varepsilon' < \varepsilon$, are linear, continuous and compact. Moreover, differential operators define continuous linear operators on this space. Then the theorem follows from the next classical lemma, which we state without proof (cf. Schwartz [1], or Houzel [1, § 1]).

Lemma 4.3.8. *Let $E^\bullet$ and $F^\bullet$ be two bounded complexes of Fréchet spaces and u a morphism from $E^\bullet$ to $F^\bullet$:*

$$\begin{array}{ccccccc} 0 \to & E_0 & \to \ldots \to & E_r & \to 0 \\ & u \downarrow & & u \downarrow & \\ 0 \to & F_0 & \to \ldots \to & F_r & \to 0, \end{array}$$

(all the maps are linear and continuous). Assume that for each j, the map $u: E_j \to F_j$ is compact, and u is a quasi-isomorphism from $E^\cdot$ to $F^\cdot$ (i.e. for each j, u induces an isomorphism $H^j(E^\cdot) \cong H^j(F^\cdot))$. Then for all j, the spaces $H^j(E^\cdot)$ are finite dimensional. ☐

Example 4.3.9. a) Let Z be a submanifold of X of codimension r. Then:

$$\mathcal{E}xt^j_{\mathscr{D}_X}(\mathscr{B}_{Z|X}, \mathscr{O}_X) = 0 \qquad j \neq r,$$

$$= \underline{\mathbb{C}}_Z \qquad j = r.$$

b) Let $X = \mathbb{C}$, t a coordinate on X, $P = t D_t - 1/2$, $\mathcal{M} = \mathscr{D}_X / \mathscr{D}_X \cdot P$. Consider the stratification $\mathbb{C} = (\mathbb{C} \setminus \{0\}) \cup \{0\}$. Then the sheaf $\mathscr{H}om_{\mathscr{D}_X}(\mathcal{M}, \mathscr{O}_X)$ is locally constant (but not constant) of rank one on $\mathbb{C} \setminus \{0\}$, and is zero on $\{0\}$. The sheaf $\mathcal{E}xt^1_{\mathscr{D}_X}(\mathcal{M}, \mathscr{O}_X)$ is zero on $\mathbb{C} \setminus \{0\}$, but is isomorphic to $\mathbb{C}$ on $\{0\}$.

Remark 4.3.10. If $\mathcal{M}$ is holonomic, one defines the analytical index of $\mathcal{M}$ at $x \in X$ by:

$$\chi(\mathcal{M}, x) = \sum_j (-1)^j \dim_{\mathbb{C}}(\mathcal{E}xt^j_{\mathscr{D}_X}(\mathcal{M}, \mathscr{O}_X)_x).$$

This index has been computed by M. Kashiwara [5], in terms of topological invariants of the characteristic cycle of $\mathcal{M}$.

Remark 4.3.11. Using Remark 1.4.7 and Theorem 4.3.3 we find that if $\mathcal{M}$ is holonomic, then the sheaves $\mathcal{E}xt^j_{\mathscr{D}_X}(\mathcal{M}, \mathscr{B}_{Z|X})$ are all weakly $\mathbb{C}$-constructible. In fact they are even $\mathbb{C}$-constructible, and from this result one deduces easily that for any pair of holonomic $\mathscr{D}_X$-modules $\mathcal{M}$ and $\mathcal{N}$, the sheaves $\mathcal{E}xt^j_{\mathscr{D}_X}(\mathcal{M}, \mathcal{N})$ are $\mathbb{C}$-constructible (M. Kashiwara [4]).

Remark 1.3.12. The correspondence which associates the complex $\mathscr{H}om_{\mathscr{D}_X}(\mathcal{M}^\cdot, \mathscr{O}_X)$ to the holonomic Module $\mathcal{M}$ on X, is an "equivalence of categories" between a subcategory of the category of (bounded complexes of) holonomic Modules on X, namely that of "regular holonomic Modules" of M. Kashiwara and T. Kawaï [1], and the category of (bounded complexes of) $\mathbb{C}$-constructible sheaves on X.

This is the Riemann-Hilbert correspondence; cf. Z. Mebkhout [1] (for the case of infinite order differential operators), and M. Kashiwara [7].

It is possible to give a "relative version" of Theorem 4.3.7.

Proposition 4.3.13. *Let f be a smooth surjective map from X to Y, $y \in Y$, $\Sigma = f^{-1}(y)$. Let $\mathcal{M}$ be a right coherent $\mathscr{D}_X$-module. Assume the set $\mathrm{char}_f(\mathcal{M}) \times_X \Sigma$ is isotropic in $T^*\Sigma$. Then for each j the sheaf $\mathscr{T}or_j^{\mathscr{D}_X}(\mathcal{M}, \mathscr{D}_{X \to Y})|_\Sigma$ is weakly $\mathbb{C}$-constructible, and for each $x \in \Sigma$, the stalk at x of this sheaf is a $\mathscr{D}_{Y,y}$-module of finite type.*

Sketch of proof. By Theorems 4.3.4 and 4.2.5 it remains only to prove the finiteness of the $\mathscr{D}_{Y,y}$-modules $\mathscr{T}\!or_j^{\mathscr{D}_X}(\mathscr{M}, \mathscr{D}_{X \to Y})_x$.

Let $\mathscr{N}_0$ be a coherent sub-$\mathscr{D}_{X/Y}$-module of $\mathscr{M}$ which generates $\mathscr{M}$ in a neighborhood of x. Let $\mathscr{M}_1$ be the coherent $\mathscr{D}_X$-module defined by the exact sequence:

$$0 \to \mathscr{M}_1 \to \mathscr{N}_0 \otimes_{\mathscr{D}_{X/Y}} \mathscr{D}_X \to \mathscr{M} \to 0.$$

Since the set $\mathrm{char}_f^1(\mathscr{M}_1)$ is contained in $\mathrm{char}_f^1(\mathscr{N}_0 \otimes_{\mathscr{D}_{X/Y}} \mathscr{D}_X) = \mathrm{char}_f^1(\mathscr{M})$, we may argue by induction on j and reduce the proof to the case where $\mathscr{M} = \mathscr{N}_0 \otimes_{\mathscr{D}_{X/Y}} \mathscr{D}_X$. Then:

$$(4.3.7) \qquad \mathscr{N}_0 \otimes_{\mathscr{D}_{X/Y}} \mathscr{D}_X \otimes_{\mathscr{D}_X} \mathscr{D}_{X \to Y} = (\mathscr{N}_0 \otimes_{\mathscr{D}_{X/Y}} \mathscr{O}_X) \otimes_{f^{-1}\mathscr{O}_Y} f^{-1}\mathscr{D}_Y$$

and it is enough to prove that the $\mathscr{O}_{Y,y}$-modules $\mathscr{T}\!or_j^{\mathscr{D}_{X/Y}}(\mathscr{N}_0, \mathscr{O}_X)$ are of finite type.

If we choose a local chart in Σ in a neighborhood of x, then the balls of center x and radius ε, $0 < \varepsilon \ll 1$, in Σ will have a non characteristic boundary with respect to $\mathscr{N}_0$. Now the proof goes as for Theorem 4.3.7, with Lemma 4.3.8 replaced by the following:

Lemma 4.3.14. *Let Y be a complex manifold, $y \in Y$, and let $X = Y \times \mathbb{C}^p$. Let $\overline{B}_\varepsilon$ be the closed ball in $\mathbb{C}^p \cong \mathbb{R}^{2p}$, with center at 0 and radius ε. Let $\mathrm{E}_\varepsilon^{\cdot}$ be a bounded complex of the type:*

$$0 \to \mathscr{O}_X^{N_0}(\{y\} \times \overline{B}_\varepsilon) \xrightarrow[P_0]{} \cdots \to \mathscr{O}_X^{N_r}(\{y\} \times \overline{B}_\varepsilon) \to 0,$$

where the P_j's are linear continuous and $\mathscr{O}_{Y,y}$-linear (e.g. P_j is a matrix with entries in $\mathscr{D}_{X/Y}$). Assume that the restriction morphisms $\mathrm{E}_\varepsilon^{\cdot} \to \mathrm{E}_{\varepsilon'}^{\cdot}$, $0 < \varepsilon' \leqslant \varepsilon$ are quasi-isomorphisms. Then the cohomology groups $H^j(\mathrm{E}_\varepsilon^{\cdot})$ are modules of finite type over $\mathscr{O}_{Y,y}$ for $0 \leqslant \varepsilon \leqslant \varepsilon_0$.

For the proof we refer to C. Houzel [1].   □

Remark 4.3.15. Using a similar method, one can prove the coherency of the direct images of $\mathscr{D}_X$-modules for non proper morphisms (cf. Houzel-Schapira [1]).

Exercises to III.4

Ex. 4.1. Let X be a real manifold of class $C^r (r \geqslant 2)$, $(y_1, \ldots, y_p, t_1, \ldots, t_q) = (y, t)$ a system of coordinates on X, $(y, t; \eta, \tau)$ the associated coordinates on $T^* X$. Let $Y = \{(y, t); \ t = 0\}$, and let S be a conic subset of $T^* X$. Prove: $(y, \eta) \in T^* Y \cap C_{T_Y^* X}(S) \Leftrightarrow$ there exists a sequence $\{(y_n, t_n; \eta_n, \tau_n)\}$ in S such that $(y_n, \eta_n) \xrightarrow[n]{} (y, \eta)$, $|t_n| \xrightarrow[n]{} 0$, $|t_n| |\tau_n| \xrightarrow[n]{} 0$. (cf. Kashiwara-Schapira [4]).

Ex. 4.2. Let $X = Y \times \mathbb{C}$ where Y is complex analytic, with coordinates (y, t), and let Y be identified with $Y \times \{0\}$. Let f be a holomorphic function on Y, Z the smooth hypersurface $Z = \{(y, t) \in X; \; t - f(y) = 0\}$. Prove: $(y, \eta) \in T^*Y \cap C_{T_Y^*X}(T_Z^*X) \Leftrightarrow$ there exists a sequence $\{(y_n, c_n)\}$ in $Y \times \mathbb{C}$ such that $y_n \xrightarrow{n} y$, $c_n\, df(y_n) \xrightarrow{n} \eta$, $c_n f(y_n) \xrightarrow{n} 0$. (Use Ex. 4.1).

Ex. 4.3. Let $X = \bigsqcup_\alpha X_\alpha$ be a Whitney $\mathbb{C}$-analytic stratification. Let $Y \subset X$ be a submanifold, and assume Y is transversal to each stratum X_α. Prove that $T^*Y \cap C_{T_Y^*X}(\bigsqcup_\alpha T_{X_\alpha}^*X) = \bigsqcup_\alpha T_{X_\alpha \cap Y}^*Y$.

Ex. 4.4. Let $X = \mathbb{R}^n$ with coordinates $(x_1, \ldots, x_n)$, and let $A = \{x \in X; \; x_1 \geqslant 0, x_2 < 0\}$. Calculate $SS(\mathbb{C}_A)$. (Hint: use formula (4.2.3)).

Ex. 4.5. On $X = \mathbb{C}^n$ with coordinates $(x_1, \ldots, x_n)$, let $\mathcal{M}$ be the holonomic $\mathcal{D}_X$-module $\mathcal{D}_X/\mathcal{J}$, where $\mathcal{J}$ is the left Ideal generated by the operators $\sum_{j=1}^{n} D_j^2$, $(x_i D_j - x_j D_i)_{1 \leqslant i, j \leqslant n}$. Let $X_0 = \{0\}$, $X_1 = \{x \in X; \; \sum_{j=1}^{n} x_j^2 = 0, \; x \neq 0\}$, $X_2 = \{x \in X; \; \sum_{j=1}^{n} x_j^2 \neq 0\}$. Calculate the sheaves $\mathcal{E}xt^j_{\mathcal{D}_X}(\mathcal{M}, \mathcal{O}_X)|_{X_\alpha}$ and the induced systems $\mathcal{T}or^{\mathcal{O}_X}_j(\mathcal{O}_{X_\alpha}, \mathcal{M})$ for $\alpha = 0, 1, 2, j \in \mathbb{N}$.

Ex. 4.6. Prove formula (4.3.4) in the particular case where $\mathcal{M}^\cdot$ is the complex $0 \to \mathcal{D}_X \xrightarrow{P} \mathcal{D}_X \to 0$ and P has simple characteristics (i.e.: $d\sigma(P) \wedge \omega \neq 0$ on $\dot{T}^*X \cap \sigma(P)^{-1}(0)$). (Hint: use the results of § 2.2, as in Ex. 3.1).

Ex. 4.7. By considering the hypersurface $Z = \{t \in \mathbb{C}; \; t = 0\}$ and the operator $P = t D_t - 1/2$, show that the first inclusion in Theorem 4.3.3 may be strict.

Ex. 4.8. Let Ω be a relatively compact open subset of $\mathbb{C}^n$. Assume Ω strictly pseudo-convex with C^1-boundary. Let P be a differential operator defined in a neighborhood of $\bar{\Omega}$, and assume $\partial\Omega$ is non characteristic for P. Prove that P, acting on $\mathcal{O}(\Omega)$, has a kernel and a cokernel of finite dimension (cf. Bony-Schapira [1]).

Ex. 4.9. Let $x = (x_1, \ldots, x_n)$ be the coordinates on $\mathbb{C}^n$. Let $P = \sum_{|\alpha| = |\beta| \leqslant m} a_{\alpha,\beta}(x) x^\alpha D_x^\beta$, and assume $\sum_{|\alpha| = |\beta| = m} a_{\alpha,\beta}(0) x^\alpha \bar{x}^\beta \neq 0$ for $x \neq 0$. Let $\mathfrak{m}$ be the maximal ideal of $O_{\mathbb{C}^n,0}$. Prove that there exists $k_0 \geqslant 0$ such that for $k \geqslant k_0$, P induces an isomorphism on $\mathfrak{m}^k$ (cf. Kashiwara-Kawaï-Sjöstrand [1] or Bengel-Gérard [1]).

Notes

The microcharacteristic variety $C_V(\mathcal{M})$ was first introduced by Bony-Schapira [2] for the case of one operator, then by Kashiwara-Schapira [1] for systems, (cf. also Kashiwara-Schapira [2] for the real case). We follow Monteiro-Fernandès [3] for the construction and the study of the variety $C_V^1(\mathcal{M})$, using the Ring $\mathscr{E}_V$ (and Proposition 1.2.2) of Kashiwara-Oshima [1]. Let us mention that this variety and many others which roughly speaking correspond to "Gevrey classes", were introduced at the same time by Laurent [3] by a different method (as mentioned in the introduction of this Chapter).

The variety $\mathrm{char}_f^1(\mathcal{M})$ is an important tool in the study of direct images (Houzel-Schapira [1]). The filtration $F^Y \mathscr{D}_X$ associated to a submanifold Y of X was introduced by M. Kashiwara [6], but the results of § 1.4 are due to Laurent-Schapira [1] as well as Theorem 1.5.2, which is a variation on a theorem of Kashiwara-Schapira [3] [4], and Theorem 1.6.6. which is related to results of Kashiwara-Kawaï-Sjöstrand [1]. The formal microcharacteristic variety $\hat{C}_{T_Y X}(\mathcal{M})$ is closely related to a variety first introduced by Monteiro-Fernandès [3] in the study of formal microdifferential operators.

The generalization of the classical Cauchy-Kowalewski theorem to differential systems (Theorem 2.1.4), was first achieved in 1970 by M. Kashiwara [1]. On the other hand the study of the Cauchy problem whose data are no more holomorphic but possess singularities along hypersurfaces, was initiated with Y. Hamada's paper in 1969 (Hamada [1]), (cf. also Hamada-Leray-Wagschal [1] and Pallu de la Barrière-Schapira [1]). In 1978, Kashiwara-Schapira [1], using the variety $C_V(\mathcal{M})$, gave a general formulation of the Cauchy problem for microdifferential systems $\mathcal{M}$ with values in $\mathscr{S}^\infty$, where $\mathscr{S}$ is another system and $\mathscr{S}^\infty$ is obtained from $\mathscr{S}$ by applying microdifferential operators of infinite order.

In this context, the theorems of § 2 are due to these authors. For example Theorem 2.2.1 is proved in Kashiwara-Schapira [1] for solutions having logarithmic or essential singularities, under the hypothesis that Y is non-microcharacteristic. As they are stated here, the results of this section were obtained by T. Monteiro-Fernandès [2] [3] and Y. Laurent [2] [3].

The results of § 3 were first stated in the case of holomorphic solutions of differential Modules. In this situation Lemma 3.1.4 is due to M. Zerner [1], its extensions (Lemma 3.1.6 and Theorem 3.1.2 in the case where the open set $\{x;\ \phi(x) < 0\}$ is pseudo-convex) are due to Bony-Schapira [1], and finally Theorem 3.2.2 is due to Kashiwara [5].

For other sheaves of solutions, the story is the same as for the Cauchy problem. The results of § 3 were first proved by Kashiwara-Schapira [2] using the variety $C_V(\mathcal{M})$, for solutions in $\mathscr{S}^\infty$, but the method was rather different since these results were obtained as particular cases of a propagation theorem for microfunction solutions of micro-hyperbolic systems. Then Theorem 3.2.1 was proved in this formulation (and by this method) by Monteiro-Fernandès [2] and also, (for the non filtered part of this theorem), using another method, by Y. Laurent [2].

In section 4.1 we have followed Kashiwara [5] (cf. also Kashiwara-Schapira [4]) for the microlocal study of Whitney's conditions on stratifications. The theory of the micro-support and the results of § 4.2 are due to Kashiwara-Schapira [4].

The important Theorem 4.3.7 was obtained by M. Kashiwara [2] in 1975. The finiteness result in Proposition 4.1.13 is a particular case of a theorem of Houzel-Schapira [1].

Finally let us mention that Kashiwara's constructibility theorem was the starting point of many important works (cf. Remarks 4.3.10, 4.3.11, 4.3.12).

Appendices

A. Symplectic Geometry

References are made to Abraham-Mardsen [1], Arnold [1], Duistermaat [1], Weinstein [1].

A.1. Symplectic Vector Spaces

Let $\mathbb{K} = \mathbb{R}$ or $\mathbb{C}$, and let E be a finite dimensional vector space over $\mathbb{K}$. A symplectic form σ on E is a non degenerate alternate bilinear form on E.

Let (E, σ) be a symplectic vector space. Then the dimension of E is even, say $2n$, and there exists a basis of E, $(e_1, \ldots, e_n; f_1, \ldots, f_n)$ such that if we denote by $(e_1^*, \ldots, e_n^*; f_1^*, \ldots, f_n^*)$ the dual basis on E^*, then:

$$(A.1.1) \qquad \sigma = \sum_{i=1}^{n} f_i^* \wedge e_i^*.$$

Such a basis is said to be symplectic. Since σ is non degenerate, it defines an isomorphism H from E^* to E by the formula:

$$(A.1.2) \qquad \langle \theta, v \rangle = \sigma(v, H(\theta)), \quad v \in E, \quad \theta \in E^*$$

and if we calculate the image by H of the dual basis of a symplectic basis we immediately find:

$$(A.1.3) \qquad \begin{cases} H(e_i^*) = -f_i & i = 1, \ldots, n, \\ H(f_j^*) = e_j & j = 1, \ldots, n. \end{cases}$$

A vector space $F \subset E$ is said to be *isotropic* (resp. *Lagrangean*, resp. *involutive*) if, $F^\perp$ denoting the orthogonal space to F in E, we have $F \subset F^\perp$ (resp. $F = F^\perp$, resp. $F^\perp \subset F$).

Some authors also use "co-isotropic" instead of "involutive".

Remark that if F is isotropic (resp. Lagrangean, resp. involutive) then $\dim F \leqslant n$ (resp. $= n$, resp. $\geqslant n$) and F is Lagrangean if and only if $\dim F = n$ and F is isotropic (resp. involutive). A line is always isotropic and a hyperplane always involutive.

A.2. Symplectic Manifolds

Now let Y and X be two manifolds (real or complex analytic), f a map from Y to X. To f are naturally associated $\overline{\omega}$ and ρ:

$$T^*Y \xleftarrow{\rho} Y \underset{X}{\times} T^*X \xrightarrow{\overline{\omega}} T^*X,$$

where $\overline{\omega}((y, x, \xi))=(x, \xi)$, $\rho((y, x, \xi))=(y, {}'f'(x)\xi)$. If we identify Y with the graph of f in $Y \times X$, and $Y \underset{X}{\times} T^*X$ with $T^*_Y(Y \times X)$ the conormal bundle to this graph, $\overline{\omega}$ and ρ are induced by the projections from $T^*(Y \times X)$ to T^*X and to T^*Y. In particular, if we choose T^*X for Y, π (the projection $T^*X \to X$) for f, we get a map $T^*X \underset{X}{\times} T^*X \to T^*T^*X$:

$$(x, \eta; \, x, \xi) \to (x, \eta; \, \xi, 0).$$

The restriction of this map to the diagonal $T^*X \underset{T^*X}{\times} T^*X$, that is to T^*X, gives a section ω of the bundle T^*T^*X above T^*X. This 1-form ω is called the canonical 1-form on T^*X, or the Liouville form on T^*X. If $(x_1, \ldots, x_n)$ is a system of local coordinates on X, $(x; \sum_j \xi_j dx_j)=(x; \xi)$ the associated coordinates on T^*X, then:

(A.2.1)
$$\omega = \sum_{j=1}^{n} \xi_j dx_j.$$

The differential $\sigma=d\omega$ is a symplectic form on T^*X (i.e.: σ induces at each $p \in T^*X$ a symplectic form on $T_p T^*X$). We still denote by H the isomorphism $T^*T^*X \cong TT^*X$ attached to σ, and call it the *symplectic isomorphism*.

Let f be a function on some open set U of T^*X, df its differential. The *Hamiltonian field* H_f of f is the vector field on U image of df by H. In the preceding coordinates (x, ξ):

$$df = \sum_{j=1}^{n} \left(\frac{\partial f}{\partial \xi_j} d\xi_j + \frac{\partial f}{\partial x_j} dx_j \right),$$

(A.2.2)
$$H_f = \sum_{j=1}^{n} \left(\frac{\partial f}{\partial \xi_j} \frac{\partial}{\partial x_j} - \frac{\partial f}{\partial x_j} \frac{\partial}{\partial \xi_j} \right).$$

The *Poisson bracket* $\{f, g\}$ of two functions f and g is defined by:

(A.2.3)
$$\{f, g\}=H_f(g).$$

One checks the relations:

(A.2.4)
$$\begin{cases} \{f, g\}= -\{g, f\}, \\ \{f, hg\}=h\{f, g\}+g\{f, h\}, \\ \{\{f, g\}, h\}+\{\{g, h\}, f\}+\{\{h, f\}, g\}=0, \end{cases}$$

and we get:

(A.2.5) $[H_f, H_g] = H_{\{f, g\}}$.

A submanifold V of T^*X is said to be isotropic (resp. Lagrangean, resp. involutive) if at each $p \in V$ the tangent space $T_p V$ has the corresponding property in $T_p T^*X$. When X is a complex manifold this definition is extended to complex analytic varieties. One say that such a set S is isotropic (resp. Lagrangean, resp. involutive) if there exists an open dense submanifold $V \subset S$ with this property, (cf. Chapter III, § 4 for more details). A manifold V is isotropic if $d\omega|_V = 0$.

A manifold V is involutive iff the Poisson bracket $\{f, g\}$ is zero on V for any functions f and g which are zero on V. In fact the bundle $(TV)^\perp$ is generated by the vector fields H_f, with $f|_V = 0$, thus $(TV)^\perp \subset TV$ is equivalent to $H_f(g) = 0$ for any f, g, with $f|_V = 0$, $g|_V = 0$. Moreover since $[H_f, H_g]$ equals $H_{\{f, g\}}$, we find that if V is involutive the sub-bundle $(TV)^\perp$ of TV satisfies the Frobenius integrability condition (i.e.: $(TV)^\perp$ is closed under bracket $[,]$). By the Frobenius theorem, an involutive manifold V admits a foliation, and the leaves of this foliation are called the *bicharacteristic leaves* of V. Remark that the dimension of the leaves is the codimension of V. In particular if V is Lagrangean, the leaves are open.

Example A.2.1. Let Z be a submanifold of X. Then $Z \underset{X}{\times} T^*X$ is involutive.

Now recall that the normal bundle $T_Z X$ and the conormal bundle $T_Z^* X$, are defined by the exact sequence of bundle over Z:

$$0 \to TZ \to Z \underset{X}{\times} TX \to T_Z X \to 0,$$

$$0 \to T_Z^* X \to Z \underset{X}{\times} T^*X \to T^*Z \to 0.$$

It is immediately verified that $T_Z^* X$ is Lagrangean in T^*X. When $Z = X$, $T_X^* X$, the zero section of T^*X, is identified to X.

A.3. Homogeneous Symplectic Structures

The 1-form ω on T^*X induces a richer structure than merely that of a symplectic manifold, and we shall now describe this "homogeneous symplectic" structure of T^*X.

Let $H(\omega)$ be the image of ω by the Hamiltonian isomorphism. If we have chosen coordinates (x, ξ):

(A.3.1) $\omega = \sum_j \xi_j dx_j$, $H(\omega) = -\sum_j \xi_j \dfrac{\partial}{\partial \xi_j}$.

Thus $-H(\omega)$ is just the radial vector field on the bundle T^*X (or what we called in Chapter I, § 1 the Euler operator e).

We shall say that a submanifold V is *conic* if $H(\omega)$ is tangent to it, and that a function f on T^*X is homogeneous (with respect to the fibre) if $H(\omega)(f)=kf$ for some $k\in\mathbb{Z}$. Thus a submanifold V is conic iff it can locally be defined by equations $\{f=0\}$, with f homogeneous. Remark that a conic manifold V is isotropic if and only if $\omega|_V=0$, since for a vector $v\in TT^*X$, $\langle\omega,v\rangle=d\omega(v,H(\omega))$.

We say that a conic involutive manifold V is *regular* if $\omega|_V$ is everywhere different from zero.

This is equivalent to the local existence of homogeneous functions $f_1,\ldots,f_r$ which vanish on V, where $r=\operatorname{codim} V$, such that:

$$(A.3.2)\qquad \begin{cases} \{f_i,f_j\}=0 \text{ on } V, \\ df_1\wedge\ldots\wedge df_r\wedge\omega\neq 0 \text{ on } V. \end{cases}$$

In particular a Lagrangean manifold is nowhere regular.

Example A.3.1. Let $V=\{(x,\xi)\in T^*\mathbb{C}^n;\ x_1+x_2^2=0\}$. Then V is involutive, as is any hypersurface, and conic (with respect to ξ).

On V, $dx_1+2x_2\,dx_2=0$. Thus $\omega|_V=(\xi_2-2x_2\xi_1)dx_2+\sum_{i>2}\xi_j\,dx_j$, and V is regular except on the set $\xi_2-2x_2\xi_1=\xi_3=\ldots=\xi_n=0$.

More generally if $V=Y\times_X T^*X$, for a submanifold $Y\subset X$, V is regular outside T_Y^*X.

Example A.3.2. Let P be a microdifferential operator on an open set $U\subset T^*X$, $\sigma(P)$ its principal symbol, $V=\{(x,\xi);\ \sigma(P)(x,\xi)=0\}$. One says that V is the characteristic variety of P. Since V is a hypersurface, V is involutive, but V is regular involutive iff $d\sigma(P)\wedge\omega\neq 0$ on V. The bicharacteristic curves of P are the leaves of V, that is the integral curves of the vector field:

$$(A.3.3)\qquad H_{\sigma(P)}=\sum_{j=1}^n\left(\partial_{\xi_j}\sigma(P)\frac{\partial}{\partial x_j}-\partial_{x_j}\sigma(P)\frac{\partial}{\partial\xi_j}\right).$$

A.4. Contact Transformations

Let X and Y be two manifolds (of the same dimension), U and V two open sets in T^*X and T^*Y respectively. Let ϕ be a diffeomorphism from U to V (a bi-holomorphic map, in the complex case). We denote by ω_X (resp. ω_Y) the canonical 1-form on T^*X (resp. T^*Y). Then ϕ is a symplectomorphism if $\phi^*(d\omega_Y)=d\omega_X$, and ϕ is a homogeneous symplectomorphism if ϕ is homogeneous and $\phi^*(\omega_Y)=\omega_X$. But we prefer to use the term "*contact transformation*" instead of homogeneous symplectomorphism, although this is not really correct, since a contact structure is locally the structure obtained on the quotient space of $\dot{T}^*X=T^*X\setminus T_X^*X$ by the integral curves of the radial field, (that is, for example in the complex case, the orbits of the action of $\mathbb{C}^{\times}$, cf. Arnold [1]).

Let ϕ be a homogeneous diffeomorphism from U to V, Λ_ϕ its graph in $U \times V$. The inverse image of a 1-form α on V is characterized by the condition $(\phi^* \alpha - \alpha)|\Lambda_\phi = 0$. Thus if we denote by Λ_ϕ^a the image of Λ_ϕ by the antipodal map of $T^* Y$, we find that ϕ is a contact transformation if and only if Λ_ϕ^a is isotropic, thus Lagrangean.

Assume we have coordinates (x, ξ) on $T^* X$, (y, η) on $T^* Y$. Then ϕ is defined by two sets of functions, $f_j(x, \xi)$ homogeneous of degree 0, $g_k(x, \xi)$ homogeneous of degree 1, $1 \leq j, k \leq n$, with:

$$y_j = f_j(x, \xi), \qquad \eta_k = g_k(x, \xi),$$

and ϕ is a contact transformation iff Λ_ϕ^a is involutive, that is iff:

(A.4.1) $\qquad \{f_j, f_k\} = 0, \qquad \{g_j, g_k\} = 0, \qquad \{f_j, g_k\} = -\delta_j^k.$

Let us recall two important results, without proofs (cf. Duistermaat [1]).

Theorem A.4.1. *Let ϕ be a contact transformation. Then locally we may find two contact transformations ϕ_1 and ϕ_2 such that $\phi = \phi_2 \circ \phi_1$, and the graphs $\Lambda_{\phi_i}^a (i = 1, 2)$ are conormal bundles to some hypersurfaces in the base spaces.*

Theorem A.4.2. *Let Λ be a conic Lagrangean manifold in a neighborhood of $p \in \dot{T}^* X$. Then there exists a system of local homogeneous symplectic coordinates $(x; \xi)$ such that $p = (0; dx_n)$ and $\Lambda = \{(x, \xi); \xi_1 = \ldots = \xi_{n-1} = x_n = 0\}$.*

We shall also need the following results.

Lemma A.4.3. *Let V be a conic regular involutive submanifold of $T^* X$ defined in a neighborhood of $p \in \dot{T}^* X$. Let λ be a Lagrangean plane of $T_p T^* X$ contained in $T_p V$. Then there exists a conic Lagrangean manifold Λ in a neighborhood of p such that:*

$$\Lambda \subset V, \qquad T_p \Lambda = \lambda.$$

Proof. Let $n = \dim X$, $r = \operatorname{codim} V$. Let $f_1, \ldots, f_r$ be a system of homogeneous functions vanishing on V and satisfying (A.3.2). By the classical theory of differential equations, we may find a function ϕ on V such that:

$$\begin{cases} (H(\omega))|_V(\phi) = (H_{f_1})|_V(\phi) = \ldots = (H_{f_r})|_V(\phi) = 0, \\ d\phi(p) \in T_\lambda^*(T_p V). \end{cases}$$

Moreover if $r < n - 1$ we may also assume:

$$df_1(p) \wedge \ldots \wedge df_r(p) \wedge d\phi(p) \wedge \omega(p) \neq 0.$$

Then we argue by induction on r, and replace V by $V' = \{q \in V; \phi(q) = 0\}$. $\square$

Theorem A.4.4. *Let $g_1, \ldots, g_r$ be homogeneous functions of degree one, $f_1, \ldots, f_d (d \leqslant r)$ homogeneous functions of degree 0, these functions being defined in a neighborhood of $p \in \dot{T}^* X$. Assume:*

- $dg_1 \wedge \ldots \wedge dg_r \wedge \omega \neq 0$,
- $\{g_i, g_j\} \equiv 0$, $\{f_l, f_k\} \equiv 0$, $\{g_i, f_l\} = \delta_i^l$ *for* $1 \leqslant i, j \leqslant r$, $1 \leqslant l, k \leqslant d$.

Then there exists a system of homogeneous symplectic coordinates $(x_1, \ldots, x_n; \xi_1, \ldots, \xi_n)$ in a neighborhood of p such that:

$$p = (0; \, dx_n), \quad g_j = \xi_j (1 \leqslant j \leqslant r), \quad f_l = x_l (1 \leqslant l \leqslant d).$$

Proof. We may assume that in a system of homogeneous symplectic coordinates (x, ξ) on $T^* X$ we have:

$$p = (0; \, dx_n); \, dg_j(p) = d\xi_j, \, (1 \leqslant j \leqslant r), df_l(p) = dx_l, \, (1 \leqslant l \leqslant d).$$

Let $T^* Y$ be another copy of $T^* X$, and let (y, η) be the coordinates on $T^* Y$.

Let $V \subset T^*(Y \times X)$ be the involutive manifold defined by the equations:

$$\begin{cases} \eta_j = -g_j(x, \xi) & 1 \leqslant j \leqslant r, \\ y_l = f_l(x, \xi) & 1 \leqslant l \leqslant d. \end{cases}$$

Then $T_p V \supset \lambda$ where λ is the conormal to the diagonal of $Y \times X$, and by Lemma A.4.3 there exists a Lagrangean manifold Λ contained in V, and such that $T_p \Lambda = \lambda$. Then Λ will be the graph of a canonical transformation in a neighborhood of p, and this proves the result.   □

Corollary A.4.5. *Let V and W be two conic involutive manifolds in a neighborhood of $p \in \dot{T}^* X$, V being regular involutive. Assume that for any function ϕ defined in a neighborhood of p, such that $\phi|_W = 0$ and $d\phi \neq 0$, the vector H_ϕ is not tangent to V. Then there exists a system of local homogeneous symplectic coordinates $(x; \xi)$ such that:*

$$p = (0; dx_n)$$
$$V = \{(x, \xi); \xi_1 = \ldots = \xi_r = 0\}, \quad r < n$$
$$W = \{(x, \xi); x_1 = \ldots = x_d = 0\}, \quad d \leqslant r.\bullet$$

Proof.   Let $V = \{(x, \xi); \quad g_1(x, \xi) = \ldots = g_r(x, \xi) = 0\}$, $W = \{(x, \xi); f_1(x, \xi) = \ldots = f_d(x, \xi) = 0\}$, where the g_j's are homogeneous of degree one, the f_j's homogeneous of degree zero, $dg_1 \wedge \ldots \wedge dg_r \wedge \omega \neq 0$, $df_1 \wedge \ldots \wedge df_d \neq 0$.

By hypothesis, for a given f_i, there exists some j, with $\{f_i, g_j\} \neq 0$. Thus we may assume $\{f_1, g_1\} \neq 0$. If $g_1 = \xi_1$ in a system of symplectic coordinates, this means $\dfrac{\partial f}{\partial x_1} \neq 0$; thus by the Weierstrass Preparation theorem, we may find a non vanishing function θ such that $\{\theta f_1, g_1\} \equiv -1$. By applying Theorem A.4.4 we get that there exist homogeneous symplectic coordinates $(x; \xi)$ such that in these new coordinates,

$$f_1 = x_1, \quad g_1 = \xi_1.$$

We write $x = (x_1, x')$, $\xi = (\xi_1, \xi')$ and we divide g_j by ξ_1. Then we may assume $g_j(x, \xi')$ does not depend on ξ_1. Let $\tilde{V}$ be the submanifold of V defined by:

$$\tilde{V} = \{(x, \xi); \; x_1 = \xi_1 = g_j(0, x', \xi') = 0, \quad j = 2, \ldots, r\}.$$

Since V is involutive, V is invariant by $\dfrac{\partial}{\partial x_1}$, the Hamiltonian flow of H_{ξ_1}. Thus:

$$V = \{(x, \xi); \; \xi_1 = g_j(0, x', \xi') = 0, \quad j = 2, \ldots, r\}.$$

Similarly:

$$W = \{(x, \xi); \; x_1 = f_j(x', 0, \xi') = 0, \quad j = 2, \ldots, d\}.$$

To complete the proof, we repeat this procedure with the functions g_j and f_k ($2 \leqslant j \leqslant r$, $2 \leqslant k \leqslant d$) in the variables (x', ξ'). $\quad\square$

B. Homological Algebra

References are made to Cartan-Eilenberg [1], Godement [1], Hartshorne [1], [2], Bourbaki [1], Freyd [1].

B.1. Categories and Derived Functors

We shall partly follow Hartshorne [2]. We do not give the precise definition of an abelian category, but we only recall that if $\mathfrak{C}$ is such a category, for each M, N $\in Ob(\mathfrak{C})$, Hom(M, N) has a structure of an abelian group, and the composition law is linear, every morphism has a kernel and a cokernel, every morphism can be factored into an epimorphism followed by a mono-morphism, a morphism which is both a monomorphism and an epimor-phism is an isomorphism. Moreover finite direct sums and finite direct products exist, and are isomorphic.

The first example we shall meet is the category we denote by $\mathfrak{A}$ of (say left) A-modules over a unitary ring A. In particular $\mathfrak{Z}$ well denote the category of abelian groups.

Let $\mathfrak{C}$ be an abelian category.

A complex $M^{\cdot}$ is a collection of objects M_j, $j \in \mathbb{Z}$, and morphisms $d_j : M_j \to M_{j+1}$, such that $d_{j+1} \circ d_j = 0 \; \forall j$. We often write d instead of d_j, and say d is the differential of the complex. It is convenient to write $M^{\cdot}$ as a line:

$$\ldots \to M_j \xrightarrow{d} \ldots \, .$$

If the objects M_j are only specified on a certain interval of $\mathbb{Z}$, we set $M_j = 0$ outside of this interval. For example an object M is identified with the complex "concentrated in degre 0"

$$\ldots \to 0 \to M \to 0 \to \ldots \, .$$

The p-shifted complex $M^{\cdot}_{[p]}$ is defined by:

(B.1.1) $$(M^{\cdot}_{[p]})_j = M_{j+p}$$

and d_j is replaced by $(-1)^p d_j$.

The j-th cohomology object of the complex $M^{\cdot}$ is defined by:

(B.1.2) $$H^j(M^{\cdot}) = \operatorname{Ker} d_j / \operatorname{Im} d_{j-1} \, .$$

A complex is bounded to the left if $M_j = 0$ for $j \ll 0$. One defines also complexes bounded to the right, and bounded complexes.

A morphism ϕ from $M^{\cdot}$ to $N^{\cdot}$ is a collection of morphisms ϕ_j from M_j to N_j such that:

$$\phi_{j+1} \circ d = d \circ \phi_j \quad \forall j \, .$$

Then ϕ induces for all j a morphism from $H^j(M^{\cdot})$ to $H^j(N^{\cdot})$. If ϕ induces an isomorphism on all the cohomology objects one says that ϕ is a quasi-isomorphism. A complex quasi-isomorphic to zero is said exact. A short exact sequence is an exact complex of the type: $0 \to L \to M \to N \to 0$. Now assume we have a short exact sequences of complexes, $0 \to L^{\cdot} \to M^{\cdot} \to N^{\cdot} \to 0$. Then there are natural maps for all j, $\delta^j: H^j(N^{\cdot}) \to H^{j+1}(L^{\cdot})$, giving rise to a long exact sequence:

(B.1.3) $$\ldots \to H^j(M^{\cdot}) \to H^j(N^{\cdot}) \to H^{j+1}(L^{\cdot}) \to \ldots \, .$$

Two morphisms ϕ and ψ from the complex $M^{\cdot}$ to the complex $N^{\cdot}$ are said homotopic, (written $\phi \sim \psi$), if there is a collection of morphisms $k_j: M_j \to N_{j-1}$, such that:

(B.1.4) $$\phi - \psi = dk + kd \, .$$

If ϕ and ψ are homotopic, then they induce the same morphisms on the cohomology objects, $H^j(M^{\cdot}) \to H^j(N^{\cdot})$, for all j.

Two complexes M and N are said to be homotopy equivalent if there exist $\phi: M^{\cdot} \to N^{\cdot}$ and $\psi: N^{\cdot} \to M^{\cdot}$ such that $\phi \circ \psi$ (resp. $\psi \circ \phi$) is homotopic to the identity of $N^{\cdot}$ (resp. $M^{\cdot}$).

A double complex $M^{\cdot\cdot}$ is a sequence of complexes $(M^{\cdot}_j, d')$ and morphisms of complexes d''_j from $M^{\cdot}_j$ to $M^{\cdot}_{j+1}$, such that $d''_{j+1} \circ d''_j = 0$. We often write d'' instead of d''_j, and represent $M^{\cdot\cdot}$ as a diagram:

$$\begin{array}{ccc}
\downarrow & & \downarrow \\
\longrightarrow \mathsf{M}_{i,j} \xrightarrow[d']{} & \mathsf{M}_{i+1,j} & \longrightarrow \cdots \\
\downarrow{\scriptstyle d''} & & \downarrow{\scriptstyle d''} \\
\longrightarrow \mathsf{M}_{i,j+1} \xrightarrow[d']{} & \mathsf{M}_{i+1,j+1} & \longrightarrow \cdots \\
\downarrow & & \downarrow \\
\vdots & & \vdots
\end{array}$$

We shall only have to consider double complexes which satisfy a bounded-ness condition, such as for example:

(B.1.5) $\exists k$ such that $\mathsf{M}_{i,j}=0 \quad \forall i<k \quad$ and $\mathsf{M}_{i,j}=0 \quad \forall j<k$

or else:

(B.1.6) $\exists k$ such that $\mathsf{M}_{i,j}=0 \quad$ for $|i|>k$.

In such a case it is possible to associate to $\mathsf{M}^{\cdot\cdot}$ a simple complex $\mathsf{M}^{\cdot}$ by setting:

$$\mathsf{M}_k = \bigoplus_{i+j=k} \mathsf{M}_{i,j}$$

and defining the differential d_k by:

$$d_k|_{\mathsf{M}_{i,j}} = d_j'' + (-1)^j d_i',$$

where $d_k|_{\mathsf{M}_{i,j}}$ stands for the restriction of d_k to $\mathsf{M}_{i,j}$.

A morphism ϕ from $\mathsf{M}^{\cdot}$ to $\mathsf{N}^{\cdot}$ may be considered as a double complex:

$$0 \to \mathsf{M}^{\cdot} \xrightarrow[\phi]{} \mathsf{N}^{\cdot} \to 0.$$

Let $\mathsf{L}^{\cdot}$ be the simple complex associated to this double complex. Then $\mathsf{L}^{\cdot}$ is called "the mapping cone" of ϕ, and denoted by $C(\phi)$. Remark that $\mathsf{L}^{\cdot} \cong \mathsf{N}^{\cdot} \oplus \mathsf{M}^{\cdot}[+1]$, and that we have an exact sequence of complexes:

$$0 \to \mathsf{N}^{\cdot} \to \mathsf{L}^{\cdot} \to \mathsf{M}^{\cdot}[+1] \to 0,$$

which gives rise to a long exact sequence on the cohomology objects. If $\mathsf{M}^{\cdot}$ and $\mathsf{N}^{\cdot}$ are two complexes, then $\mathrm{Hom}(\mathsf{M}^{\cdot},\mathsf{N}^{\cdot})$ is a double complex whose i-th-line is the complex $\mathrm{Hom}(\mathsf{M}_i,\mathsf{N}^{\cdot})$. One defines in an obvious way morphisms of double complexes. We shall need a special case of a general result on spectral sequences.

Let $\mathsf{M}^{\cdot\cdot}$ be a double complex with horizontal differential d' and vertical differential d''. One defines the new complex $H_I(\mathsf{M}^{\cdot\cdot})=\tilde{\mathsf{M}}^{\cdot\cdot}$, where $\tilde{\mathsf{M}}_{i,j}$ is the i-th cohomology object with respect to d' of the j-th line. The vertical differential on $H_I(\mathsf{M}^{\cdot\cdot})$ is the one induced by d'', and the horizontal one is now zero. Similarly one defines $H_{II}(\mathsf{M}^{\cdot\cdot})$ by interchanging the roles of d' and d''.

Proposition B.1.1. *Let* $\phi\colon \mathsf{M}^{\cdot\cdot}\to\mathsf{N}^{\cdot\cdot}$ *be a morphism of double complexes. Assume* ϕ *induces an isomorphism from* $H_{II}(H_I(\mathsf{M}^{\cdot\cdot}))$ *to* $H_{II}(H_I(\mathsf{N}^{\cdot\cdot}))$, *and assume* $\mathsf{M}^{\cdot\cdot}$ *and* $\mathsf{N}^{\cdot\cdot}$ *satisfy both condition (B.1.5) or (B.1.6). Then* ϕ *induces a quasi-isomorphism from the simple complex associated to* $\mathsf{M}^{\cdot\cdot}$ *to the simple complex associated to* $\mathsf{N}^{\cdot\cdot}$.

For the proof we refer to Godement [1, Th. 4.3.1].

Now let $\mathfrak{C}$ and $\mathfrak{C}'$ be two abelian categories, F a functor from $\mathfrak{C}$ to $\mathfrak{C}'$ (we only consider additive functors). One says that F is left exact (resp. right exact) if for any short exact sequence in $\mathfrak{C}$:

$$0 \to \mathsf{L} \to \mathsf{M} \to \mathsf{N} \to 0,$$

the sequence:

$$0 \to F(\mathsf{L}) \to F(\mathsf{M}) \to F(\mathsf{N}),$$

(resp. the sequence:

$$F(\mathsf{L}) \to F(\mathsf{M}) \to F(\mathsf{N}) \to 0),$$

is exact.

If F is both right and left exact, one says F is exact. If F is a contravariant functor there is a similar definition. For example F is left exact if for a short exact sequence as above, the sequence below is exact:

$$0 \to F(\mathsf{N}) \to F(\mathsf{L}) \to F(\mathsf{M}).$$

For a fixed object M in $\mathfrak{C}$, the functor $\mathrm{Hom}(\mathsf{M}, \cdot)$ is covariant left exact, and the functor $\mathrm{Hom}(\cdot, \mathsf{M})$ is contravariant left exact, from $\mathfrak{C}$ to $\mathfrak{Z}$.

Definition B.1.2. An object I (resp. P) is injective (resp. projective) if $\mathrm{Hom}(\cdot, \mathsf{I})$ (resp. $\mathrm{Hom}(\mathsf{P}, \cdot)$) is exact. One says that $\mathfrak{C}$ has enough injectives (resp. projectives), if for any object M in $\mathfrak{C}$ there is an exact sequence $0\to\mathsf{M}\to\mathsf{I}$ (resp. $\mathsf{P}\to\mathsf{M}\to 0$), with I injective (resp. P projective).

An injective resolution of M is a complex $\mathsf{I}^{\cdot}$ and a morphism $\varepsilon\colon\mathsf{M}\to\mathsf{I}^{\cdot}$ such that the objects of $\mathsf{I}^{\cdot}$ are injective and zero for $j<0$, and ε is a quasi-isomorphism, that is $H^j(\mathsf{I}^{\cdot})=0$ for $j\neq 0$ and $\mathsf{M}\xrightarrow{\sim}H^0(\mathsf{I}^{\cdot})$. Similarly a projective resolution of M is a complex $\mathsf{P}^{\cdot}$ and a morphism $\varepsilon\colon\mathsf{P}^{\cdot}\to\mathsf{M}$ such that the objects of $\mathsf{P}^{\cdot}$ are projective and zero for $j>0$, and ε is a quasi-isomorphism.

We shall say that $\mathsf{I}^{\cdot}$ is of length n (resp. of finite length) if $\mathsf{I}_j=0$ for $j>n$ (resp. $j\gg 0$), and similarly for projective resolutions. If $\mathfrak{C}$ has enough injectives (resp. projectives), any object has an injective resolution (resp. projective resolution). Moreover two such resolutions are always homotopy equivalent. This allows to state the following:

Definition B.1.3. Assume $\mathfrak{C}$ has enough injectives, and let F be a covariant left exact functor. The right derived functors $R^j F$ are defined as follows. For $M \in 0b(\mathfrak{C})$, let $M^{\bullet}$ be an injective resolution of M. Then:

$$R^j F(M) = H^j(F(M^{\bullet})).$$

There is an analogous definition for the left derived functor $L^j F(M)$ of a covariant right exact functor F (replace injective by projective), and for a contravariant right exact functor (use injective resolutions) or left exact functors (use projective resolutions), cf. Hartshorne [2].

Theorem B.1.4 (cf. Hartshorne [p. 204]). *Assume $\mathfrak{C}$ has enough injectives and let F be a covariant left exact functor from $\mathfrak{C}$ to another abelian category $\mathfrak{C}'$. Then:*
a) *For each j, $R^j F$ is an additive functor and $R^j F = 0$ for $j < 0$.*
b) *There is a natural isomorphism $F \cong R^0 F$.*
c) *For each short exact sequence $0 \to L \to M \to N \to 0$ there are natural morphisms $\delta^i: R^i F(N) \to R^{i+1} F(L)$ giving rise to long exact sequences:*

$$\ldots \to R^i F(L) \to R^i F(M) \to R^i F(N) \xrightarrow[\delta^i]{} R^{i+1} F(L) \to \ldots,$$

with δ^i functorial with respect to short exact sequences for all i. In other words, for any morphism from $0 \to L \to M \to N \to 0$ to another exact sequence $0 \to L' \to M' \to N' \to 0$, the diagram below is commutative:

$$
\begin{array}{ccc}
\to R^i F(N) & \to & R^{i+1} F(L) \to \\
\downarrow & & \downarrow \\
\to R^i F(N') & \to & R^{i+1} F(L') \to .
\end{array}
$$

d) *For each injective object I of $\mathfrak{C}$, $R^j F(I) = 0$ for $j \neq 0$.*

An object M in $\mathfrak{C}$ such that $R^j F(M) = 0$ for $j \neq 0$ is said to be acyclic for F, or F-acyclic.

We do not state the corresponding results for contravariant or right exact functors.

By definition an F-acyclic resolution of an object M is a complex $J^{\bullet}$ and a morphism $\varepsilon: M \to J^{\bullet}$ such that the objects of $J^{\bullet}$ are F acyclic and zero for $j < 0$, and ε is a quasi-isomorphism.

Proposition B.1.5. *Let $J^{\bullet}$ be an F-acyclic resolution of M. Then there are natural isomorphisms for all j:*

$$R^j F(M) \cong H^j(F(J^{\bullet})).$$

Remark that $R^j F(M) = 0$ for $j > n$, iff M admits an F-acyclic resolution of length n.

Finally let us recall that a functor F is faithful if $F(M) = 0$ implies $M = 0$.

B.2. Rings and Modules

Now we particularize the preceding study to the category $\mathfrak{A}$ of A-modules over a unitary ring A. (We do not require the ring A to be commutative. When not otherwise specified, we consider left A-modules. Thus A-module means left A-module).

Then we have the notions of injective modules and projective modules. Remark that projective modules are direct summands of free modules, but injective modules are not easy to characterize. The category $\mathfrak{A}$ has enough injectives and enough projectives.

Let M and N be two modules. We have:

$$R^i \operatorname{Hom}_A(\cdot, M)(N) = R^i \operatorname{Hom}_A(N, \cdot)(M).$$

This group is denoted by $\operatorname{Ext}_A^i(N, M)$. Thus it can be calculated by taking the i-th cohomology either of $\operatorname{Hom}_A(N, I^\cdot)$ or of $\operatorname{Hom}_A(P^\cdot, M)$, where $I^\cdot$ (resp. $P^\cdot$) is an injective (resp. projective) resolution of M (resp. N). Remark that $\operatorname{Ext}_A^i(N, M)$ is an A-module if A is commutative.

Applying Theorem B.1.4, we may associate long exact sequences to short exact ones. For example if $0 \to N' \to N \to N'' \to 0$ is exact, we get the long exact sequence:

(B.2.1) $0 \to \operatorname{Ext}_A^0(N'', M) \to \operatorname{Ext}_A^0(N, M) \to \operatorname{Ext}_A^0(N', M) \to \operatorname{Ext}_A^1(N'', M) \to \ldots$

Of course $\operatorname{Ext}_A^0(N, M) = \operatorname{Hom}_A(N, M)$.

Now recall that if N is a right A-module, the group $N \otimes_A M$ is well defined. If A is commutative, then $N \otimes_A M$ is even an A-module. The module M is flat if the covariant functor $\cdot \otimes_A M$ (on the category of right A-modules) is exact. Projective A-modules are flat.

We may also define flat right A-modules, and we remark:

$$L^{-j}(N \otimes_A \cdot)(M) = L^{-j}(\cdot \otimes_A M)(N).$$

This group is denoted by $\operatorname{Tor}_j^A(N, M)$. It can be calculated either by replacing M by a flat resolution $M^\cdot$ or by replacing N by a flat resolution $N^\cdot$. of course $\operatorname{Tor}_0^A(N, M) = N \otimes_A M$.

To a short exact sequence $0 \to M' \to M \to M'' \to 0$, one functorially associates a long exact sequence.

(B.2.2) $\ldots \to \operatorname{Tor}_1^A(N, M'') \to \operatorname{Tor}_0^A(N, M') \to \operatorname{Tor}_0^A(N, M) \to \operatorname{Tor}_0^A(N, M'') \to 0$

and similarly with N an M interchanged. Recall that M is faithfully flat if M is flat and $N \otimes_A M = 0$ implies $N = 0$. An s-presentation of a module M is an exact sequence:

$$M_{-s} \to M_{-s+1} \to \ldots \to M_0 \to M \to 0.$$

This presentation is free, or finite free, or projective, etc. … if the M_j's are free or free of finite type, or projective, etc. … A module M is of finite type iff it admits a finite free 0-presentation.

The ring A is (left) noetherian iff any finite free 0-presentation of a module extends as a finite free 1-presentation (thus as a finite free ∞-presentation).

The category $\mathfrak{A}^f$ of (left) A-modules of finite type over a (left) noetherian ring A is an abelian category.

Now for the reader's convenience we give here the proofs of some easy results that we need (cf. Malgrange [2] for Lemma B.2.1).

Lemma B.2.1. *We assume* A *noetherian, and such that any module of finite type admits a projective resolution of finite length by projective modules of finite type. Let* M *be a module of finite type which satisfies* $\mathrm{Ext}_A^j(M, A)=0$, *for* $j>n$. *Then:*

a) $\mathrm{Ext}_A^j(M, N)=0$ *for* $j>n$ *for any module* N.

b) M *admits a projective resolution of length* n *by projective modules of finite type.*

Proof. a) Let P be a projective module. There exists a module Q such that $P \oplus Q$ is free. Thus:

$$\mathrm{Ext}_A^j(M, P)=0 \quad \forall j>n.$$

Now let N be a module of finite type, and consider a finite free projective resolution of N of length r:

$$N^\bullet: 0 \to P_{-r} \to \ldots \to P_0 \to 0.$$

We argue by induction on r, the result being proved for $r=0$. Let N_1 be the image of P_{-1} in P_0. Then by the induction hypothesis, $\mathrm{Ext}_A^j(M, N_1)=0$ $\forall j>n$. Applying the functor $\mathrm{Hom}_A(M, \cdot)$ to the short exact sequence $0 \to N_1 \to P_0 \to N \to 0$ we get the long exact sequence:

$$\ldots \to \mathrm{Ext}_A^{j-1}(M, N) \to \mathrm{Ext}_A^j(M, N_1) \to \mathrm{Ext}_A^j(M, P_0) \to \ldots$$

Thus $\mathrm{Ext}_A^j(M, N)=0$ $\quad \forall j>n$.

Now any module N is an inductive limit of submodules of finite type, $N = \varinjlim_\alpha N_\alpha$. Since M is of finite type, we have:

$$\mathrm{Ext}_A^j(M, N) = \varinjlim_\alpha \mathrm{Ext}_A^j(M, N_\alpha)$$

and these groups are zero for $j>n$.

b) Let us consider a projective resolution of M of length s by projective modules of finite type:

$$M^\bullet: 0 \to Q_{-s} \to \ldots \to Q_0 \to 0$$

and replace Q_{-n} by R_{-n}, the kernel of the map $Q_{-n} \to Q_{-n+1}$. Then $\mathrm{Ext}_A^j(R_{-n}, N)=0$ $\forall j>0$ for any module N, and thus R_{-n} is projective. $\square$

Lemma B.2.2. *We assume* A *noetherian, and such that any module of finite type admits a projective resolution of finite length by free modules of finite type. Then:*

a) *Projective modules of finite type are stably finite-free (i.e.: if P is such a module, there exist L_0 and L_1 free of finite type such that $P \oplus L_1 = L_0$).*

b) *Let M be a module of finite type such that $Ext_A^j(M, A) = 0 \quad \forall j > n$. Then M admits a projective resolution of length $\sup(1, n)$ by free modules of finite type.*

Proof. a) Let P be a projective module of finite type, and consider a free presentation of P:

$$0 \to L_{-n} \to \ldots \to L_0 \to P \to 0.$$

If $n = 1$, we get $P \oplus L_{-1} = L_0$ since P is projective. We argue by induction on n. Let P_1 be the kernel of the map $L_0 \to P$. Then P_1 is projective and stable free by the induction hypothesis. Let L'_{-1} and L'_0 be free modules of finite type such that $P_1 \oplus L'_{-1} = L'_0$. We get: $P \oplus L'_0 = L_0 \oplus L'_{-1}$.

b) Consider a projective presentation of finite length of M by finite free modules:

$$0 \to L_{-r} \to \ldots \to L_0 \to M \to 0$$

and replace L_{-n} by P_{-n} the kernel of the map $L_{-n} \to L_{-n+1}$. Then P_{-n} is projective, and we may find a finite free module L such that $P_{-n} \oplus L$ is free. We get a free presentation of length n of M:

$$0 \to P_{-n} \oplus L \to L_{-n+1} \oplus L \to L_{-n+2} \to \ldots \to L_0 \to M \to 0. \quad \square$$

B.3. Graded Rings and Modules

Now we shall briefly describe another abelian category, which is important in our study.

A graded ring (over $\mathbb{Z}$), GA, is a unitary ring A endowed with a family of subgroups $(A_k)_{k \in \mathbb{Z}}$ such that:

(B.3.1)
$$\begin{cases} A = \bigoplus_k A_k, \\ 1 \in A_0, \quad A_k \cdot A_l \subset A_{k+l}. \end{cases}$$

(We do not ask A to be commutative and consider left A-modules).

A graded module GM over GA, or a GA-module, is an A-module M endowed with a family of subgroups $(M_k)_{k \in \mathbb{Z}}$, such that:

(B.3.2)
$$M = \bigoplus_k M_k, \quad A_l M_k \subset M_{k+l}.$$

A morphism of graded modules, or a GA-linear map, from GM to GM, is an A-linear map from M to N such that:

$$u(M_k) \subset N_k \quad \forall k.$$

Thus u is homogeneous of degree 0. The notions of graded sub-module, quotient, kernel, cokernel, ... are clear.

Remark for example that a graded ideal G I of G A satisfies:

$$G I = \bigoplus_k (G I \cap A_k).$$

Thus G I is generated by homogeneous elements.

Example B.3.1.: The ring $\mathbb{C}[x]$ of polynomials over $\mathbb{C}$ is graded by $\mathbb{C}[x] = \bigoplus_i \mathbb{C} \cdot x^i$. The ideal generated by $1 + x$ is not a graded ideal.

The notions of finite type G A-modules, s-presentation of a G A-module, finite free G A-module, etc. ... are clear, as well as the notion of noetherian graded ring, (cf. II, Remark 1.1.10).

Example B.3.2. The G A-module A endowed with the graduation $A'_k = A_{k-p}$, is free of rank one over G A. We shall denote it by $G A_{[p]}$. Remark that $G A_{[p]}$ is generated by the element 1 of degre p.

Proposition B.3.3. *The category $\mathfrak{G}\mathfrak{A}$ of (left) graded modules over G A is abelian.*

If G A is noetherian, the category $\mathfrak{G}\mathfrak{A}^f$ of graded modules of finite type over G A is abelian.

As the category of graded modules over a graded ring is very similar to the category of modules over a ring, we shall not always repeat the definitions or the proofs in this category. On the contrary the category of filtered modules over a filtered ring is not so easy to manipulate (it is not abelian), and this will be dealt with in (II, § 1).

B.4. Koszul Complexes

We shall study a rather general situation in which we may calculate explicitly the cohomology groups $\text{Ext}_A^j(N, M)$ or $\text{Tor}_j^A(N, M)$, by means of the so-called Koszul complexes.

Let M be an A-module (not necessarily of finite type) and let $\phi_1, \ldots, \phi_p$ be a family of commuting endomorphisms of M:

$$(B.4.1) \qquad \begin{cases} \phi_j \in \text{End}_A(M), \\ [\phi_i, \phi_j] \equiv 0 \quad 1 \leq i, j \leq p. \end{cases}$$

Let $e_1, \ldots, e_p$ be the canonical basis of $\mathbb{Z}^p$. For a subset I of $\{1, \ldots, p\}$, $I = (i_1, \ldots, i_k)$, we define the elements of $\overset{j}{\Lambda}(\mathbb{Z}^p)$:

$$e_I = e_{i_1} \wedge \ldots \wedge e_{i_k}.$$

Thus we have the relations, for a permutation π of $\{1, \ldots, k\}$ with signature ε_π:

$$e_{i_{\pi(1)}} \wedge \ldots \wedge e_{i_{\pi(k)}} = \varepsilon_\pi e_{i_1} \wedge \ldots \wedge e_{i_k}$$

and the $\mathbb{Z}$-module $\overset{k}{\Lambda}(\mathbb{Z}^p)$ is the free module with basis $(e_I)_I$, where I runs over the family of ordered subsets of $\{1, \ldots, p\}$ with $|I| = k$, (i.e. $I = (i_1, \ldots, i_k), i_1 < i_2 < \ldots < i_k$).

We set:

$$\mathsf{M}^{(k)} = \mathsf{M} \otimes_{\mathbb{Z}} \overset{k}{\Lambda}(\mathbb{Z}^p)$$

and we define the differential d from $\mathsf{M}^{(k)}$ to $\mathsf{M}^{(k+1)}$ by letting, for an element $x e_I$ in $\mathsf{M}^{(k)}$:

$$d(x e_I) = \sum_{j=1}^{p} \phi_j(x) \, e_j \wedge e_I.$$

The commutativity of the operators ϕ_j clearly implies $d \circ d = 0$. Hence we obtain a complex:

(B.4.2) $0 \longrightarrow \mathsf{M}^{(0)} \underset{d}{\longrightarrow} \ldots \longrightarrow \mathsf{M}^{(p)} \longrightarrow 0$.

Let us denote this complex by $\mathsf{M}^{\cdot}$.

Definition B.4.1. We shall say that $\mathsf{M}^{\cdot}$ is the Koszul complex associated to the sequence $(\phi_1, \ldots, \phi_p)$.

Proposition B.4.2. a) *Assume for each j, $1 \leqslant j \leqslant p$, ϕ_j is surjective as an endomorphism of the sub-module $\operatorname{Ker}\phi_1 \cap \ldots \cap \operatorname{Ker}\phi_{j-1}$ of M. Then $H^j(\mathsf{M}^{\cdot}) = 0$ for $j \neq 0$, and $H^0(\mathsf{M}^{\cdot}) \simeq \operatorname{Ker}\phi_1 \cap \ldots \cap \operatorname{Ker}\phi_p$.*

b) *Assume for each j, $1 \leqslant j \leqslant p$, ϕ_j is injective as an endomorphism of the module $\mathsf{M}/(\phi_1(\mathsf{M}) + \ldots + \phi_{j-1}(\mathsf{M}))$. Then $H^j(\mathsf{M}^{\cdot}) = 0$ for $j \neq p$, and $H^p(\mathsf{M}^{\cdot}) \simeq \mathsf{M}/(\phi_1(\mathsf{M}) + \ldots + \phi_p(\mathsf{M}))$.*

Sketch of proof. We argue by induction on p, the case where $p = 1$ being obvious. Let M'' be the Koszul complex associated to the sequence $(\phi_1, \ldots, \phi_{p-1})$. Then one checks that $\mathsf{M}^{\cdot}$ is the simple complex associated to the double complex:

$$0 \to \mathsf{M}'' \underset{\phi_p}{\longrightarrow} \mathsf{M}'' \to 0.$$

Let us study case a) for example. In that case we have a quasi-isomorphism:

$$H^0(\mathsf{M}'') \overset{\sim}{\longrightarrow} \mathsf{M}''.$$

Then $\mathsf{M}^{\cdot}$ is quasi-isomorphic to the complex:

$$0 \to H^0(\mathsf{M}'') \underset{\phi_p}{\longrightarrow} H^0(\mathsf{M}'') \to 0,$$

which completes the proof. $\square$

Remark B.4.3. A sequence $(\phi_1, \ldots, \phi_p)$ satisfying one of the hypotheses of Proposition B.4.2 (and also (B.4.1) of course), is said to be a *regular sequence.*

Example B.4.4. Let us particularize the preceding Proposition to the case where $p = 2$. We get the complex:

$$0 \to M \xrightarrow{d} M \times M \xrightarrow{d} M \to 0,$$

where $d(x) = (\phi_1(x), \phi_2(x))$, and $d(y, z) = \phi_2(y) - \phi_1(z)$.

Let us assume ϕ_1 surjective on M, and ϕ_2 surjective on $\mathrm{Ker}\,\phi_1$. Let $(y, z) \in M \times M$, with $\phi_2(y) = \phi_1(z)$. Choose $x \in M$ with $\phi_1(x) = y$. Then $\phi_2 \circ \phi_1(x) = \phi_2(y) = \phi_1(z) = \phi_1 \circ \phi_2(x)$. Thus $\phi_1(z - \phi_2(x)) = 0$, and we may find $t \in M$, with $\phi_1(t) = 0$, $\phi_2(t) = z - \phi_2(x)$. Finally $y = \phi_1(t + x)$, $z = \phi_2(t + x)$.

B.5. The Mittag-Leffler Condition

Let $\{E_n, \rho_{n,p}\}_{n \in \mathbb{N}}$ be a projective system of abelian groups. One says that the Mittag-Leffler (M-L for short) condition is satisfied, if for each n, the decreasing sequence of subgroups of E_n, $\{\rho_{n,p}(E_p)\}_{p \geqslant n}$ is stationary.

Let $\{E_n^{\cdot}\}_n$ be a complex of projective systems of abelian groups, $E_\infty^{\cdot}$ the complex for which $E_\infty^i = \varprojlim_n E_n^i$. The natural maps from $\varprojlim_n E_n^i$ to E_p^i defines the morphisms:

$$\phi_i : H^i(E_\infty^{\cdot}) \to \varprojlim_n H^i(E_n^{\cdot}).$$

Proposition B.5.1 (Grothendieck [1]). a) *Assume that for any i the projective system $\{E_n^i\}_n$ satisfies the M-L condition. Then for any i, ϕ_i is surjective.*

b) *Assume moreover that $\{H^{i-1}(E_n)\}_n$ satisfies the M-L condition. Then ϕ_i is bijective.*

Corollary B.5.2. *Let $\psi : \{E_n^{\cdot}\}_n \to \{F_n^{\cdot}\}_n$ be a morphism of complex of projective systems of abelian groups. Assume:*

i) *For each i, the projective systems $\{E_n^i\}_n$ and $\{F_n^i\}_n$ satisfy the M-L condition.*

ii) *For each n, each j, ψ induces an isomorphism $H^j(E_n^{\cdot}) \cong H^j(F_n^{\cdot})$. Then for each j, ψ induces an isomorphism:*

$$H^j(\varprojlim_n E_n^{\cdot}) \cong H^j(\varprojlim_n F_n^{\cdot})$$

Proof. Let $\{G_n^{\cdot}\}_n$ be the mapping cone of ψ. By hypothesis, $H^j(G_n^{\cdot}) = 0$. Since $\{G_n^i\}_n$ satisfies the M-L condition for each i, we may apply Proposition B.5.1 and we obtain:

$$H^j(\varprojlim_n G_n^{\cdot}) = 0.$$

That is, $\varprojlim_n E_n^{\cdot} \to \varprojlim_n F_n^{\cdot}$ is a quasi-isomorphism.

C. Sheaves

References are made to Godement [1].

C.1. Presheaves and Sheaves

Let X be a topological space. We denote by $\mathfrak{O}\mathfrak{P}(X)$ the category whose objects are the open sets of X, and morphisms are inclusions of open sets. A presheaf $\mathscr{F}$ (of abelian groups) is a contravariant functor from $\mathfrak{O}\mathfrak{P}(X)$ to $\mathfrak{Z}$, the category of abelian groups, and a morphism of presheaves is a natural transformation of such functors.

Thus a presheaf $\mathscr{F}$ assigns to each open set U an abelian group $\mathscr{F}(U)$, and to each open inclusion $V \subset U$ a group homomorphism $\rho_V^U: \mathscr{F}(U) \to \mathscr{F}(V)$, called the restriction morphism, with the evident compatibility conditions $\rho_U^U = I$, $\rho_W^U = \rho_W^V \circ \rho_V^U$ for all triplets $W \subset V \subset U$ of open sets.

A morphism ϕ of presheaves, from $\mathscr{F}$ to $\mathscr{G}$, is a family of group homomorphisms ϕ_U (we shall often simply write ϕ instead of ϕ_U) from $\mathscr{F}(U)$ to $\mathscr{G}(U)$ compatible with the restrictions. If we denote by the same letters ρ_V^U the restriction morphisms of $\mathscr{G}$, it means that the diagram below is commutative for all $V \subset U$:

$$
\begin{array}{ccc}
\mathscr{F}(U) & \xrightarrow{\phi_U} & \mathscr{G}(U) \\
{\scriptstyle \rho_V^U} \downarrow & & \downarrow {\scriptstyle \rho_V^U} \\
\mathscr{F}(V) & \xrightarrow{\phi_V} & \mathscr{G}(V)
\end{array}
$$

The presheaf kernel of ϕ is the presheaf $U \mapsto \mathrm{Ker}\,\phi_U$. One defines similarly the image presheaf of ϕ, or the cokernel presheaf of ϕ. We may also regard a presheaf $\mathscr{F}$ as a projective system $(\mathscr{F}(U), \rho_V^U)$ indexed by the ordered family of open sets of X. A sheaf $\mathscr{F}$ is a presheaf $\mathscr{F}$ which satisfies the following condition:

(C.1.1)   For any open set $U \subset X$ and any open covering stable by finite intersection $U = \bigcup_{i \in I} U_i$, the morphism $\mathscr{F}(U) \to \varprojlim_{i \in I} \mathscr{F}(U_i)$ is an isomorphism.

Then we define a morphism of sheaves as a morphism of the underlying presheaves.

If ϕ is a morphism of sheaves, then the kernel presheaf of ϕ is a sheaf, but the image presheaf, and the cokernel presheaf, are not sheaves in general. To a presheaf $\mathscr{F}$ we associate its stalk $\mathscr{F}_x$ at $x \in X$:

$$\mathscr{F}_x = \varinjlim_{U \ni x} \mathscr{F}(U).$$

If s is an element of $\mathscr{F}(U)$, we denote by s_x its image in $\mathscr{F}_x$.

Remark that if ϕ is a morphism of sheaves from $\mathscr{F}$ to $\mathscr{G}$ which induces an isomorphism $\mathscr{F}_x \overset{\sim}{\longrightarrow} \mathscr{G}_x$ for all $x \in X$, then ϕ is an isomorphism. If $\mathscr{F}$ and $\mathscr{G}$ are only presheaves, this is no longer true.

Proposition C.1.1 (cf. Hartshorne [2, p. 64]). *Given a presheaf $\mathscr{F}$ there is a sheaf $\mathscr{F}^+$ and a morphism $\theta: \mathscr{F} \to \mathscr{F}^+$ such that for any sheaf $\mathscr{G}$ and morphism ϕ from $\mathscr{F}$ to $\mathscr{G}$, there is a unique morphism ϕ^+ from $\mathscr{F}^+$ to $\mathscr{G}$, with $\phi = \phi^+ \circ \theta$. Moreover $(\mathscr{F}^+, \theta)$ is unique up to isomorphism.*

The sheaf $\mathscr{F}^+$ is called the sheaf associated to the presheaf $\mathscr{F}$. It may be constructed using projective limits, or more simply as follows: to each open set $U \subset X$, $\mathscr{F}^+(U)$ is the set of functions from U to $\prod_{x \in U} \mathscr{F}_x$ such that $s(x) \in \mathscr{F}_x \quad \forall x \in U$, and such that each $y \in U$ has an open neighborhood V on which there exists $t \in \mathscr{F}(V)$ with $t_z = s(z) \quad \forall z \in V$.

Example C.1.2. Let G be an abelian group. The constant sheaf $\underline{G}_X$ on X is the sheaf associated to the presheaf $U \mapsto G$, the restriction morphisms being the identity. Let Z be a locally closed subset of X. The constant sheaf $\underline{G}_Z$ on Z extends naturally as a sheaf on X supported by Z. We still denote by $\underline{G}_Z$ this sheaf on X. Thus $(\underline{G}_Z)_x = G$ if $x \in Z$, $(\underline{G}_Z)_x = 0$ if $x \notin Z$.

Notation C.1.3. We sometimes improperly write $s \in \mathscr{F}$ to mention that s belongs to $\mathscr{F}(U)$ for some open set $U \subset X$.

Let $\phi: \mathscr{F} \to \mathscr{G}$ be a morphism of sheaves. We define $\mathrm{Ker}\,\phi$, $\mathrm{Coker}\,\phi$, $\mathrm{Im}\,\phi$ as the sheaves associated to the presheaves kernel, cokernel, and image of ϕ. (In fact the kernel of ϕ was already a sheaf). It is immediately verified that $\mathrm{Im}\,\phi$ is a subsheaf of $\mathscr{G}$.

Proposition C.1.4. *The category of sheaves of abelian groups on X is an abelian category.*

We denote it by $\mathfrak{Z}_X$.

Remark that if ϕ is a morphism of sheaves from $\mathscr{F}$ to $\mathscr{G}$, $\mathrm{Ker}\,\phi = 0$ iff ϕ is injective on each stalk $\mathscr{F}_x$, or iff ϕ is injective on $\mathscr{F}(U)$ for each U. Now $\mathrm{Im}\,\phi = \mathscr{G}$ iff ϕ is surjective from $\mathscr{F}_x$ to $\mathscr{G}_x$ at each $x \in X$, but it does not imply that ϕ is surjective on $\mathscr{G}(U)$.

Example C.1.5. Let $X = \mathbb{C} \setminus \{0\}$, and let $\mathcal{O}_X$ be the sheaf of holomorphic functions on X. Let t be a holomorphic coordinate on $\mathbb{C}$, and consider the morphism ϕ from $\mathcal{O}_X$ to $\mathcal{O}_X$ defined by:

$$\phi(f) = \frac{\partial}{\partial t} f.$$

Then $\operatorname{Ker} \phi = \mathbb{C}_X$, the constant sheaf on X, and for any $x \in X$, ϕ is surjective on $\mathcal{O}_{X,x}$. Thus we have an exact sequence of sheaves on X:

$$0 \to \mathbb{C}_X \to \mathcal{O}_X \xrightarrow{\phi} \mathcal{O}_X \to 0.$$

However the sequence:

$$0 \to \mathbb{C} \to \mathcal{O}(\mathbb{C} \setminus \{0\}) \xrightarrow{\frac{\partial}{\partial t}} \mathcal{O}(\mathbb{C} \setminus \{0\}) \to 0$$

is not exact, since $1/t$ does not have any primitive on $\mathbb{C} \setminus \{0\}$.

C.2. Cohomology of Sheaves

We denote by $\Gamma(U, \cdot)$ the functor from $\mathfrak{Z}_X$ to $\mathfrak{Z}$, $\mathcal{F} \mapsto \mathcal{F}(U)$ (U is open in X). An element of $\mathcal{F}(U)$ is called a section of $\mathcal{F}$ on U. We say that a section s of $\mathcal{F}$ on U is zero on an open set $V \subset U$, if $\rho_V^U(s) = 0$.

Definition C.2.1. Let s be a section of $\mathcal{F}$ on U. The support of s, denoted $\operatorname{supp}(s)$, is the complement in U of the union of all open sets $V \subset U$ such that s is zero on V.

Let Z be a closed subset of U. One puts:

$$\Gamma_Z(U, \mathcal{F}) = \{s \in \mathcal{F}(U); \operatorname{supp}(s) \subset Z\}.$$

If V is open in U and Z is contained in V, then the natural morphism $\Gamma_Z(U, \mathcal{F}) \to \Gamma_Z(V, \mathcal{F})$ is clearly an isomorphism. Thus it is possible to define $\Gamma_Z(X, \mathcal{F})$ for a locally closed set Z in X as $\Gamma_Z(U, \mathcal{F})$, for any open set U containing Z as a closed subset.

Let $\Gamma_Z(X, \cdot)$ be the functor $\mathcal{F} \mapsto \Gamma_Z(X, \mathcal{F})$ from $\mathfrak{Z}_X$ to $\mathfrak{Z}$. This functor is covariant left exact, but in general not right exact (cf. Ex. C.1.5). The class of acyclic objects for all the functors $\Gamma_Z(X, \cdot)$, (Z locally closed in X), is the class of flabby sheaves.

Definition C.2.2. A sheaf $\mathcal{F}$ is flabby if the restriction morphisms $\mathcal{F}(X) \to \mathcal{F}(U)$ are all surjective.

The category $\mathfrak{Z}_X$ has enough injectives (cf. Godement [1]) (and injective sheaves are flabby). Thus the derived functors of $\Gamma_Z(X, \cdot)$ are well defined. They are denoted by $H_Z^i(X, \cdot)$. Of course $H_Z^0(X, \cdot) = \Gamma_Z(X, \cdot)$.

To an exact sequence of sheaves $0 \to \mathscr{F} \to \mathscr{G} \to \mathscr{H} \to 0$ is associated a long exact sequence:

(C.2.1) $0 \to H_Z^0(X, \mathscr{F}) \to H_Z^0(X, \mathscr{G}) \to H_Z^0(X, \mathscr{H}) \to H_Z^1(X, \mathscr{F}) \to \dots .$

To calculate the groups $H_Z^i(X, \mathscr{F})$ it is sufficient to replace $\mathscr{F}$ by a flabby resolution of $\mathscr{F}$.

Example C.2.3. Let X be a complex manifold of dimension n. The Dolbeault resolution of $\mathscr{O}_X$ with hyperfunctions as coefficients is the complex:

(C.2.2) $$0 \to \mathscr{O}_X \to \mathscr{B}_{X^{\mathbb{R}}}^{(0,0)} \xrightarrow{\bar{\partial}} \dots \to \mathscr{B}_{X^{\mathbb{R}}}^{(0,n)} \to 0$$

It is a flabby resolution of $\mathscr{O}_X$ (cf. Komatsu [1], Schapira [1]).

When Z is open in X, we write $H^i(Z, \cdot)$ instead of $H_Z^i(X, \cdot)$. If Z is locally closed in X, we may consider the presheaf $U \mapsto \Gamma_{Z \cap U}(U, \mathscr{F})$. This is a sheaf which is denoted by $\Gamma_Z(\mathscr{F})$ and its derived functors by $\mathscr{H}_Z^i(\mathscr{F})$.

Now let Y be another topological space, f a continuous map from Y to X. The direct image $f_*\mathscr{G}$ of a sheaf $\mathscr{G}$ on Y, is the sheaf on X defined by:

(C.2.3) $f_*(\mathscr{G})(U) = \mathscr{G}(f^{-1}(U)), \quad U$ open in X.

The functor f_* from $\mathfrak{Z}_Y$ to $\mathfrak{Z}_X$ is covariant left exact. Its derived functors are denoted by $R^i f_*(\cdot)$.

Remark that if X is a single point, then $R^i f_*(\mathscr{G}) = H^i(Y, \mathscr{G})$.

The inverse image $f^{-1}\mathscr{F}$ of a sheaf $\mathscr{F}$ on X is the sheaf on Y associated to the presheaf:

$$V \mapsto \varinjlim_{U \supset f(V)} \mathscr{F}(U),$$

(V is open in Y, U open in X). Then for $y \in Y$:

(C.2.4) $(f^{-1}\mathscr{F})_y = \mathscr{F}_{f(y)}$

and the functor f^{-1} from $\mathfrak{Z}_X$ to $\mathfrak{Z}_Y$ is exact.

If Y is a subspace of X and f the injection $Y \hookrightarrow X$, we write $\mathscr{F}|_Y$ instead of $f^{-1}\mathscr{F}$. If X is paracompact, and Y is closed in X, then $\Gamma(Y, \mathscr{F}|_Y) = \varinjlim_{U \supset Y} \Gamma(U, \mathscr{F})$ (the limit is taken over all open sets U containing Y). We say that a sheaf $\mathscr{F}$ on X is zero on an open set $U \subset X$, if $\mathscr{F}|_U = 0$. The complement of the union of all open sets on which $\mathscr{F}$ is zero is called the support of $\mathscr{F}$, and denoted by $\mathrm{supp}(\mathscr{F})$.

Remark that if $0 \to \mathscr{F} \to \mathscr{G} \to \mathscr{H} \to 0$ is an exact sequence of sheaves, then:

(C.2.5) $\mathrm{supp}(\mathscr{G}) = \mathrm{supp}(\mathscr{F}) \cup \mathrm{supp}(\mathscr{H})$

Example C.2.4. Let $X = \mathbb{C}$, with coordinate t, $\mathscr{F}$ the sheaf of holomorphic solutions of the equation $(t D_t - 1/2) f = 0$. Then $\mathscr{F}_x$ is a vector space of dimension one for each $x \in \mathbb{C} \setminus \{0\}$, but $\mathscr{F}_0 = \{0\}$. Remark that $\mathrm{supp}(\mathscr{F}) = \mathbb{C}$.

The sheaf $\mathscr{F}$ is a locally constant sheaf on $\mathbb{C}\setminus\{0\}$, but not a constant sheaf on this space.

We shall mainly consider sheaves of rings, or Rings, and sheaves of modules, or Modules, as in the following definition.

A Ring $\mathscr{A}$ on X is a sheaf on X such that for each open set $U \subset X$, $\mathscr{A}(U)$ is a unitary ring and the operations on $\mathscr{A}(U)$ are compatible with the restriction morphisms. We define in an obvious way the notion of Module over the Ring $\mathscr{A}$, and we shall write "a Module over $\mathscr{A}$" or "an $\mathscr{A}$-module".

The pair $(X, \mathscr{A})$ is called a ringed space.

Let $\mathscr{M}$ and $\mathscr{N}$ be two left $\mathscr{A}$-modules. The presheaf:

$$U \mapsto \mathrm{Hom}_{\mathscr{A}|_U}(\mathscr{N}|_U, \mathscr{M}|_U)$$

is a sheaf, denoted $\mathscr{H}om_{\mathscr{A}}(\mathscr{N}, \mathscr{M})$. Remark that in general $(\mathscr{H}om_{\mathscr{A}}(\mathscr{N}, \mathscr{M}))_x \neq \mathrm{Hom}_{\mathscr{A}_x}(\mathscr{N}_x, \mathscr{M}_x)$. We shall not identify the sheaf $\mathscr{H}om_{\mathscr{A}}(\mathscr{N}, \mathscr{M})$ with the group

(C.2.6) $\mathrm{Hom}_{\mathscr{A}}(\mathscr{N}, \mathscr{M}) = \Gamma(X, \mathscr{H}om_{\mathscr{A}}(\mathscr{N}, \mathscr{M}))$.

If $\mathscr{N}$ is a right $\mathscr{A}$-module, then the sheaf $\mathscr{N} \otimes_{\mathscr{A}} \mathscr{M}$ is by definition the sheaf associated to the presheaf:

$$U \mapsto \mathscr{N}(U) \otimes_{\mathscr{A}(U)} \mathscr{M}(U)$$

Then we have:

(C.2.7) $(\mathscr{N} \otimes_{\mathscr{A}} \mathscr{M})_x = \mathscr{N}_x \otimes_{\mathscr{A}_x} \mathscr{M}_x$.

We shall denote by $\mathfrak{A}_X$ the category of sheaves of $\mathscr{A}$-modules over X. Then it can be proved (cf. Godement [1]) that the category $\mathfrak{A}_X$ has enough injectives.

Thus it is possible to define the right derived functors of $\mathscr{H}om_{\mathscr{A}}(\mathscr{M}, \cdot)$ and the right derived functors of $\mathrm{Hom}_{\mathscr{A}}(\mathscr{M}, \cdot)$. They are denoted by $\mathscr{E}xt^j_{\mathscr{A}}(\mathscr{M}, \cdot)$ and $\mathrm{Ext}^j_{\mathscr{A}}(\mathscr{M}, \cdot)$ respectively.

In general it is not possible to find projective resolutions of $\mathscr{A}$-modules, but only flat resolutions of $\mathscr{A}$-modules. This is enough to define the left derived functors of $\cdot \otimes_{\mathscr{A}} \mathscr{M}$. They are denoted by $\mathscr{T}or_j^{\mathscr{A}}(\cdot, \mathscr{M})$.

C.3. Čech Cohomology

We shall not review here the whole theory of Čech cohomology, but only the "Leray's acyclic covering theorem" which we shall need in Chapter III.

Let $\mathscr{F}$ be a sheaf on X, and let $\mathfrak{U} = (U_i)_{i \in I}$ be an open covering of X. For a finite sequence J of I, we set $U_J = \bigcap_{i \in J} U_i$.

Definition C.3.1. One says that the covering $\mathfrak{U}$ is $\mathscr{F}$-acyclic if for any finite sequence J of I and any $k > 0$, we have $H^k(U_J, \mathscr{F}) = 0$

Now for a (not necessarly acyclic) covering $\mathfrak{U}$ of X, we define the group $C^q(\mathfrak{U}, X, \mathscr{F})$, $(q \geqslant 0)$ as the subgroup of $\prod\limits_{|J|=q+1} \mathscr{F}(U_J)$ consisting of alternating elements. Thus an element f of $C^q(\mathfrak{U}, X, \mathscr{F})$ is a family $(f_J)_{|J|=q+1}$, such that if $J=(i_0, \dots, i_q)$,

$$f_{i_0, \dots, i_q} \in \mathscr{F}(U_{i_0} \cap \dots \cap U_{i_q})$$

and if π is a permutation of the set $\{0, \dots, q\}$ with signature ε_π then:

$$f_{i_{\pi(0)}, \dots, i_{\pi(q)}} = \varepsilon_\pi f_{i_0, \dots, i_q}$$

One constructs a differential δ from $C^q(\mathfrak{U}, X, \mathscr{F})$ to $C^{q+1}(\mathfrak{U}, X, \mathscr{F})$ by setting:

$$(\delta f)_{i_0, \dots, i_{q+1}} = \sum_{j=0}^{q} (-1)^j f'_{i_0, \dots, \hat{i}_j, \dots, i_q},$$

where $\hat{i}_j$ means that the index i_j must be omitted, and where $f'_{i_0, \dots, \hat{i}_j, \dots, i_q}$ is the restriction to $U_{i_0} \cap \dots \cap U_{i_q}$ of $f_{i_0, \dots, \hat{i}_j, \dots, i_q}$. One checks immediately that $\delta \circ \delta = 0$, and thus we get a complex:

(C.3.1) $\qquad 0 \to C^0(\mathfrak{U}, X, \mathscr{F}) \xrightarrow[\delta]{} C^1(\mathfrak{U}, X, \mathscr{F}) \to \dots$.

We may also consider the sheaves $\check{C}^q(\mathfrak{U}, \mathscr{F})$ associated to the presheaves $\Omega \mapsto C^q(\mathfrak{U} \cap \Omega, \Omega, \mathscr{F})$ (where $\mathfrak{U} \cap \Omega$ is the covering $(U_i \cap \Omega)_{i \in I}$ of the open set Ω), and we obtain a complex of sheaves:

(C.3.2) $\qquad 0 \to \check{C}^0(\mathfrak{U}, \mathscr{F}) \xrightarrow[\delta]{} \check{C}^1(\mathfrak{U}, \mathscr{F}) \to \dots$

Now replacing in (C.3.2), the sheaf $\mathscr{F}$ by a flabby resolution $\mathscr{F}^{\cdot}$ of $\mathscr{F}$, we get a double complexe $\check{C}^{\cdot}(\mathfrak{U}, \mathscr{F}^{\cdot})$, and using this double complex it is easy to prove:

Proposition C.3.2. *Assume the covering $\mathfrak{U}$ is $\mathscr{F}$-acyclic. Then for each $k \geqslant 0$, the group $H^k(X, \mathscr{F})$ is canonically isomorphic to the k-th cohomology group of the sequence (C.3.1).*

Example C.3.3. Let X be a complex manifold, $\mathscr{F}$ a coherent $\mathscr{O}_X$-module. Then an open covering by Stein open subsets is $\mathscr{F}$-acyclic (cf. Hörmander [2], Gunning-Rossi [1]). In particular a convex open covering of an open set X of $\mathbb{C}^n$ is $\mathscr{O}_X$-acyclic.

Explicit calculations using Čech cohomology may be found in Schapira [1].

C.4. An Extension Theorem

Let X be a topological space, $\mathscr{F}^{\cdot}$ a complex of sheaves (of abelian groups) on X. We shall only consider complexes bounded from below.

By replacing each $\mathscr{F}^i$ by an injective resolution and considering the simple complex associated to the double complex so obtained, one sees (using Proposition B.1.1) that $\mathscr{F}^{\cdot}$ is quasi-isomorphic to a complex $\mathscr{I}^{\cdot}$ of injective sheaves, $\mathscr{I}^{\cdot}$ being bounded from below.

Let U be an open subset of X, Z a closed subset of U. We set:

(C.4.1) $$H_Z^k(U, \mathscr{F}^{\cdot}) = H^k(\Gamma_Z(U, \mathscr{I}^{\cdot})),$$

where $\Gamma_Z(U, \mathscr{I}^{\cdot})$ is the complex of groups obtained by applying the functor $\Gamma_Z(U, \cdot)$ to $\mathscr{I}^{\cdot}$.

Recall that $H_Z^k(U, \mathscr{F}^{\cdot})$ is uniquely defined, up to isomorphism, by (C.4.1), and that if $\mathscr{F}^{\cdot}$ is a complex of flabby sheaves, then:

$$H_Z^k(U, \mathscr{F}^{\cdot}) \cong H^k(\Gamma_Z(U, \mathscr{F}^{\cdot})).$$

If $Z = U$, we write $H^k(U, \mathscr{F}^{\cdot})$ instead of $H_U^k(U, \mathscr{F}^{\cdot})$.

For Z a locally closed subset of X we also introduce the sheaves $\mathscr{H}_Z^k(\mathscr{F}^{\cdot})$ associated to the presheaves $U \to H_Z^k(U, \mathscr{F}^{\cdot})$. Then:

(C.4.2) $$\mathscr{H}_Z^k(\mathscr{F}^{\cdot}) \cong \mathscr{H}^k(\Gamma_Z(\mathscr{I}^{\cdot})),$$

where $\Gamma_Z(\mathscr{I}^{\cdot})$ is the complex of sheaves $U \mapsto \Gamma_Z(U, \mathscr{I}^{\cdot})$.

The support of $\mathscr{F}^{\cdot}$, denoted by $\operatorname{supp}(\mathscr{F}^{\cdot})$, is the closed set:

(C.4.3) $$\operatorname{supp}(\mathscr{F}^{\cdot}) = \overline{\bigcup_j \operatorname{supp}(H^j(\mathscr{F}^{\cdot}))}.$$

We shall need the following Lemma that we recall without proof.

Lemma C.4.1 (Kashiwara [5]). *Let $(E_t, \rho_{s,\,t})_{t \in \mathbb{R}}$ be a projective system of sets. Assume that the canonical maps $E_t \to \varprojlim_{s < t} E_s$ and $\varprojlim_{t > s} E_t \to E_s$ are both injective (resp. surjective) for all t and s. Then the maps $\rho_{s,\,t}$ are all injective (resp. surjective) for all s and t, with $s \leqslant t$.*

Proposition C.4.2. *Let $\mathscr{F}^{\cdot}$ be a complex bounded from below of sheaves on X, $\{U_n\}$ an increasing sequence of open subsets of X and $\{Z_n\}$ a decreasing sequence of closed subsets of X.*

Set $U = \bigcup_n U_n$, $Z = \bigcap_n Z_n$.

a) For any k, $\phi_k: H_Z^k(U, \mathscr{F}^{\cdot}) \to \varprojlim_n H_{Z_n}^k(U_n, \mathscr{F}^{\cdot})$ is surjective.

b) If $\{H_{Z_n}^{k-1}(U_n, \mathscr{F}^{\cdot})\}$ satisfies the M-L condition, then ϕ_k is bijective.

Proof. We may assume all the sheaves $\mathscr{F}^k$ flabby. Let $E_n^{\cdot}$ be the simple complex associated to the double complex:

$$
\begin{array}{ccccccc}
\cdots \to & \Gamma(U_n, \mathscr{F}^{k-1}) & \to & \Gamma(U_n, \mathscr{F}^k) & \to \cdots \\
& \downarrow & & \downarrow & \\
\cdots \to & \Gamma(U_n \backslash Z_n, \mathscr{F}^{k-1}) & \to & \Gamma(U_n \backslash Z_n, \mathscr{F}^k) & \to \cdots .
\end{array}
$$

Then $H_{Z_n}^k(U_n, \mathscr{F}^{\cdot}) = H^k(E_n^{\cdot})$ and $H_Z^k(U, \mathscr{F}^{\cdot}) = H^k(\varprojlim_n E_n^{\cdot})$.

Since $\{E_n^k\}_n$ satisfies the M-L condition, the Proposition follows from Proposition B.5.1.

Theorem C.4.3 (cf. Kashiwara [5], Kashiwara-Schapira [4]). *Assume X is a Hausdorff space, let $\{\Omega_t\}_{t \in \mathbb{R}}$ be a family of open subsets of X and let $\mathscr{F}^{\cdot}$ be a complex bounded from below of sheaves on X. Assume:*

i) $\Omega_t = \bigcup_{s<t} \Omega_s$.

ii) *For $t \geq s$, $\overline{\Omega_t \backslash \Omega_s} \cap \operatorname{supp}(\mathscr{F}^{\cdot})$ is compact.*

iii) *Setting $Z_s = \bigcap_{t>s} \overline{\Omega_t \backslash \Omega_s}$, we have for any s, t with $s \leq t$, any $x \in Z_s \backslash \Omega_t$:*

$$(\mathscr{H}^i_{X \backslash \Omega_t}(\mathscr{F}^{\cdot}))_x = 0 \quad \forall i.$$

Then we have the isomorphisms:

$$H^i(\bigcup_s \Omega_s, \mathscr{F}^{\cdot}) \xrightarrow{\sim} H^i(\Omega_t, \mathscr{F}^{\cdot})$$

for any i, any t.

Proof. We may assume that all the $\mathscr{F}^k$ are flabby, and, replacing X by $\operatorname{supp}(\mathscr{F}^{\cdot})$, that $\overline{\Omega_t \backslash \Omega_s}$ is compact for $t \geq s$.

Consider the assertions:

$$(\text{C.4.4})^k_{t_0} \qquad \varinjlim_{t>t_0} H^k(\Omega_t, \mathscr{F}^{\cdot}) \xrightarrow{\sim} H^k(\Omega_{t_0}, \mathscr{F}^{\cdot}),$$

$$(\text{C.4.5})^k_{t_0} \qquad \varprojlim_{s<t_0} H^k(\Omega_s, \mathscr{F}^{\cdot}) \xleftarrow{\sim} H^k(\Omega_{t_0}, \mathscr{F}^{\cdot}).$$

By Lemma C.4.1 it is enough to prove $(\text{C.4.4})^k_{t_0}$ and $(\text{C.4.5})^k_{t_0}$ for all k, all t_0.

We argue by induction on k, and assume we have proved $(\text{C.4.4})^k_{t_0}$ for all k, all t_0, and $(\text{C.4.5})^k_{t_0}$ for all $k < k_0$, all t_0. Then for each t_0, the sequence $\{H^{k_0-1}(\Omega_{t_0-1/n}, \mathscr{F}^{\cdot})\}_n$ satisfies the M-L condition by the induction hypothesis, and by applying Proposition C.4.2 we find $(\text{C.4.5})^{k_0}_{t_0}$. Since $\mathscr{F}^{\cdot}$ is bounded from below we can start the induction, and it is thus enough to prove $(\text{C.4.4})^k_{t_0}$ for all k, all t_0.

By iii) we have the quasi-isomorphisms:

$$(\text{C.4.6}) \quad (\Gamma_{X \backslash \Omega_t}(\mathscr{F}^{\cdot}))_x \cong (\Gamma_{X \backslash \Omega_{t_0}}(\mathscr{F}^{\cdot}))_x \cong 0 \quad \text{for any } x \in Z_{t_0}, \text{ any } t \geq t_0.$$

Let j_t be the inclusion map $\Omega_t \hookrightarrow X$.

If $\mathscr{G}$ is a flabby sheaf on X, the sequence of sheaves:

$$0 \to \Gamma_{X \backslash \Omega_t}(\mathscr{G}) \to \mathscr{G} \to j_{t*} j_t^{-1} \mathscr{G} \to 0$$

is exact. Applying this result to the sheaves $\Gamma_{X \backslash \Omega_0}(\mathscr{F}^i)$, we find by (C.4.6):

$$(j_{t*} j_i^{-1} \Gamma_{X \setminus \Omega_{t_0}}(\mathscr{F}\,'))|_{Z_0} \simeq 0.$$

Thus:

$$\varinjlim_{U \supset Z_{t_0}} \Gamma(U \cap \Omega_t, \Gamma_{X \setminus \Omega_{t_0}}(\mathscr{F}\,')) \simeq 0,$$

hence:

$$\varinjlim_{U \supset Z_{t_0}} \Gamma((U \cup \Omega_{t_0}) \cap \Omega_t, \mathscr{F}\,') \simeq \Gamma(\Omega_{t_0}, \mathscr{F}\,').$$

For any open set U containing Z_{t_0}, there exists $t > t_0$ such that U contains $\Omega_t \setminus \Omega_{t_0}$. Therefore $U \cup \Omega_{t_0}$ contains Ω_t, which completes the proof. □

C.5. Coherent Sheaves

Let $(X, \mathscr{A})$ be a ringed space. An $\mathscr{A}$-module $\mathscr{M}$ is finite free if it is iso-morphic to $\mathscr{A}^N$ for some $N \in \mathbb{N}$. One says that $\mathscr{M}$ is locally finite free if each $x \in X$ has an open neighborhood U such that $\mathscr{M}|_U$ is a finite free $\mathscr{A}|_U$-module.

An s-presentation of $\mathscr{M}$ is an exact sequence of $\mathscr{A}$-modules:

(C.5.1) $$\mathscr{M}_{-s} \to \ldots \to \mathscr{M}_0 \to \mathscr{M} \to 0$$

One says that this presentation is finite free, or locally finite free, etc. ... if the $\mathscr{M}_j$'s have the corresponding property. A Module $\mathscr{M}$ is locally of fin-ite type if locally it admits a free 0-presentation, that is if locally on X there exists an exact sequence: $\mathscr{A}^N \to \mathscr{M} \to 0$.

Definition C.5.1. a) We shall say that the Ring $\mathscr{A}$ is coherent on X if for any open set $U \subset X$, any finite free 1-presentation of an $\mathscr{A}|_U$-module on U, extends locally on U to a finite free 2-presentation.

b) We shall say that a Module over a coherent Ring $\mathscr{A}$ is coherent if it admits, locally, a finite free 1-presentation.

c) The Ring $\mathscr{A}$ is said to be noetherian if:

    i) $\mathscr{A}$ is coherent.

    ii) $\forall x \in X$, the ring $\mathscr{A}_x$ is noetherian.

    iii) For any open set $U \subset X$, any increasing family of coherent sub-Modules of a finite free $\mathscr{A}|_U$-module on U is locally stationary.

d) If $\mathscr{A}$ is coherent, a pseudo-coherent $\mathscr{A}$-module $\mathscr{M}$ is an $\mathscr{A}$-module such that for any open set $U \subset X$, any sub-Module $\mathscr{N}$ of $\mathscr{M}$ on U which is of finite type is coherent (on U).

One proves easily that coherent Modules are pseudo-coherent, and that if we have an exact sequence of $\mathscr{A}$-modules, ($\mathscr{A}$ being coherent), $0 \to \mathscr{L} \to \mathscr{M} \to \mathscr{N} \to 0$, such that two Modules are coherent, then the third one is also coherent.

When $\mathscr{A}$ is coherent, we say that $(X, \mathscr{A})$ is a coherent ringed space.

Example C.5.2. Let I be an ideal of the algebra $\mathbb{Z}[t_1, \ldots, t_n]$, and let $\mathscr{A}$ be a noetherian Ring on X. Then $\mathscr{A} \otimes_{\mathbb{Z}} (\mathbb{Z}[t_1, \ldots, t_n]/I)$ is a noetherian Ring.

Example C.5.3. Let X be a complex manifold. Then the Ring $\mathscr{O}_X$ is noetherian (in particular it is coherent) (cf. Serre [2]).

Proposition C.5.4. *The category* $\mathfrak{A}_X^C$ *of left coherent* $\mathscr{A}$-*modules on a coherent ringed space is an abelian category.*

Remark C.5.5. We could state an analogous definition to Definition C.5.1 for graded Rings, and obtain for example the notion of coherent graded Modules. We leave the exact formulation to the reader.

Assume $\mathscr{A}$ is coherent, and let $\mathscr{M}$ be a coherent $\mathscr{A}$-module. Then for each $s \geqslant 0$ and each $x \in X$, there exist an open neighborhood U of x and a finite free s-presentation of $\mathscr{M}$ on U. Hence on U there exists a complex $\mathscr{M}^{\cdot}$ of free Modules of finite type, and a morphism $\varepsilon: \mathscr{M}^{\cdot} \to \mathscr{M}$ such that:

$$\mathscr{H}^0(\mathscr{M}^{\cdot}) \cong \mathscr{M},$$
$$\mathscr{H}^j(\mathscr{M}^{\cdot}) = 0 \quad \text{for} \ -s < j < 0.$$

If $\mathscr{N}$ is an $\mathscr{A}$-module we have for $j < s$:

$$\mathscr{E}xt_{\mathscr{A}}^j(\mathscr{M}, \mathscr{N})_x = \mathscr{E}xt_{\mathscr{A}}^j(\mathscr{M}^{\cdot}, \mathscr{N})_x$$
$$= \mathrm{Ext}_{\mathscr{A}_x}^j(\mathscr{M}_x^{\cdot}, \mathscr{N}_x).$$

Thus we get for a coherent $\mathscr{A}$-module $\mathscr{M}$ and any $\mathscr{A}$-module $\mathscr{N}$:

$$\mathscr{E}xt_{\mathscr{A}}^j(\mathscr{M}, \mathscr{N})_x = \mathrm{Ext}_{\mathscr{A}_x}^j(\mathscr{M}_x, \mathscr{N}_x) \quad \forall j.$$

Remark that if $\mathscr{N}$ is a right $\mathscr{A}$-module, then for $-s < -j$:

$$\mathscr{T}or_j^{\mathscr{A}}(\mathscr{N}, \mathscr{M}) = H^{-j}(\mathscr{N} \otimes_{\mathscr{A}} \mathscr{M}^{\cdot}).$$

D. $\mathscr{O}_X$-modules

References are made to Banica-Stanasila [1], Serre [1], Gunning-Rossi [1], Narasimhan [1], Hervé [1], Hörmander [2], Cartan [1], Lejeune [1].

D.1. Support and Multiplicities

Let X be a complex manifold of dimension n, $\mathscr{O}_X$ the sheaf of rings of holomorphic functions on X.

Recall (Serre [2]), that $\mathcal{O}_X$ is a noetherian Ring (hence coherent).

Let S be a closed analytic subset of X. We denote S_{reg} the regular part of S, that is the set of points of S on a neighborhood of which S is a (smooth) manifold. We also set $S_{\mathrm{sing}} = S \backslash S_{\mathrm{reg}}$.

Recall that S may be written as a locally finite union $S = \bigcup_j S_j$, where the S_j's are closed irreducible analytic subsets of X. In many cases the study of S reduces to the study of its irreducible components.

If S is irreducible, S_{reg} is a dense connected open subset of S, and we define the dimension of S, $\dim(S)$, to be the dimension of S_{reg}. In the general case, the dimension of S is the biggest dimension of its irreducible components.

We shall make a constant use of the "Hilbert Nullstellensatz".

Let J_x be an ideal of $\mathcal{O}_{X,x}$ and let S_x be the germ at x of the set of common zeros of elements of J_x (remark that J_x is finitely generated, and we may take for S_x the germ of the set of common zeros of a finite set of generators of J_x). Then if an element f of $\mathcal{O}_{X,x}$ vanishes on S_x, there exists $k > 0$ such that $f^k \in J_x$. We may also formulate this theorem using sheaves. Let $\mathcal{J}$ be a coherent Ideal of $\mathcal{O}_X$, $S = \mathrm{supp}(\mathcal{O}_X/\mathcal{J})$. Then if f is a section of $\mathcal{O}_X$ which vanishes on S, locally there exists $k > 0$ such that f^k belongs to $\mathcal{J}$.

Let us recall the construction of the graded group Γ_X of cycles of X. First we introduce Γ_X^d, the free abelian group generated by the closed irreducible analytic sets of X of codimension d. An element of Γ_X^d is a finite sum $\sum_j n_j S_j$, where S_j is irreducible with codimension d. The group Γ_X is by definition the direct sum of the groups Γ_X^d:

$$\Gamma_X = \bigoplus_{d=0}^{n} \Gamma_X^d .$$

A cycle on X is an element of Γ_X. A cycle $\sum_j n_j S_j$ is said to be positive if it is a linear combination with positive coefficients of irreducible analytic sets. (i.e. $n_j \geqslant 0 \ \forall j$).

Now let $\mathcal{F}$ be a coherent $\mathcal{O}_X$-module, $S = \mathrm{supp}(\mathcal{F})$ its support. Let U be an open set in X, W an irreducible closed analytic subset of U which contains $S \cap U$. It is possible to define the multiplicity of $\mathcal{F}$ on W as follows.

Let $\mathcal{J}_W$ be the Ideal of definition of W, that is the Ideal of $\mathcal{O}_X$ of sections vanishing on W, and let $\mathcal{O}_W = \mathcal{O}_X/\mathcal{J}_W$, the structural sheaf of W. Assume first $\mathcal{J}_W \mathcal{F} = 0$. Then $\mathcal{F}$ is an $\mathcal{O}_W$-module and it can be proved that generically, i.e. outside a proper closed analytic set V of W, the sheaf $\mathcal{F}$ is locally free of finite rank on $\mathcal{O}_W$. Since $W \backslash V$ is connected, this rank is constant, say r, and one sets $\mathrm{mult}_W(\mathcal{F}) = r$. In the general case one sets:

Definition D.1.1. $\mathrm{mult}_W(\mathcal{F}) = \displaystyle\sum_{j=0}^{\infty} \mathrm{mult}_W(\mathcal{J}_W^j \mathcal{F} / \mathcal{J}_W^{j+1} \mathcal{F})$.

(We have assumed $\mathrm{supp}\,\mathscr{F} \subset W$, hence this sum is finite by the Hilbert Nullstellensatz).

Remark that $\mathrm{mult}_W(\mathscr{F}) = 0$ iff $\mathrm{supp}(\mathscr{F}) \cap U \subsetneqq W$.

If $S = \bigcup_j S_j$ is the decomposition of $S = \mathrm{supp}(\mathscr{F})$ into irreducible components, and n_j the multiplicity of $\mathscr{F}$ on S_j, we define the cycle of $\mathscr{F}$ by:

(D.1.1)
$$[\mathrm{supp}\,\mathscr{F}] = \sum_j n_j S_j ,$$

and we denote by $[\mathrm{supp}\,\mathscr{F}]_d$ the homogeneous component of $[\mathrm{supp}\,\mathscr{F}]$ of degree d. Now recall (Serre [1]) that an additive function χ from an abelian category $\mathfrak{C}$ to an abelian group Γ is a map from $0b(\mathfrak{C})$ to Γ such that for any short exact sequence $0 \to \mathsf{L} \to \mathsf{M} \to \mathsf{N} \to 0$ in $\mathfrak{C}$, we have:

(D.1.2)
$$\chi(\mathsf{M}) = \chi(\mathsf{L}) + \chi(\mathsf{N}) .$$

Let $\mathfrak{O}_X^C$ be the abelian category of coherent $\mathcal{O}_X$-modules, and $\mathfrak{O}_{X,d}^c$ be the subcategory of $\mathcal{O}_X$-modules whose support has codimension at least d.

Proposition D.1.2. *The function from $\mathfrak{O}_{X,d}^C$ to Γ_X which assignes the cycle $[\mathrm{supp}\,\mathscr{F}]_d$ to the coherent sheaf $\mathscr{F}$, is additive.*

In particular if W is a locally closed irreducible analytic set in X, and if $0 \to \mathscr{F} \to \mathscr{G} \to \mathscr{H} \to 0$ is an exact sequence of coherent $\mathcal{O}_X$-modules, with $\mathrm{supp}(\mathscr{G}) \subset W$, then we have:

(D.1.3)
$$\mathrm{mult}_W(\mathscr{G}) = \mathrm{mult}_W(\mathscr{F}) + \mathrm{mult}_W(\mathscr{H}) .$$

Now we recall the restriction formula for multiplicities (cf. Serre [1]).

Let Y and X be two complex manifolds, ϕ a holomorphic map from Y to X.

a) Let Z be a closed analytic subset of Y, and assume ϕ is finite on Z. For a cycle $[S] = \sum_j n_j S_j$ contained in Z (i.e.: $S_j \subset Z \;\forall j$), we set:

$$\phi_*[S] = \sum_j n_j [\mathrm{supp}(\phi_* \mathcal{O}_{S_j})] .$$

(Remark that if $\mathscr{G}$ is a coherent $\mathcal{O}_Y$-module and ϕ is finite on $\mathrm{supp}\,\mathscr{G}$, then $\phi_* \mathscr{G}$ is $\mathcal{O}_X$-coherent).

b) Now let Z be a closed irreducible analytic subset of X. Assume Z and $\phi^{-1}(Z)$ of pure codimension d. Then one sets:

$$\phi^*[Z]_d = \sum_j (-1)^j [\mathrm{supp}\,\mathscr{T}\!or_j^{\phi^{-1}\mathcal{O}_X}(\mathcal{O}_Y, \phi^{-1}\mathcal{O}_Z)]_d$$

and one extends this formula by linearity.

(Remark that if $\mathscr{F}$ is a coherent $\mathcal{O}_X$-module, then the Modules $\mathscr{T}\!or_j^{\phi^{-1}\mathcal{O}_X}(\mathcal{O}_Y, \phi^{-1}\mathscr{F})$ are $\mathcal{O}_Y$-coherent).

Proposition D.1.3. a) *Let $\mathscr{G}$ be a coherent $\mathcal{O}_Y$-module. Assume ϕ is finite on* supp $\mathscr{G}$. *Then:*

$$[\operatorname{supp}(\phi_*\mathscr{G})] = \phi_*[\operatorname{supp}\mathscr{G}].$$

b) *Let $\mathscr{F}$ be a coherent $\mathcal{O}_X$-module. Let $S = \operatorname{supp}(\mathscr{F})$, and assume S and $\phi^{-1}(S)$ have pure codimension d. Then:*

$$\sum_j (-1)^j [\operatorname{supp} \mathscr{T}\!\mathit{or}_j^{\phi^{-1}\mathcal{O}_X}(\mathcal{O}_Y, \phi^{-1}\mathscr{F})]_d = \phi^*[\operatorname{supp}\mathscr{F}]_d.$$

Finally remark that the property of the Ring $\mathcal{O}_X$ of being noetherian may be easily deduced from Proposition D.1.2 and from the fact that any decreasing family of analytic subsets is locally stationary (cf. Narasimhan [1, § 4, Th. 2]).

D.2. Homological Dimension

We shall also need some results relating the dimension of supp $(\mathscr{F})$ and the vanishing of the sheaves $\mathscr{E}\!\mathit{xt}^j_{\mathcal{O}_X}(\mathscr{F}, \mathcal{O}_X)$.

We refer to Banica-Stanasila [1, Ch. 0], Siu-Trautman [1, § 1].

Proposition D.2.1. *Let $\mathscr{F}$ be a coherent $\mathcal{O}_X$-module. Then:*
a) $\mathscr{E}\!\mathit{xt}^j_{\mathcal{O}_X}(\mathscr{F}, \mathcal{O}_X) = 0$ *for* $j < \operatorname{codim}(\operatorname{supp}(\mathscr{F}))$
b) $\operatorname{codim}(\operatorname{supp}(\mathscr{E}\!\mathit{xt}^j_{\mathcal{O}_X}(\mathscr{F}, \mathcal{O}_X))) \geqslant j$

Finally recall the classical Hilbert **syzygy** theorem:

Proposition D.2.2. *Let $\mathscr{F}$ be a coherent $\mathcal{O}_X$-module. Locally on X, $\mathscr{F}$ admits a projective resolution of length at most n, by free $\mathcal{O}_X$-modules of finite rank.*

Bibliography

Abraham, R. and Marsden, J.E. [1]: Foundation of Mechanics – Cummings Publ. (1978).

Adjamagbo, K. [1]: Réseaux sur des anneaux filtrés – C.R. Acad. Paris, t. *294*, I, 681–684, (1982).

Andronikof, E. [1]: Systèmes déterminées et systèmes normaux d'équations aux dérivées partielles – C.R. Acad. Sci. Paris, t. *293*, I, 257–260, (1981).

— [2]: Systèmes déterminés d'équations aux dérivées partielles – Sem. J. Vaillant – Univ. Paris VI, (1982).

Arnold, V. [1]: Méthodes mathématiques de la mécanique classique – Editions MIR – Moscou – trad. française (1976).

Atiyah, M.F. and Macdonald, I.G. [1]: Introduction to commutative Algebra – Addison-Wesley, Reading, Mass. (1969).

Banica, C. and Stanasila, O. [1]: Méthodes algébriques dans la théorie globale des espaces complexes – Vol. I et II – Gauthier-Villars-Bordas (1977).

Bengel, G. and Gérard, R. [1]: Formal and convergent solutions of singular partial differential equations – Manuscripta Math. *38*, 343–373, (1982).

Bernstein, I.N. [1]: Modules over a ring of differential operators – Study of fundamental solutions of equations with constant coefficients – Funct. Anal. and its Appl. Vol. *5*, 89–101, (1971).

— [2]: The analytic continuation of generalized functions with respect to a parameter – Funct. Anal. and its Appl. Vol. *6*, 272–285, (1972).

Björk, J.E. [1]: Rings of differential operators – North-Holland Math. Library, (1979).

Bony, J.M. [1]: Propagation des singularités différentiables pour des opérateurs à coefficients analytiques – Astérisque. Soc. Math. de France. 34–35, 43–91 (1976).

Bony, J.M. and Schapira, P. [1]: Existence et prolongement des solutions holomorphes des équations aux dérivées partielles – Inventiones Math. *17*, 95–105, (1972).

— [2]: Propagation des singularités analytiques pour des solutions des équations aux dérivées partielles – Ann. Inst. Fourier Grenoble, *26-1*, 81–140, (1976).

Bourbaki, N. [1]: Algèbre commutative – Vol. XXVII et XXVIII, Hermann, Paris, (1961–65).

Boutet de Monvel, L. [1]: Opérateurs pseudo-différentiels analytiques – Sem. Grenoble (1975–76).

Boutet de Monvel, L. and Kree, P. [1]: Pseudo-differential operators and Gevrey classes – Ann. Inst. Fourier Grenoble, *17-1*, 295–323, (1967).

Cartan, H. [1]: Séminaire E.N.S. Exposés de C. Houzel (1960–61).

Cartan, H. and Eilenberg, S. [1]: Homological algebra – University Press, Princeton, (1956).

Duistermaat, J.J. [1]: Fourier integral operators – Lecture Notes Courant Institute – New York, (1973).

Ehrenpreis, L. [1]: Fourier Analysis in several complex variables – Wiley-Interscience, New York (1970).

Freyd, P. [1]: Abelian categories, an introduction to the theory of functors – Harper and Row, New York, (1964).

Gabber, O. [1]: The integrability of the characteristic variety - American Journ. of Math. Vol. *103*, No 3, 445–468, (1981).

Godement, R. [1]: Théorie des faisceaux - Hermann. Paris, (1964).

Grothendieck, A. [1]: Local cohomology - Notes by R. Hartshorne. Lecture Notes in Math. *41*, Springer-Verlag, (1967).

— [2]: Eléments de géométrie algébrique III - Publ. I.H.E.S. n° 11, (1961).

Guillemin, V., Quillen, D. and Sternberg, S. [1]: The integrability of characteristics - Comm. Pure Appl. Math. *23*, 39–77, (1970).

Gunning, R. and Rossi, H. [1]: Analytic functions of several complex variables - Prentice-Hall (1965).

Hamada, Y. [1]: The singularities of the solutions of the Cauchy problem - Publ. R.I.M.S., Kyoto Univ. *5*, 20–40, (1969).

Hamada, Y., Leray, J. and Wagschal, C. [1]: Systèmes d'équations aux dérivées partielles à caractéristiques mutliples: problème de Cauchy ramifié, hyperbolicité partielle - J. Math. Pure Appl. *55*, 297–352, (1976).

Hartshorne, R. [1]: Residues and duality - Lecture Notes in Math. *20*, Springer-Verlag, (1966).

— [2]: Algebraic geometry - Graduate Texts in Math. 52, Springer-Verlag, (1977).

Hervé, M. [1]: Several complex variables - Tata Institute. Oxford Univ. Press, (1963).

Hironaka, H. [1]: Subanalytic sets - Number theory, algebraic geometry and commutative algebra. In honour to Y. Akizuki - Kinokumiya - Tokyo, 453–493, (1973).

Hörmander, L. [1]: Linear partial differential operators - Grundlehren der Math. 116, Springer-Verlag, (1963).

— [2]: An introduction to complex analysis in several variables - Van Norstrand. Princeton, (1966).

— [3]: Fourier integral operators I - Acta Math. *127*, 79–183, (1971).

— [4]: The analysis of linear partial differential operators I - Grundlehren der Math. 256, Springer-Verlag, (1983).

Houzel, C. [1]: Espaces analytiques relatifs et théorème de finitude - Math. Ann. *205*, 13–54, (1973).

— [2]: Algèbre filtrée - Preprint.

Houzel, C. and Schapira, P. [1]: Images directes des Modules différentiels - C.R. Acad. Sci. Paris t. *298*, I, 461–464, (1984).

Kashiwara, M. [1]: Algebraic study of systems of partial differential equations - Thesis Univ. Tokyo, (1971). (In japenese).

— [2]: On the maximally overdetermined systems of linear differential equations, I - Publ. R.I.M.S. Kyoto Univ. *10*, 563–579, (1975).

— [3]: b-functions and holonomic systems - Inventiones Math. *38*, 33–53, (1976).

— [4]: On the holonomic systems of linear differential equation II - Inventiones Math. *49*, 121–135, (1979).

— [5]: Systems of microdifferential equations - Cours Université Paris-Nord. Rédigé par T. Monteiro-Fernandès. Progress in Math. Birkhäuser, (1983).

— [6]: Vanishing cycle sheaves and holonomic systems of differential equations - Lecture Notes in Math. *1016*, Springer-Verlag, 134–142, (1983).

— [7]: The Riemann-Hilbert problem for holonomic systems - Publ. R.I.M.S., Kyoto Univ. *20*, 319–365, (1984).

Kashiwara, M. and Kawaï, T. [1]: On the holonomic systems of microdifferential equations III - Publ. R.I.M.S. Kyoto Univ. *17*, 813–979, (1981).

— [2]: Some applications of boundary value problems for elliptic systems of linear differential operators. Ann. Math. Studies n° 93 - Princeton Univ. Press. (1980).

Kashiwara, M., Kawaï, T. and Sjöstrand, J. [1]: On a class of linear partial differential equations whose formal solutions always converge - Arkiv för Math. *17*, n° 1, 83–91, (1979).

Kashiwara, M. and Oshima, T. [1]: Systems of differential equations with regular singularities and their boundary value problems - Annals of Math. *106*, 145–200, (1977).

Kashiwara, M. and Schapira, P. [1]: Problème de Cauchy pour les systèmes microdifferentiels dans le domaine complexe - Inventiones Math. *46*, 17–38, (1978).
— [2]: Micro-hyperbolic systems - Acta Math. *142*, 1–55, (1979).
— [3]: Variété caractéristique de la restriction d'un Module différentiel - Journée E.D.P. St-Jean-de-Monts. Publ. Ecole Polytechnique Palaiseau, (1981).
— [4]: Microlocal study of sheaves - RIMS Prepubl *469* Kyoto Univ. 1984, and C.R. Acad. Sci Paris, t. *295*, 487–490, (1982), Proc. Japan Acad. Vol. *59*, A, 349–359, (1983) - Asterisque - Soc. Math. France, (1985).
Komatsu, H. [1]: Resolutions by hyperfunctions of sheaves of solutions of differential equations with constant coefficients - Math. Annalen *176*, 77–86, (1968).
Laurent, Y. [1]: Deuxième microlocalisation: Operateurs 2-microdifferentials - C.R. Acad. Sci. t. *290*, I, 79–82, (1980).
— [2]: Deuxième microlocalisation: Système d'équations 2-microdifferentials - C.R. Acad. Sci. t. *290*, I, 147–150, (1980).
— [3]: Thèse Univ. Paris-Sud (1982).
Laurent, Y. and Schapira, P. [1]: Images inverses des Modules différentiels - A paraître.
Lebeau, G. [1]: Fonctions harmoniques et spectre singulier - Ann. Sci. Ec. Norm. Sup. 4ème série. *13*, 269–291, (1980).
Lejeune, M. [1]: Operateurs différentiels et pseudo-différentiels - Sem. Grenoble Exposé 3, (1975–76).
Leray, J. [1]: Problème de Cauchy I - Bull. Soc. Math. de France *85*, 389–430, (1957).
Malgrange, B. [1]: Sur les systèmes différentiels à coefficients constants - Sem. J. Leray. Collège de France, Paris, (1961–62).
— [2]: Opérateurs différentiels et pseudo-différentiels - Sem. Grenoble. Exposés 1, 2, 4, (1975–76).
— [3]: L'involutivité des caractéristiques des systèmes différentiels et microdifférentiels - Sem. Bourbaki. N° 522, (1977–78).
Maslov, V.P. [1]: Theory of perturbations and asymptotic methods - Moscow University Ed., (1965).
Matsuura, S. [1]: On general systems of partial differential operators with constant coefficients - J. Math. Soc. Japan *13*, n° 1, 94–103, (1961).
Mebkhout, Z. [1]: Sur le problème de Riemann-Hilbert. C.R. Acad. Sci. t. *290*, I, 415–417, (1980).
Monteiro-Fernandès, T. [1]: Variété 1-microcaractéristique pour les $\mathscr{E}_X$-modules coherents. C.R. Acad. Sci., t. *290*, I, 787–790, (1980).
— [2]: Problèmes de Cauchy microdifférentiels et théorèmes de propagation - C.R. Acad. Sci., t. *290*, I, 833–836, (1980).
— [3]: Thèse Univ. Paris-Nord (1982). Asterisque. Soc. Math. France (1984).
Nagumo, M. [1]: Über das Anfangswertproblem partieller differentialgleichungen - Japan Journ. Math. *18*, 41–47, (1941).
Narasimhan, R. [1]: Introduction to the theory of analytic spaces - Lecture Notes in Math. *25*, Springer-Verlag (1966).
Nirenberg, L. [1]: An abstract form of the non linear Cauchy-Kowalewski theorem - J. Diff. Geometry *6*, 561–576, (1972).
Palamodov, V.P. [1]: Linear differential operators with constant coefficients - Grundlehren der Math. 168, Springer-Verlag, (1970).
Pallu de la Barriere, P. and Schapira, P. [1]: Application de la théorie des microfonctions holomorphes au problème de Cauchy à données singulières - Sem. Goulaouic-Schwartz, exposé 23 (1975–76).
Pham, F. [1]: Singularités des systèmes différentiels de Gauss-Manin - Progress in Math. n° 2, Birkhäuser-Boston, (1979).
Quillen, D.: Formal properties of overdetermined systems of linear partial differential equations - Thesis Harvard (1964).
Sato, M. [1]: Hyperfunctions and partial differential equations - Proc. Int. Conf. on functional analysis and related topics 91–94, Tokyo Univ. Press., Tokyo (1969).

— [2]: Regularity of hyperfunction solutions of partial differential equations - Actes Congr. Int. Math. Nice 785-794, (1970).

Sato, M. and Kashiwara, M. [1]: The determinant of matrices of pseudodifferential operators - Proc. Japan Acad. *51*, 17-19, (1975).

Sato, M., Kashiwara, M. and Kawaï, T. [1]: Hyperfunctions and pseudodifferential equations - Lecture Notes in Math. *287*, Springer-Verlag 265-529, (1973).

Sato, M., Kashiwara, M., Kimura, T. and Oshima, T. [1]: Microlocal analysis of prehomogeneous vector spaces - Inventiones Math. *62*, 117-179, (1980).

Schapira, P. [1]: Théorie des hyperfonctions - Lecture Notes in Math. *126* - Springer-Verlag, (1970).

— [2]: Conditions de positivité dans une variété symplectique complexe. Application à l'étude des microfonctions - Ann. Sci. Ec. Norm. Sup. 4ème série, t. *14*, 121-139, (1981).

— [3]: Une introduction à l'étude des systèmes microdifférentiels - Astérisque *89-90*, Soc. Math. France 45-83, (1981).

Schwartz, L. [1]: Homomorphismes et applications complètement continues - C.R. Acad. Sci. *236*, 2472-2473, (1953).

Serre, J-P. [1]: Algèbre locale-Multiplicités - Lecture Notes in Math. *11* - Springer-Verlag, (1965).

— [2]: Prolongements de faisceaux analytiques coherents - Ann. Inst. Fourier Grenoble, *16*, n° 1, 363-374, (1966).

Siu, Y.T. [1]: Every Stein subvariety admits a Stein neighborhood - Inventiones Math. *38*, 89-100, (1976).

Siu, Y.T. and Trautman, G. [1]: Gap sheaves and extension of coherent analytic subsheaves - Lecture Notes in Math. *172*, Springer-Verlag, (1971).

Sjöstrand, J. [1]: Singularités analytiques microlocales - Astérisque *95*, Soc. Math. France, (1982).

Treves, F. [1]: Basic linear partial differential equations - Academic press, (1975).

Verdier, J.L. [1]: Stratifications de Whitney et théorème de Bertini-Sard - Inv. Math. *36*, 295-312, (1976).

Weinstein, A. [1]: Symplectic manifolds and their Lagrangean submanifolds - Advances in Math. *6*, 329-346, (1971).

Wells, R.O.Jr. [1]: Differential analysis on complex manifolds - Graduate Texts in Math., *65*, Springer-Verlag, (1980).

Whitney, H. [1]: Tangents to an analytic variety - Ann. of Math. *81*, 496-549, (1965).

— [2]: Complex analytic varieties - Addison Wesley, (1972).

Zerner, M. [1]: Domaine d'holomorphie des fonctions vérifiant une équation aux dérivées partielles - C.R. Acad. Sci. Paris Série A, t. *272*,1646-1648, (1971).

List of Notations and Conventions

General Notations

$\mathbb{N}$:   set of non negative integers

$\mathbb{Z}$:   ring of integers

$\mathbb{Q}$:   field of rational numbers

$\mathbb{R}$:   field of real numbers

$\mathbb{C}$:   field of complex numbers

$\mathbb{R}^+$: multiplicative group $\{c \in \mathbb{R}; c > 0\}$

$\mathbb{C}^\times$: multiplicative group $\{c \in \mathbb{C}; c \neq 0\}$

$\#A$: number of elements of a finite set A

δ_j^i: Kronecker symbol, $\delta_j^i = 0$ for $i \neq j$ and $\delta_i^i = 1$

$\dfrac{\partial}{\partial x_i}$ or ∂_{x_i} or ∂_i: partial derivation with respect to x_i (in a system of local coordinates $(x_1, \ldots, x_n)$)

$|\alpha| = \alpha_1 + \ldots + \alpha_p$, where $\alpha = (\alpha_1, \ldots, \alpha_p) \in \mathbb{N}^p$

$\alpha! = \alpha_1! \ldots \alpha_p!$

$y^\alpha = y_1^{\alpha_1} \ldots y_p^{\alpha_p}$, where $y = (y_1, \ldots, y_p)$ and the y_j's belong to some ring

$|\cdot|$: a norm on a vector space E

     When $E = \mathbb{C}^n$, $|x| = \langle x, \bar{x} \rangle^{1/2}$, where $\langle x, \bar{x} \rangle = \displaystyle\sum_{j=1}^n x_j \bar{x}_j$

$d(x, y) = |x - y|$

$K_\varepsilon = \{x \in E; d(x, K) < \varepsilon\}$

$\bar{A} = $ the closure of a subset A of E

$\{x_n\}_{n \in \mathbb{N}}$, or $\{x_n\}_n$: a sequence indexed by $n \in \mathbb{N}$

$x_n \xrightarrow{n} x$: x_n tends to x when n tends to infinity

$|f|_K = \sup\limits_{x \in K} |f(x)|$, for f a continuous function on a compact set K

$A \times B$: cartesian Product

$\mathbb{R}^n$:   Euclidian n-space

$\wedge$:   wedge product

$\sqcup$:   disjoint union

$\varinjlim$: inductive limit

$\varprojlim$: projective limit

ker: kernel

coker: cokernel

im: image

Manifold-Bundles

X, Y, Z: real or complex manifolds

$X \underset{Z}{\times} Y$: the fiber product of X and Y over Z, associated to two
maps, $f: X \to Z$, $g: Y \to Z$, the set $\{(x, y) \in X \times Y; f(x) = g(y)\}$

TX or $\tau: TX \to X$: the tangent bundle to X

T^*X or $\pi: T^*X \to X$: the cotangent bundle to X

$T_Y X$: the normal bundle to a submanifold Y of X, defined by the exact sequence:

$$0 \to TY \to Y \underset{X}{\times} TX \to T_Y X \to 0$$

$T_Y^* X$: the conormal bundle to Y in X, defined by the exact sequence:

$$0 \to T_Y^* X \to Y \underset{X}{\times} T^*X \to T^*Y \to 0$$

$T^*(X/Y)$: the relative contangent bundle, associated to a submersion $f: X \to Y$ (III, § 1.3)

e_A: the Euler vector field on a vector bundle A. If (y, z) is a system of coordinates on A such that $z = (z_1, \ldots, z_p)$ are the linear coordinates of the fibers, then $e_A = \sum_{j=1}^{p} z_j \dfrac{\partial}{\partial z_j}$ (I, § 1.2 and III, § 1.4).

$\bar{\omega}, \rho$: the natural maps associated to a map $f: Y \to X$; (I, § 4.3).

$$\bar{\omega}: Y \underset{X}{\times} T^*X \to T^*X$$

$$\rho: Y \underset{X}{\times} T^*X \to T^*Y$$

$T_Y^* X$: the kernel of ρ.

$T_X^* X$: the zero section of T^*X, often identified to X.

$\dot{T}^* X$: $T^*X \setminus T_X^* X$.

P^*X: the projective cotangent bundle.

$\dot{\pi}$: the restriction of $\pi: T^*X \to X$ to $\dot{T}^* X$.

γ: the projection $\dot{T}^* X \to P^*X = \dot{T}^* X / \mathbb{C}^\times$.

a: the antipodal map on T^*X, $a((x; \xi)) = (x; -\xi)$.

ω or ω_X: the canonical 1-form on T^*X. If $x = (x_1, \ldots, x_n)$ is a system of local coordinates on X, $(x, \xi) = (x_1, \ldots, x_n; \xi_1 dx_1, \ldots, \xi_n dx_n)$ the associated coordinates on T^*X, then $\omega = \sum_{j=1}^{n} \xi_j dx_j$ (Appendix A.2).

H: the Hamiltonian (or "symplectic") isomorphism from T^*T^*X to TT^*X $\langle \theta, v \rangle = \langle d\omega, v \wedge H(\theta) \rangle$, $v \in TT^*X$, $\theta \in T^*T^*X$ (Appendix A.1).

$H_f = H(df)$: the Hamiltonian vector field of f. With the coordinates (x, ξ) as precedingly:

$$H_f = \sum_{j=1}^{n} \left(\frac{\partial f}{\partial \xi_j} \frac{\partial}{\partial x_j} - \frac{\partial f}{\partial x_j} \frac{\partial}{\partial \xi_j} \right) \text{ (Appendix A.2).}$$

$\{f, g\} = H_f(g)$: the Poisson bracket of f and g (Appendix A.2).

$Z \prec Y$: Y dominates Z (III, § 4.1).

$X^{\mathbb{R}}$: real underlying manifold of the complex manifold X (III, § 4.1).

$\overline{X}$: the anti-holomorphic manifold associated to the complex manifold X (III, § 4.1).

Algebra

$F^{\cdot}$: a complex (Appendix B).

$H^j(F^{\cdot})$: the j-th cohomology object of $F^{\cdot}$.

$\text{Ext}^j(\cdot, \cdot)$: the j-th derived functor of $\text{Hom}(\cdot, \cdot)$.

$\text{Tor}_j(\cdot, \cdot)$: the $-j$-th derived functor of $\cdot \otimes \cdot$.

GM or $gr(M)$: the graded module associated to a filtered module FM.

$\text{Icar}(M)$: the radical of the annihilator of GM (II, 1.3.2).

Sheaves

Ring, Module, etc. ...: sheaf of rings, sheaf of modules, etc. ... (Appendix C.2).

$\text{supp}(\mathscr{F})$: the support of the sheaf $\mathscr{F}$.

$\text{supp}(\mathscr{F}^{\cdot})$: $\bigcup_j \text{supp}(H^j(\mathscr{F}^{\cdot}))$, where $\mathscr{F}^{\cdot}$ is a complex of sheaves (III, 4.2.3).

$\mathscr{E}\!xt^j(\cdot, \cdot)$: the j-th derived functor of $\mathscr{H}\!om(\cdot, \cdot)$ (in the category of sheaves).

$\mathscr{T}\!or_j(\cdot, \cdot)$: the $-j$-th derived functor of $\cdot \otimes \cdot$ (in the category of sheaves).

$\Gamma_Z(U, \mathscr{F})$: the group of sections of $\mathscr{F}$ on the open set U, supported by the closed subset Z of U.

$\Gamma_Z(X, \mathscr{F})$: for Z a locally closed subset of X, the group $\Gamma_Z(U, \mathscr{F})$, where U is any open subset of X containing Z as a closed subset.

$\Gamma_Z(\mathscr{F})$: the sheaf $U \mapsto \Gamma_Z(U, \mathscr{F})$.

$H^j_Z(U, \cdot)$: the j-th derived functor of $\Gamma_Z(U, \cdot)$.

$H^j(U, \cdot)$: $H^j_U(U, \cdot)$.

$\mathscr{H}^j_Z(\mathscr{F})$: the sheaf associated to the presheaf $U \mapsto H^j_{Z \cap U}(U, \mathscr{F})$, or equivalently the j-the derived functor of $\Gamma_Z(\cdot)$ applied to $\mathscr{F}$.

$SS(\mathscr{F}^{\cdot})$: the micro-support of a complex of sheaves $\mathscr{F}^{\cdot}$ (III, 4.2.5).

$\mathscr{F}|_Y$: the restriction of the sheaf $\mathscr{F}$ to the subspace Y.

$f_*(\mathscr{F})$: the direct image of $\mathscr{F}$ by f.

$f^{-1}(\mathscr{F})$: the inverse image of $\mathscr{F}$ by f.

Sheaves on a Complex Manifold

$\mathcal{O}_X$: the sheaf of holomorphic functions on the complex manifold X.

$\Omega_X^{(p)}$: the sheaf of holomorphic p-forms on X.

Ω_X: $\Omega_X^{(\dim X)}$.

$\Omega_X^{\otimes-1}$: $\mathcal{H}om_{\mathcal{O}_X}(\Omega_X, \mathcal{O}_X)$. Thus $\Omega_X^{\otimes-1}\otimes_{\mathcal{O}_X}\Omega_X=\mathcal{O}_X$.

$\mathcal{O}_\Lambda(j)$: on a complex vector bundle $\tau\colon\Lambda\to Y$ the subsheaf of $\mathcal{O}_\Lambda$ of sections homogeneous of degree j in the fibers (i.e.: the sheaf of solutions of the equation $(e_\Lambda-j)f=0\}$.

$\mathcal{O}_\Lambda^h$: $\bigoplus_{j\in\mathbb{Z}}\mathcal{O}_\Lambda(j)$.

$\mathcal{O}_{[\Lambda]}$: $\tau^{-1}\tau_*(\mathcal{O}_\Lambda^h)$, that is the sub-Ring of $\mathcal{O}_\Lambda$ of sections which are polynomials in the fibers.

$\mathcal{J}_Y$: the defining Ideal of an analytic subset Y of X.

$\mathcal{S}_Y$: $\bigoplus_{k\geqslant0}\mathcal{J}_Y^k/\mathcal{J}_Y^{k+1}$ (III, §1.1).

$[\mathrm{supp}(\mathcal{F})]$: the analytic cycle of a coherent $\mathcal{O}_X$-module $\mathcal{F}$ (Appendix D.1).

$\mathrm{mult}_W(\mathcal{F})$: the multiplicity of $\mathcal{F}$ along W (Appendix D.1).

Modules of Microdifferential Operators

$\hat{\mathcal{E}}_X$: the Ring on T^*X of formal microdifferential operators (I, §1.2).

$\mathcal{E}_X$: the Ring on T^*X of microdifferential operators (I, §1.3).

$\hat{\mathcal{E}}_X(k)$: the subsheaf of $\hat{\mathcal{E}}_X$ of sections of order at most k.

$\mathcal{E}_X(k)=\hat{\mathcal{E}}_X(k)\cap\mathcal{E}_X$.

$\mathcal{D}_X$: the Ring on X of differential operators, $\mathcal{D}_X=\mathcal{E}_X|_{T_X^*X}$.

$\mathcal{D}_X(k)=\mathcal{E}_X(k)|_{T_X^*X}$.

$\mathcal{E}_X(P_1,\ldots,P_r)$: the left Ideal of $\mathcal{E}_X$ generated by $P_1,\ldots,P_r$.

$(P_1,\ldots,P_r)\mathcal{E}_X$: the right Ideal of $\mathcal{E}_X$ generated by $(P_1,\ldots,P_r)$.

$\mathcal{C}_{Z|X}$: the sheaf of holomorphic microfunctions along a submanifold Z of X (I, §4.1).

$\mathcal{B}_{Z|X}$: the sheaf of holomorphic hyperfunctions along Z; $\mathcal{B}_{Z|X}=\mathcal{C}_{Z|X}|_{T_X^*X}$.

$\mathcal{E}_{Y\to X}$: the $(\rho^{-1}\mathcal{E}_Y,\overline{\omega}^{-1}\mathcal{E}_X)$-bi-Module associated to $f\colon Y\to X$ (I, 4.3.1).

$\mathcal{E}_{X\leftarrow Y}$: the $(\overline{\omega}^{-1}\mathcal{E}_X,\rho^{-1}\mathcal{E}_Y)$-bi-Module associated to $f\colon Y\to X$ (II, §3.1).

$\mathcal{D}_{Y\to X}=\mathcal{E}_{Y\to X}|_{Y\times_X T_X^*X}$ (I, 4.3.1 and II, 3.1.10).

$\mathcal{D}_{X\leftarrow Y}=\mathcal{E}_{X\leftarrow Y}|_{Y\times_X T_X^*X}$ (II, §3.1).

δ_Z: the fundamental class of Z in X (I, 4.1.2).

$1_{Y\to X}$: the canonical section of $\mathcal{E}_{Y\to X}$ (I, 4.3.1)

Y_Z: the canonical section of $\mathcal{C}_{Z|X}|_{\dot{T}^*X}$ for Z a hypersurface (I, 4.2.1).

$\mathcal{O}_{Z|X}^1$: sheaf of holomorphic functions with meromorphic or logarithmic singularities along Z (III, §2.2).

$\mathscr{E}_V$: the sub-Algebra of $\mathscr{E}_X|_V$ generated by $\{P\in\mathscr{E}_X(1);\ \sigma_1(P)|_V=0\}$ (I, § 3.1 and III, 1.2.1).

$\mathscr{D}_{X/Y}$: The sub-Algebra of $\mathscr{D}_X$ of relative differential operators associated to a submersion $f: X\to Y$ (II, § 1.5 and III, § 1.3).

$F^Y\mathscr{D}_X$: the filtration on $\mathscr{D}_X$ associated to an immersion $Y\hookrightarrow X$ (II, § 1.5 and III, § 1.4).

W_n: the Weyl algebra on an n-dimensional vector space (II, § 1.5).

$\mathscr{D}_{[\Lambda]}$: on a vector bundle Λ, the sub-Ring of $\mathscr{D}_\Lambda$ of operators with coefficients in $\mathscr{O}_{[\Lambda]}$ (III, § 1.4).

$\mathscr{D}_{\Lambda,\,k}$: the subsheaf of $\mathscr{D}_{[\Lambda]}$ of homogeneous operators of degree k (III, § 1.4).

$\mathscr{M}_Y$: the inverse image of $\mathscr{M}$ (II, 3.1.2).

$\int_f\mathscr{N}$: the direct image of $\mathscr{N}$ (II, 3.1.2).

$\mathscr{N}\hat{\otimes}\mathscr{M}$ – the external product of $\mathscr{N}$ and $\mathscr{M}$ (II, 3.1.1).

Principal Symbols and Characteristic Varieties

$\sigma(P)$: the principal symbol of a section P of $\mathscr{E}_X$ (I, § 1.2 and § 1.3).

ord(P): the order of a section P of $\mathscr{E}_X$ (I, § 1.2 and § 1.3).

$\sigma_m(P)$: the symbol of order m of a section of $\mathscr{E}_X(m)$.

char$(\mathscr{M})$: the characteristic variety of a coherent $\mathscr{E}_X$-module $\mathscr{M}$ (II, § 2.3) or even of a pseudo-coherent $\mathscr{E}_X$-module (III, § 1.5).

[char$(\mathscr{M})$]: the characteristic cycle of a coherent $\mathscr{E}_X$-module (II, § 2.3).

$\sigma_Y(f)$: the principal symbol of f along Y (III, § 1.1).

$C_Y(\mathscr{M})$: the normal cone of $\mathscr{M}$ along Y (III, 1.1.2).

$C_Y(S)$: the normal cone of the subset S along Y (III, 1.1.2 and III, § 4.1).

$C(S, V)$: the normal cone of S along V (III, § 4.1).

$C_V(\mathscr{M})=C_V(\text{char}(\mathscr{M}))$: the microcharacteristic variety of $\mathscr{M}$ along V (III, 1.1.7).

$\sigma_V^1(\sigma(P))$: associated to a section P of $\mathscr{E}_V$ (III, 1.2.4).

$C_V^1(\mathscr{M})$: the 1-microcharacteristic variety of $\mathscr{M}$ along V (III, 1.2.8).

char$_{X/Y}(\mathscr{M})$: the characteristic variety of a coherent $\mathscr{D}_{X/Y}$-module (III, § 1.3).

char$_f^1(\mathscr{M})$: the relative 1-characteristic variety of a coherent $\mathscr{D}_X$-module (f smooth) (III, 1.3.4).

char$_f(\mathscr{M})$: the relative characteristic variety of a coherent $\mathscr{D}_X$-module (f smooth) (III, 1.3.6).

$\hat{\sigma}_Y(P)$: the principal symbol of P (in $\mathscr{D}_{[T_YX]}$) for the filtration $F^Y\mathscr{D}_X$ on $\mathscr{D}_X$ (Y is a submanifold of X) (III, 1.4.2).

$\hat{C}_{T_YX}(\mathscr{M})$: the formal microcharacteristic variety of $\mathscr{M}$ along Y (III, 1.4.4).

Index

G. de Rham

Differentiable Manifolds

Translated from the French by F. R. Smith
Introduction to the English Edition by S. S. Chern
1984. Approx. 180 pages. (Grundlehren der
mathematischen Wissenschaften, Band 266)
ISBN 3-540-13463-8

Georges de Rham's classic work on Differentiable
Manifolds is at last available in English translation. As
S. S. Chern writes in his foreword (which also was on
to describes the development of the field since the
appearance of the French edition):
"Professor de Rham's book is an introduction to differentiable manifolds. Its main objective seems to be the
first detailed proof, different from Hodge-Weyl, of
Hodge's fundamental theorem. It must have given
him great pleasure in writing the book, for Hodge
theory is a natural culmination of the de Rham
theory...

Modern developments in the general area of "elliptic
operators on manifolds", such as the index theory and
the spectral theory, have raced way beyond the
content of this book. I believe, however that in this
enthusiasm for new results a mathematician will be
well-advised to stop at this landmark, where he will
have a lot to learn both on the mathematics and on
the mathematical style."

For all mathematicians, from the beginning graduate
student to the active researcher this coherent exposi-
tion of the theory of 'differential forms' on a 'manifold' and 'harmonic forms' on a 'Riemannian space'
will be a welcome and frequently utilized reference.

Springer-Verlag
Berlin
Heidelberg
New York
Tokyo

D.H.Luecking, L.A.Rubel

Complex Analysis
A Functional
Analysis Approach

Universitext

1984. 7 figures. VII, 176 pages
ISBN 3-540-90993-1

Contents: Introduction. – Preliminaries: Set
Theory and Topology. – Preliminaries: Vector
Spaces and Complex Variables. – Properties of
C(G) and H(G). – More About C(G) and H(G).
– Duality. – Duality of H(G) – The Case of the
Unit Disc. – The Hahn-Banach Theorem, and
Applications. – More Applications. – The Dual
of H(G). – Runge's Theorem. – The Cauchy
Theorem. – Constructive Function Theory. –
Ideals in H(G). – The Riemann Mapping
Theorem. – Carathéodory Kernels and Farrel's
Theorem. – Ring (not Algebra) Isomorphisms of
H(G). – Dual Space Topologies. – Interpolation.
– Gap-Interpolation Theorems. – First-Order
Conformal Invariants. – References.

Springer-Verlag
Berlin
Heidelberg
New York
Tokyo

The authors of this book address themselves to
mathematicians and graduate students of mathematics with at least one semester of "conven-
tional" complex variables. They present the
theory of complex variables in a unified new
approach based on the identification of the dual
of the space of analytic functions on a region as a
space of germs of holomorphic functions on the
complement. Once this has been shown, many
of the standard results follow easily, and the
reader obtains an efficient and stimulating introduction to complex analysis in a spirit that is
very close to Cauchy's original ideas.